Solutions Manual for Molecular Quantum Mechanics

Solutions Manual for Molecular Quantum Mechanics

P. W. ATKINS

Oxford New York
OXFORD UNIVERSITY PRESS

Oxford University Press, Walton Street, Oxford OX2 6DP
Oxford New York Toronto
Delhi Bombay Calcutta Madras Karachi
Petaling Jaya Singapore Hong Kong Tokyo
Nairobi Dar es Salaam Cape Town
Melbourne Auckland
and associated companies in
Berlin Ibadan

Oxford is a trade mark of Oxford University Press

First published 1983
Reprinted 1983, 1984, 1986, 1988, 1989

British Library Cataloguing in Publication Data
Atkins, P. W.
Solutions manual for molecular quantum mechanics.
1. Quantum chemistry—Examinations, questions, etc.
I. Title
541.2'8'076 QD462
ISBN 0–19–855180–0

Library of Congress Cataloging in Publication Data
Atkins, P. W. (Peter Williams), 1940–
Solutions manual for molecular quantum mechanics.
1. Quantum chemistry—Problems, exercises, etc.
I. Atkins, P. W. (Peter Williams), 1940–
Molecular quantum mechanics. II. Title.
QD462.A85 1983 541.2'8'076 83–5645
ISBN 0–19–855180–0 (pbk.)

Printed in Great Britain by
Thomson Litho Ltd, East Kilbride, Scotland

Preface

The following pages contain the detailed solutions to the problems at the ends of the chapters in the second edition of *Molecular Quantum Mechanics* (1983). The references in square brackets and comments such as 'of the text' refer to that edition. The illustrations are my own rough sketches and graphs: I apologize for splodges and scruffiness, but we tried to minimize costs.

In order to combat the criticism that with full details of the solutions available there is nothing left for either instructor or student to do, I have added at the end of each solution an *Exercise*. These are along the same lines as the preceding problem. Quite often the *Exercise* invites the reader to generalise the result (e.g. from helium to atoms of general values of Z).

I have been fortunate in having the services of an impeccable typist: I acknowledge gratefully the meticulous care of Ina Godwin in the preparation of the camera-ready copy.

Oxford, 1983 P.W.A.

Contents

1 Historical introduction

1.1 Use $E = h\nu$; $h = 6.626 \times 10^{-34}$ J s ; $\nu = 1/T$ [T: period] .

(a) $E = (6.626 \times 10^{-34}\ \text{J s})/(10^{-15}\ \text{s}) = 6.626 \times 10^{-19}\ \text{J} \triangleq \underline{399\ \text{kJ mol}^{-1}}$.

(b) $E = h/(10^{-14}\ \text{s}) = 6.626 \times 10^{-20}\ \text{J} \triangleq \underline{39.9\ \text{kJ mol}^{-1}}$.

(c) $E = h/(1\ \text{s}) = \underline{6.626 \times 10^{-34}\ \text{J}}$.

1.2 Use eqn (1.1.9): $\varepsilon = h\nu e^{-h\nu/kT} \Big/ \left\{1 - e^{-h\nu/kT}\right\} = h\nu \Big/ \left\{e^{h\nu/kT} - 1\right\}$. $h/k = 4.7993 \times 10^{-11}$ s K. We have :

$h\nu/kT$	(a) $\nu = 1$ Hz	(b) $\nu = 10^{14}$ Hz
(i) $T = 100$ K	4.80×10^{-13}	48.0
(ii) $T = 1000$ K	4.80×10^{-14}	4.80

Hence, from eqn (1.1.9) :

ε/J	(a) $\nu = 1$ Hz	(b) $\nu = 10^{14}$ Hz
(i) $T = 100$ K	1.38×10^{-21}	9.51×10^{-41}
(ii) $T = 1000$ K	1.38×10^{-20}	5.50×10^{-22}

If equipartition were applicable, $\varepsilon = kT$. Then (i) $\varepsilon = 1.38 \times 10^{-21}$ J, (ii) $\varepsilon = 1.38 \times 10^{-20}$ J for both frequencies.

1.3 Write $\varepsilon_v = (v + \tfrac{1}{2})h\nu$ in eqns (1.1.7) and (1.1.8) :

$$\varepsilon = \left\{\frac{\sum_v (v+\tfrac{1}{2})\, h\nu\, e^{-(v+\frac{1}{2})h\nu/kT}}{\sum_v e^{-(v+\frac{1}{2})h\nu/kT}}\right\} = \tfrac{1}{2}h\nu + \left\{\frac{e^{-h\nu/2kT}\sum_v e^{-vh\nu/kT}}{e^{-h\nu/2kT}\sum_v e^{-vh\nu/kT}}\right\}$$

$$= \tfrac{1}{2}h\nu + h\nu e^{-h\nu/kT} \Big/ \left\{1 - e^{-h\nu/kT}\right\} .$$

[For the sums, use eqn (1.1.9).] Hence, this ε differs from that in eqn (1.1.9) by the addition of $\tfrac{1}{2}h\nu$, the zero-point energy.

Since $dU = \varepsilon dN$ [eqn (1.1.10)], the Planck distribution is

$$dU = \varepsilon(8\pi\nu^3/c^3)d\nu \ \text{[eqn (1.1.4)]}$$

$$= \varepsilon(8\pi\nu^3/c^3)\left\{\tfrac{1}{2} + \left(\frac{e^{-h\nu/kT}}{1 - e^{-h\nu/kT}}\right)\right\} d\nu.$$

1.4 Use Wien's law, eqn (1.1.2): $\lambda_{\mathbf{max}} T = \text{const.}$; const. $= hc/5k$ [equation (1.1.13)] $= 2.878\,\text{mm K}$ [End paper 1]. Hence

$$T = (2.878\ \text{mm K})/(480 \times 10^{-9}\ \text{m}) = \underline{6000\ \text{K}}\ .$$

1.5 Use eqn (1.2.2): $c^{\mathbf{E}}(T) = (\theta_{\mathbf{E}}/T)^2\, \mathrm{e}^{\theta_{\mathbf{E}}/T} \Big/ \left(1 - \mathrm{e}^{\theta_{\mathbf{E}}/T}\right)^2$.

(a) When $T \ll \theta_{\mathbf{E}}$, $\mathrm{e}^{\theta_{\mathbf{E}}/T} \gg 1$; hence $\left(1 - \mathrm{e}^{\theta_{\mathbf{E}}/T}\right)^2 \approx \mathrm{e}^{2\theta_{\mathbf{E}}/T}$. Then,

$$c^{\mathbf{E}}(T) \approx (\theta_{\mathbf{E}}/T)^2 \left(\mathrm{e}^{\theta_{\mathbf{E}}/T} \big/ \mathrm{e}^{2\theta_{\mathbf{E}}/T}\right) = (\theta_{\mathbf{E}}/T)^2\, \mathrm{e}^{-\theta_{\mathbf{E}}/T} .$$

$$\lim_{T \to 0} c^{\mathbf{E}}(T) = 0 \quad \left[\text{as} \quad \lim_{x \to \infty} x^2\, \mathrm{e}^{-x} = 0\right] .$$

(b) When $T \gg \theta_{\mathbf{E}}$, $\mathrm{e}^{\theta_{\mathbf{E}}/T} \approx 1 + \theta_{\mathbf{E}}/T$. Then,

$$c^{\mathbf{E}}(T) \approx (\theta_{\mathbf{E}}/T)^2 \left\{ \frac{1 + (\theta_{\mathbf{E}}/T)}{[1 - 1 - (\theta_{\mathbf{E}}/T)]^2} \right\} \approx (\theta_{\mathbf{E}}/T)^2\ (T/\theta_{\mathbf{E}})^2 = 1 .$$

1.6 For sodium $\theta_{\mathbf{D}}/T = 0.50$; for diamond $\theta_{\mathbf{D}}/T = 6.20$. If we use the Einstein formula (with $\theta_{\mathbf{E}} \approx \theta_{\mathbf{D}}$), then

(a) Na(s): $c^{\mathbf{E}} = 0.979$; hence $\underline{C_{V,\mathrm{m}}/R = 2.93}$

(b) C(d) : $c^{\mathbf{E}} = 0.078$; hence $\underline{C_{V,\mathrm{m}}/R = 0.24}$.

The Debye formula can be evaluated by numerical integration (e.g. using a library program, or using Simpson's rule), but it is simpler to use tabulated values. See the *American Institute of Physics Handbook*, D.E. Gray (ed.), McGraw-Hill (1972), p.4.113. Then

(a) Na(s): $c^{\mathbf{D}}(\theta_{\mathbf{D}}/T = 0.50) = 0.988$; hence $\underline{C_{V,\mathrm{m}}/R = 2.96}$

(b) C(d) : $c^{\mathbf{D}}(\theta_{\mathbf{D}}/T = 6.20) = 0.249$; hence $\underline{C_{V,\mathrm{m}}/R = 0.747}$.

EXERCISE: Evaluate C_V at 300 K for the Group 1 metals.

1.7 $$S = \int_0^T (C_V/T)\,\mathrm{d}T = 3R \int_0^T (c^{\mathbf{E}}/T)\,\mathrm{d}T$$

$$= 3R \int_0^T (\theta_{\mathbf{E}}/T)^2\ (1/T) \left\{ \frac{\mathrm{e}^{\theta_{\mathbf{E}}/T}}{[1 - \mathrm{e}^{\theta_{\mathbf{E}}/T}]^2} \right\} \mathrm{d}T .$$

Set $x = \theta_{\mathbf{E}}/T$, $\mathrm{d}T = -T^2\,\mathrm{d}x/\theta_{\mathbf{E}}$; hence

$$S/3R = \int_{\theta_{\mathbf{E}}/T}^{\infty} \left\{ x\,\mathrm{e}^x / (1 - \mathrm{e}^x)^2 \right\} \mathrm{d}x .$$

Therefore, when $T = \theta_{\mathbf{E}}$,

$$S/3R = \int_1^\infty \left\{x\,e^x/(1-e^x)^2\right\} dx = 1.048 \qquad \text{[evaluated numerically]}.$$

Hence, $\underline{S/R = 3.144}$.

EXERCISE: Evaluate S for metallic sodium and for diamond at 300 K.

1.8 Each photon carries an energy $E = h\nu = hc/\lambda = 3.37 \times 10^{-19}$ J at $\lambda = 589$ nm. 100 W corresponds to $100\ \mathrm{J\,s^{-1}}$; hence the number of photons emitted is $(100\ \mathrm{J\,s^{-1}})/(3.37 \times 10^{-19}\ \mathrm{J}) = \underline{2.97 \times 10^{20}\ \mathrm{s^{-1}}}$. ($2.97 \times 10^{20}$ photons corresponds to 4.93×10^{-4} einstein, and so 1 einstein will be generated in about 34 minutes.)

1.9 2.3 eV corresponds to $2.3 \times (1.602 \times 10^{-19}\ \mathrm{C})\ \mathrm{V} = 3.7 \times 10^{-19}$ J. Then use eqn (1.3.1) in the form

$$v = \sqrt{\left\{(2/m_e)(h\nu - \Phi)\right\}} \quad \text{so long as} \quad h\nu \geqslant \Phi$$

(a) $h\nu = hc/\lambda = 6.62 \times 10^{-19}$ J when $\lambda = 300$ nm :

$$v = \sqrt{\left\{\left[2/(9.10963 \times 10^{-31}\ \mathrm{kg})\right] \times (2.9 \times 10^{-19}\ \mathrm{J})\right\}} = \underline{8.0 \times 10^5\ \mathrm{m\,s^{-1}}}\,.$$

(b) $h\nu = 3.31 \times 10^{-19}\ \mathrm{J} < \Phi$; hence no electrons are emitted.

EXERCISE: Examine the case where the ejection speed is so great that it must be treated relativistically.

1.10 Refer to Fig. 1.1.

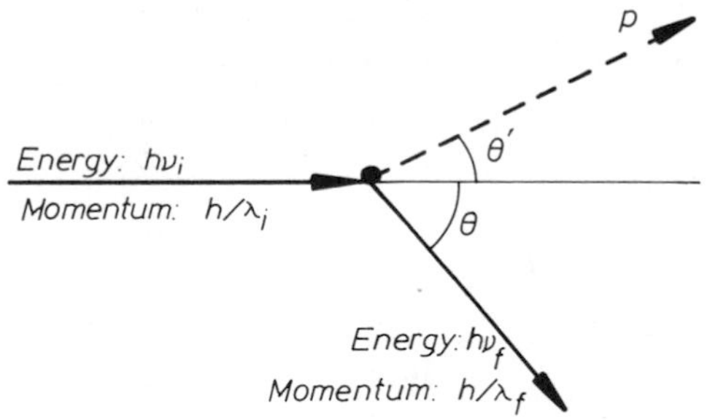

Fig. 1.1

Energy conservation: $h\nu_i + m_e c^2 = h\nu_f + \left\{p^2c^2 + m_e^2c^4\right\}^{\frac{1}{2}}$. (1)

Momentum conservation:

Parallel: $h/\lambda_i = (h/\lambda_f)\cos\theta + p\cos\theta'$ (2)

Perpendicular: $0 = (h/\lambda_f)\sin\theta + p\sin\theta'$. (3)

From (2) and (3):

$$p^2\cos^2\theta' + p^2\sin^2\theta' = [(h/\lambda_i) - (h/\lambda_f)\cos\theta]^2 + [(h/\lambda_f)\sin\theta]^2$$

$$p^2 = h^2\left\{(1/\lambda_i)^2 + (1/\lambda_f)^2\cos^2\theta - (2/\lambda_i\lambda_f)\cos\theta + (h/\lambda_f)^2\sin^2\theta\right\}$$
$$= h^2\left\{(1/\lambda_i)^2 + (1/\lambda_f)^2 - (2/\lambda_i\lambda_f)\cos\theta\right\}. \qquad (4)$$

But from (1) :

$$p^2c^2 = (h\nu_i + m_e c^2 - h\nu_f)^2 - m_e^2c^4,$$
$$= \left\{(hc/\lambda_i) + m_e c^2 - (hc/\lambda_f)\right\}^2 - m_e^2c^4. \qquad (5)$$

Therefore, combining (4) and (5) :

$$(1/\lambda_i)^2 + (1/\lambda_f)^2 - (2/\lambda_i\lambda_f)\cos\theta$$
$$= (1/\lambda_i)^2 + (1/\lambda_f)^2 + 2(m_e c/h)\left\{(1/\lambda_i) - (1/\lambda_f)\right\} - (2/\lambda_i\lambda_f).$$

Consequently,

$$(1/\lambda_i\lambda_f)(1-\cos\theta) = 2(m\ c/h)\left\{(1/\lambda_i) - (1/\lambda_f)\right\}$$
$$= 2(m_e c/h)(\lambda_f - \lambda_i)/\lambda_i\lambda_f.$$

Hence,

$$\delta\lambda = \lambda_f - \lambda_i = (h/m_e c)(1-\cos\theta) = (2h/m_e c)\sin^2\tfrac{1}{2}\theta.$$

EXERCISE : Examine the non-relativistic formulation of the effect.

1.11 From eqn (1.5.1), $1/\lambda = R_H\{(1/2^2) - (1/n^2)\}$, $n = 3, 4, \ldots$ Hence, plot $1/\lambda$ against $1/n^2$, and find R_H from the intercept at $n=\infty$ (since then $1/\lambda_\infty = R_H/4$). The data (Fig. 1.2) extrapolate to

$$1/\lambda_\infty = 2.743 \times 10^6\ \text{m}^{-1} = 2.743 \times 10^4\ \text{cm}^{-1};$$

hence

$$R_H = 4 \times (2.743 \times 10^4\ \text{cm}^{-1}) = \underline{1.097 \times 10^5\ \text{cm}^{-1}}.$$

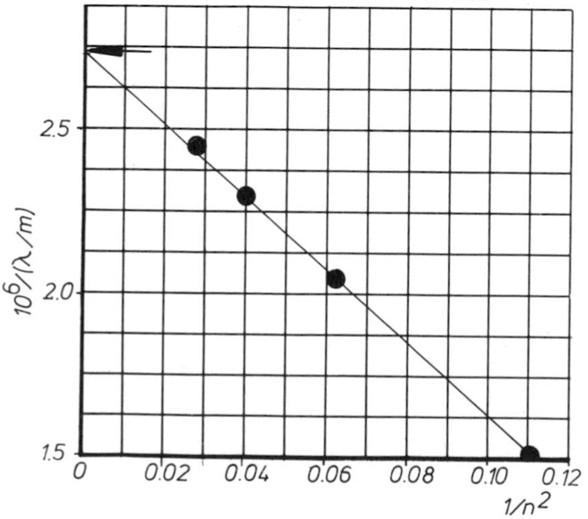

Fig 1.2

The ionization energy (I) is the energy required for the transition $n_2 = \infty \leftarrow n_1 = 1$; hence $\underline{I = hcR_H}$ = 2.179×10^{-18} J. Since 1 eV $\hat{=}$ 1.602×10^{-19} J, $I \hat{=}$ $\underline{13.6\ \text{eV}}$.

1.12 Use eqn (1.6.1) in the form $\lambda = h/p$.

(a) $\lambda = (6.626 \times 10^{-34}\ \mathrm{J\,s})/(1.0 \times 10^{-3}\ \mathrm{kg}) \times (1.0 \times 10^{-2}\ \mathrm{m\,s^{-1}}) = \underline{6.6 \times 10^{-29}\ \mathrm{m}}.$

(b) $p = m_e v/(1-v^2/c^2)^{\frac{1}{2}} = (0.95 \times 10^{-3}\ \mathrm{kg})\,c/(1-0.903)^{\frac{1}{2}} = 9.1 \times 10^{-5}\ \mathrm{kg\,m\,s^{-1}}$,
$\lambda = h/p = \underline{7.3 \times 10^{-40}\ \mathrm{m}}$.

(c) $\frac{1}{2}m\langle v^2\rangle = \frac{3}{2}kT$ [equipartition] ; $p \approx m\langle v^2\rangle^{\frac{1}{2}} = (3mkT)^{\frac{1}{2}}$.

$\lambda \approx h/(3mkT)^{\frac{1}{2}}$; $m = m_p + m_e = 1.6736 \times 10^{-27}\ \mathrm{kg}$.

$\lambda \approx 1.45 \times 10^{-10}\ \mathrm{m} = \underline{0.145\ \mathrm{nm}}$

(d) $\frac{1}{2}m_e v^2 = e\Delta\phi$; $p = m_e v = m_e(2e\Delta\phi/m_e)^{\frac{1}{2}} = (2m_e e\Delta\phi)^{\frac{1}{2}}$.

$\lambda = h/(2m_e e\Delta\phi)^{\frac{1}{2}} = (1.226 \times 10^{-9}\ \mathrm{m})/(\Delta\phi/\mathrm{V})^{\frac{1}{2}} = 1.226\ \mathrm{nm}/(\Delta\phi/\mathrm{V})^{\frac{1}{2}}$.

(i) $\Delta\phi = 1\ \mathrm{V}$; $\underline{\lambda = 1.23\ \mathrm{nm}}$; (ii) $\Delta\phi = 10\ \mathrm{kV}$; $\underline{\lambda = 12.3\ \mathrm{pm}}$.

1.13 Use eqn (1.7.1) in the form $\delta p \geqslant h/4\pi\delta x$. For simplicity take $\delta x \approx 0.10\ \mathrm{nm}$. ($\delta x$ is actually the root mean square deviation of the location: for a uniform distribution in the container $\delta x = L/\sqrt{12}$.)
Hence, with $\delta p = m_e\delta v$,

$$\delta v \geqslant h/4\pi \times (0.01\ \mathrm{nm}) \times (9.11 \times 10^{-31}\ \mathrm{kg}) = \underline{5.8 \times 10^5\ \mathrm{m\,s^{-1}}}$$

$$\delta(\mathrm{K.E.}) = \tfrac{1}{2}m_e(\delta v)^2 \geqslant 1.53 \times 10^{-19}\ \mathrm{J} \mathrel{\hat{=}} \underline{0.95\ \mathrm{eV}}.$$

Note that in a box $\delta p = \langle p^2\rangle^{\frac{1}{2}}$ because $\langle p\rangle = 0$; hence $(\delta p)^2 = \langle p^2\rangle$ and $(\delta v)^2 = \langle v^2\rangle$.

EXERCISE : Evaluate the quantities on the basis of $\delta x = L/\sqrt{12}$.

1.14 Total energy : $E \approx \langle p^2\rangle/2m_e - (e^2/4\pi\varepsilon_0)(1/\delta x)$.
Then, since $\langle p^2\rangle = (\delta p)^2$ [Problem 1.13] $\approx (h/4\pi\delta x)^2$, we have

$$E \approx h^2/2m_e(4\pi\delta x)^2 - (e^2/4\pi\varepsilon_0)(1/\delta x).$$

This has a minimum when $\mathrm{d}E/\mathrm{d}(\delta x) = 0$, or when

$$-h^2/16\pi^2 m_e(\delta x)^3 + (e^2/4\pi\varepsilon_0)(1/\delta x)^2 = 0,$$

which is when

$$\delta x \approx \varepsilon_0 h^2/4\pi e^2 m_e = \underline{5.3 \times 10^{-11}\ \mathrm{m}}.$$

The result is the Bohr radius (a_0), the radius of the first Bohr orbit, and the most probable distance of the electron from the nucleus in the quantum mechanical model of the atom.

EXERCISE: Assess how the size depends on atomic number in one-electron ions of nuclear charge Ze.

2 The Schrödinger equation

2.1 Find x such that $xp - px = \mathrm{i}\hbar$ where $p = p\times$. Try $x = \mathrm{i}\hbar\,(\mathrm{d}/\mathrm{d}p)$:

$$xp - px = \mathrm{i}\hbar\,(\mathrm{d}/\mathrm{d}p)p - p\mathrm{i}\hbar\,(\mathrm{d}/\mathrm{d}p)$$
$$= \mathrm{i}\hbar\,\{1 + p(\mathrm{d}/\mathrm{d}p)\} - \mathrm{i}\hbar p(\mathrm{d}/\mathrm{d}p) = \mathrm{i}\hbar$$

as required. Hence, use $\underline{x = \mathrm{i}\hbar\,(\mathrm{d}/\mathrm{d}p)}$.

EXERCISE: Find alternative realizations of the operators.

2.2 In each case $H\psi = E\psi$ with

(a) $H = -(\hbar^2/2m_\mathrm{e})\,\nabla_\mathrm{e}^2 - (\hbar^2/2m_\mathrm{p})\,\nabla_\mathrm{p}^2 - (e^2/4\pi\varepsilon_0 r)$.

This can be expressed in terms of the motion of the centre of mass and of the relative motion of the electron and nucleus [Appendix 6]:

$H = H_{\mathrm{c.m.}} + H_{\mathrm{relative}}$

$H_{\mathrm{c.m.}} = -(\hbar^2/2m_\mathrm{H})\,\nabla_{\mathrm{c.m.}}^2$; $H_{\mathrm{relative}} = -(\hbar^2/2\mu)\,\nabla^2 - (e^2/4\pi\varepsilon_0 r)$

with $m_\mathrm{H} = m_\mathrm{e} + m_\mathrm{p}$, $\mu = m_\mathrm{e} m_\mathrm{p}/(m_\mathrm{e} + m_\mathrm{p})$.

(b) $H = -(\hbar^2/2m_\mathrm{e})\,\nabla_\mathrm{N}^2 - (\hbar^2/2m_\mathrm{e})\,\nabla_{\mathrm{e1}}^2 - (\hbar^2/2m_\mathrm{e})\,\nabla_{\mathrm{e2}}^2$

$$- (2e^2/4\pi\varepsilon_0 r_1) - (2e^2/4\pi\varepsilon_0 r_2) + (e^2/4\pi\varepsilon_0 r_{12})$$

$H_{\mathrm{electronic}} = -(\hbar^2/2m\;)(\nabla_{\mathrm{e1}}^2 + \nabla_{\mathrm{e2}}^2) + (e^2/4\pi\varepsilon_0)\{(1/r_{12}) - (2/r_1) - (2/r_2)\}$

(c) $H_{\mathrm{electronic}} = -(\hbar^2/2m_\mathrm{e})(\nabla_{\mathrm{e1}}^2 + \nabla_{\mathrm{e2}}^2)$

$$+ (e^2/4\pi\varepsilon_0)\{-(1/r_{1\mathrm{a}}) - (1/r_{1\mathrm{b}}) - (1/r_{2\mathrm{a}}) - (1/r_{2\mathrm{b}}) + (1/r_{12}) + (1/r_{\mathrm{ab}})\}.$$

(d) $H = -(\hbar^2/2m)\,\nabla^2$.

(e) Since $F = -\mathrm{d}V/\mathrm{d}x$, $V = -\int F\,\mathrm{d}x = -Fx$ if F is uniform (independent of x). Therefore, $H = -(\hbar^2/2m)(\mathrm{d}^2/\mathrm{d}x^2) - Fx$.

EXERCISE : Write down an expression for the Hamiltonian of a general Z-electron atom.

2.3 Refer to Fig. 2.1(a). From eqn (2.2.2),

$$\phi = 2\pi L/\lambda = (2\pi/\lambda)2\sqrt{\{(\tfrac{1}{2}l)^2 + d^2\}} = (2\pi l/\lambda)\,\{1 + (2d/l)^2\}^{\frac{1}{2}}.$$

The dependence of ϕ on d is sketched in Fig. 2.1(b), where we have written $l = \alpha\lambda$ and $d = \beta\lambda$ so that

$$\phi = 2\pi\alpha\{1+(2\beta/\alpha)^2\}^{\frac{1}{2}} .$$

Since

$$\mathrm{d}\phi/\mathrm{d}\beta = \tfrac{1}{2}(2\pi\alpha)(8\beta/\alpha^2)/\{1+(2\beta/\alpha^2)\}^{\frac{1}{2}} = (8\pi\beta/\alpha)\{1+(2\beta/\alpha)^2\}^{\frac{1}{2}} ,$$

we have $\mathrm{d}\phi/\mathrm{d}\beta = 0$ when $\beta = 0$ (i.e. when $d = 0$), as was to be shown.

Fig. 2.1

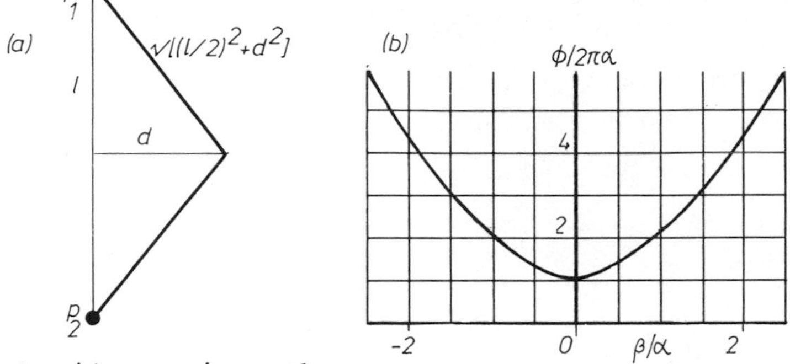

EXERCISE: Consider curving paths.

2.4 Refer to Fig. 2.2. For simplicity consider a symmetrical location of p_1 and p_2. The *true path length* when the ray passes through p' is

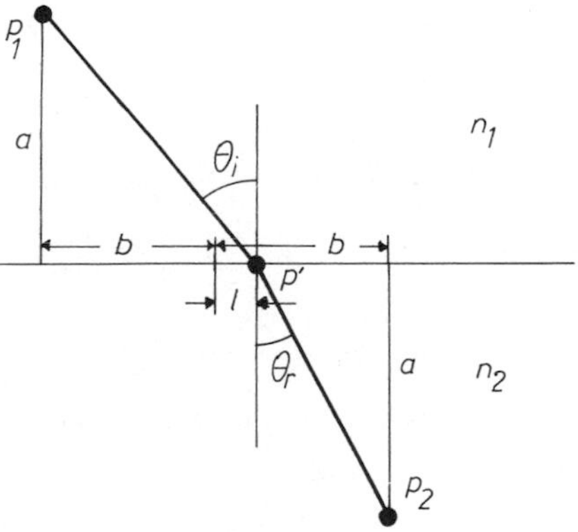

Fig. 2.2

$$\{(b+l)^2\}^{\frac{1}{2}} + \{(b-l)^2+a^2\}^{\frac{1}{2}} .$$

The *phase length* [eqn (2.2.3)] is

$$\phi = (2\pi\nu/c)\,\{n_1[(b+l)^2+a^2]^{\frac{1}{2}} + n_2[(b-l)^2+a^2]^{\frac{1}{2}}\} .$$

The minimum of ϕ occurs as $\mathrm{d}\phi/\mathrm{d}l = 0$:

$$\mathrm{d}\phi/\mathrm{d}l = (2\pi\nu/c)\left\{\frac{n_1(b+l)}{[(b+l)^2+a^2]^{\frac{1}{2}}} - \frac{n_2(b-l)}{[(b-l)^2+a^2]^{\frac{1}{2}}}\right\} = 0 .$$

Since $\qquad (b+l)/[(b+l)^2+a^2]^{\frac{1}{2}} = \sin\theta_i$

and $\qquad (b-l)/[(b-l)^2+a^2]^{\frac{1}{2}} = \sin\theta_r ,$

the minimum phase length occurs when θ_i and θ_r are related by

$$n_1 \sin\theta_i = n_2 \sin\theta_r ,$$

which is Snell's law.

EXERCISE: Consider an unsymmetrical disposition of p_1 and p_2.

2.5 $(-\hbar^2/2m)(d^2\Psi/dx^2) + V(t)\Psi = i\hbar(\partial\Psi/\partial t)$.

Try $\Psi = \psi(x)\,\theta(t)$, then

$$(-\hbar^2/2m)\,\psi''\,\theta + V(t)\,\psi\theta = i\hbar\psi\dot{\theta} ,$$

$$-(\hbar^2/2m)(\psi''/\psi) + V(t) - i\hbar\,(\dot{\theta}/\theta) = 0 .$$

By the same argument as that following eqn (2.3.4), $(-\hbar^2/2m)(\psi''/\psi) = \varepsilon$, a constant; hence

$$\psi'' = -(2m\varepsilon/\hbar^2)\,\psi . \tag{1}$$

$i\hbar(\dot{\theta}/\theta) - V(t) = \varepsilon$, the same constant; hence

$$(d/dt)\,\ln\theta = \{\varepsilon + V(t)\}/i\hbar . \tag{2}$$

Equation (1) has the solution $\psi = Ae^{ikx} + Be^{-ikx}$, $k = (2m\varepsilon/\hbar^2)^{\frac{1}{2}}$.

Equation (2) has the solution $\ln\theta(t) = \ln\theta(0) - (i/\hbar)\int_0^t \{\varepsilon + V(t)\}\,dt$.

Therefore, an absorbing $\ln\theta(0)$ into A and B,

$$\Psi = \psi(x)\exp\left\{-i\,(\varepsilon/\hbar)\,t - (i/\hbar)\int_0^t V(t)\,dt\right\}.$$

Note that $|\Psi|^2 = |\psi(x)|^2$, and so it is stationary.

Let $V(t) = V\cos\omega t$, then $\int_0^t V(t)\,dt = (V/\omega)\sin\omega t$, and so

$$\Psi = \psi(x)\exp\{-i\,(\varepsilon/\hbar)t - i(V/\hbar\omega)\sin\omega t\}$$

$$= \psi(x)(\cos\phi - i\sin\phi) , \quad \phi = \varepsilon t/\hbar + (V/\hbar\omega)\sin\omega t .$$

The behaviour of the real and imaginary parts of Ψ (essentially the functions $\cos(\tau + \sin\tau)$ and $\sin(\tau + \sin\tau)$) is shown in Fig. 2.3: the full line is $\cos(\tau + \sin\tau)$, the broken line $\sin(\tau + \sin\tau)$.

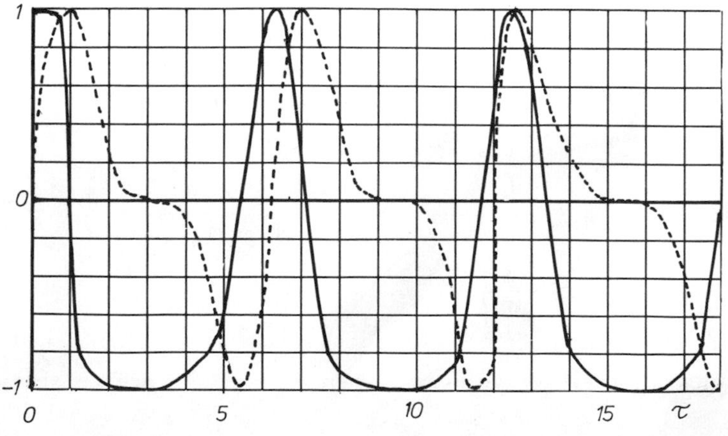

Fig. 2.3

EXERCISE: Consider the form of Ψ for an exponentially-switched cosine potential energy, $V(t) = V(1 - e^{-t/T}) \cos \omega t$, for various switching rates.

2.6 Take $H\Psi = \kappa(\partial^2\Psi/\partial t^2)$. Since H has the dimensions of energy, κ must have the dimensions of energy × time^2, or ML^2. Try $\Psi = \psi\theta$, with H an operator on x, not t. The the equation separates into $H\psi = E\psi$, $\ddot{\theta} = (E/\kappa)\theta$.

The latter admits solutions of the form $\theta \propto \cos(E/\kappa)^{\frac{1}{2}}t$. Then

$$\int |\Psi|^2 \, d\tau \propto \int |\psi|^2 \, d\tau \cos^2 (E/\kappa)^{\frac{1}{2}} t ,$$

which oscillates in time between 0 and 1; hence the total probability is not conserved.

2.7 Find a normalization constant N such that eqn (2.4.1) is satisfied.

$$\int |\psi|^2 d\tau = N^2 \int_0^\infty r^2 \, dr \int_0^{2\pi} d\phi \int_0^\pi \sin\theta \, d\theta \left\{e^{-2ar}\right\}$$

$$= N^2\{2\pi\}\{2\}\int_0^\infty r^2 e^{-2ar} \, dr = 4\pi N^2\{2!/(2a)^3\}$$

$$= N^2\pi/a^3 .$$

Hence $\underline{N = (a^3/\pi)^{\frac{1}{2}} = 1.5 \times 10^{15} \text{ m}^{-3/2}}$.

Consequently $\psi = (a^3/\pi)^{\frac{1}{2}} e^{-ar}$.

EXERCISE: ψ depends on Z as e^{-Zar}. Find N for general Z.

2.8 $$\int |\psi|^2 d\tau = N^2 \int_{-\infty}^{\infty} e^{-x^2/\Gamma^2} dx = N^2\Gamma \int_{-\infty}^{\infty} e^{-z^2} dz \, [z = x/\Gamma]$$

$$= N^2\Gamma\sqrt{\pi} .$$

Hence, $\underline{N = (1/\Gamma\sqrt{\pi})^{\frac{1}{2}}}$ and $\psi = (1/\Gamma\sqrt{\pi})^{\frac{1}{2}} e^{-x^2/2\Gamma^2}$.

$$P(-\Gamma \leqslant x \leqslant \Gamma) = \int_{-\Gamma}^{\Gamma} \psi^2 \, d\tau = (1/\Gamma\sqrt{\pi}) \int_{-\Gamma}^{\Gamma} e^{-x^2/\Gamma^2} dx$$

$$= (1/\sqrt{\pi}) \int_{-1}^{1} e^{-z^2} dz = (2/\sqrt{\pi}) \int_0^1 e^{-z^2} dz .$$

The *error function* is defined [A & S,† § 7.1.1] as

---

† A & S will denote *Handbook of Mathematical Functions*, M. Abramowitz and I.A. Stegun, Dover (1965).

$$\operatorname{erf} x = (2/\sqrt{\pi}) \int_0^x e^{-z^2}\, dz .$$

Therefore, in the present case

$$P(-\Gamma \leqslant x \leqslant \Gamma) = \operatorname{erf} 1 = \underline{0.8427 \ldots}$$

EXERCISE: Evaluate $P(x \geqslant 0)$ and $P(\Gamma \leqslant x \leqslant 2\Gamma)$.

2.9 The most probable location is given by the value of x corresponding the maximum (or maxima) of $|\psi|^2$; write this location x_*. In the present case

$$|\psi|^2 = N^2 x^2 e^{-x^2/\Gamma^2}$$

$$(d/dx)\,|\psi|^2 = N^2 \left\{2x\, e^{-x^2/\Gamma^2} - 2(x^3/\Gamma^2)\, e^{-x^2/\Gamma^2}\right\} = 0 \quad \text{at} \quad x = x_* .$$

Hence, $1 - x_*^2/\Gamma^2 = 0$, so that $\underline{x_* = \pm\,\Gamma}$.

EXERCISE: Evaluate N for the wavefunction. Consider then another excited state wavefunction $\{2(x/\Gamma)^2 - 1\}\, e^{-x^2/2\Gamma^2}$, and locate x_*.

2.10 Base the answer on $|\psi|^2 = (a^3/\pi)\, e^{-2ar}$. The probability densities are

(a) $|\psi(0)|^2 = a^3/\pi = 1/(53\,\text{pm})^3\,\pi = \underline{2.1 \times 10^{-6}\ \text{pm}^{-3}}$,

(b) $|\psi(r = 1/a, \theta, \phi)|^2 = (a^3/\pi)\, e^{-2} = \underline{2.9 \times 10^{-7}\ \text{pm}^{-3}}$.

[The values of θ and ϕ do not matter because ψ is spherically symmetrical.] The probabilities are given by

$$P = \int_{\text{volume}} \psi^2\, d\tau \approx |\psi|^2\, \delta V$$

because $|\psi|^2$ is virtually constant over the small volume of integration $\delta V = 1\ \text{pm}^3$. Hence:

(a) $P = |\psi(0)|^2\, \delta V = \underline{2.1 \times 10^{-6}}$. (b) $P = |\psi(1/a, \theta, \phi)|^2\, \delta V = \underline{2.9 \times 10^{-7}}$.

For the probability of being in a sphere of radius R centred on the nucleus, allow for the r-dependence of ψ:

$$P = \int_{\text{volume}} \psi^2\, d\tau = (a^3/\pi) \int_0^{2\pi} d\phi \int_0^{\pi} \sin\theta\, d\theta \int_0^R r^2\, dr \left\{e^{-2ar}\right\}$$

$$= (a^3/\pi)\{2\pi\}\{2\} \int_0^R r^2\, e^{-2ar}\, dr = 4 \int_0^{aR} z^2\, e^{-2z}\, dz$$

$$= 1 - \left\{1 + aR + 2(aR)^2\right\} e^{-2aR} .$$

Therefore, for $R = 1/a$,

$$P = 1 - \left\{1 + 1 + 2\right\} e^{-2} = 1 - 4\,e^{-2} = \underline{0.459}\,.$$

Find the 90 percent boundary surface by solving

$$0.90 = 1 - \left\{1 + aR + 2(aR)^2\right\} e^{-2aR};$$

That is, solve

$$\left\{1 + aR + 2(aR)^2\right\} e^{-2aR} = 0.10.$$

Do this numerically (e.g., by succesive approximation, graphically, or by a library program). The outcome is $aR = 2.557$, so that $R = 135$ pm.

EXERCISE: Find the expression for the Z-dependence of the location of the boundary surface. Repeat the calculation for a hydrogenic 2s-orbital (Tables 3.2 and 4.1).

3 Exact solutions: linear motion

3.1 Use eqn (3.1.2); for the energy in (a) use $E = e\Delta\phi$.

(a) $k = (2m_e e\Delta\phi/\hbar^2)^{\frac{1}{2}} = (5.123\times 10^9\ \text{m}^{-1})\times(\Delta\phi/\text{V})^{\frac{1}{2}}$.

(i) $\Delta\phi = 1\ \text{V}$; $k = 5.123\times 10^9\ \text{m}^{-1} = 5.123\ \text{nm}^{-1}$;

$\underline{\psi(x) = A\exp\{5.123\,\text{i}(x/\text{nm})\}}$, $A = 1/L^{\frac{1}{2}}$, $L\to\infty$.

(ii) $\Delta\phi = 10\ \text{kV}$; $k = 5.123\times 10^7\ \text{m}^{-1} = 512.3\ \text{nm}^{-1}$;

$\underline{\psi(x) = A\exp\{512.3\,\text{i}(x/\text{nm})\}}$.

(b) Since $p = (1\ \text{g})\times(10\ \text{m s}^{-1}) = 10^{-2}\ \text{kg m s}^{-1}$,
$k = p/\hbar$ [eqn (3.1.4)] $= 9.48\times 10^{31}\ \text{m}^{-1}$; hence

$$\underline{\psi(x) = A\exp\{9.48\,\text{i}\times 10^{31}(x/\text{m})\}}.$$

EXERCISE: What value of $\Delta\phi$ is needed to accelerate an electron so that its wavelength is equal to its Compton wavelength?

3.2 In each case $\underline{|\psi(x)|^2 = A^2}$, a constant ($A^2 = 1/L$; $L\to\infty$)

3.3 From eqns (3.1.7) and (3.1.6),

$$\Psi(x,t) = \int g(k)\Psi_k(x,t)\,\text{d}k = AB\int_{k-\frac{1}{2}\delta k}^{k+\frac{1}{2}\delta k}\exp\left\{\text{i}kx - \text{i}k^2\hbar t/2m\right\}\text{d}k$$

$$\Psi(x,0) = AB\int_{k-\frac{1}{2}\delta k}^{k+\frac{1}{2}\delta k}\exp\{\text{i}kz\}\,\text{d}k$$

$$= (AB/\text{i}x)\left\{\text{e}^{\text{i}(k+\frac{1}{2}\delta k)x} - \text{e}^{\text{i}(k-\frac{1}{2}\delta k)x}\right\}$$

$$= (AB\,\text{e}^{\text{i}kx}/\text{i}x)\left\{\text{e}^{\frac{1}{2}\text{i}\delta kx} - \text{e}^{-\frac{1}{2}\text{i}\delta kx}\right\} = 2AB(\text{e}^{\text{i}kx}\sin\tfrac{1}{2}\delta kx)/x.$$

$$|\Psi(x,0)|^2 = 4A^2B^2(\sin\tfrac{1}{2}\delta kx/x)^2.$$

For normalization (to unity), write $AB = N$; then

$$\int|\Psi|^2\,\text{d}\tau = 4N^2\int_{-\infty}^{\infty}(\sin\tfrac{1}{2}\delta kx/x)^2\,\text{d}x = 2N^2\delta k\int_{-\infty}^{\infty}(\sin z/z)^2\,\text{d}z \qquad [z = \tfrac{1}{2}\delta kx]$$

$$= 2N^2\delta k\sqrt{\pi} = 1;\quad \text{hence } N = (1/2\delta k\sqrt{\pi})^{\frac{1}{2}}.$$

Therefore, $\underline{\Psi(x,0) = (2/\delta k\sqrt{\pi})^{\frac{1}{2}}(\text{e}^{\text{i}kx}\sin\tfrac{1}{2}\delta kx)/x}$.

$$|\Psi(x,0)|^2 = (2/\delta k\sqrt{\pi})\,(\sin\tfrac{1}{2}\delta k x/x)^2$$

$$|\Psi(0,0)|^2 = (2/\delta k\sqrt{\pi})\lim_{x\to\infty}(\sin\tfrac{1}{2}\delta k x/x)^2 = (2/\delta k\sqrt{\pi})(\tfrac{1}{2}\delta k x/x)^2$$

we seek the value of x for which $|\Psi(x,0)|^2/|\Psi(0,0)|^2 = \frac{1}{2}$; that is

$$(\sin\tfrac{1}{2}\delta k x/x)^2/(\tfrac{1}{2}\delta k)^2 = \frac{1}{2},$$

or

$$(\sin\tfrac{1}{2}\delta k x)/(\tfrac{1}{2}\delta k x) = 1/\sqrt{2},$$

which is satisfied by $\frac{1}{2}\delta k x = \pm 1.392$ [solve numerically]. Hence the probability density falls to one half its value at $x = 0$ when $x = \pm 2.784/\delta k$.

From the uncertainty principle $\delta p\,\delta x \geqslant \frac{1}{2}\hbar$, so that $\delta k\,\delta x \geqslant \frac{1}{2}$, and hence $\delta x \geqslant 0.5/\delta k$ which is in accord with $\delta x \approx 2\times 2.784/\delta k$.

EXERCISE: Examine the properties of a Gaussian wavepacket in the same way.

3.4 $|\Psi(x,0)|^2$ is plotted in Fig. 3.1(a). For $t > 0$

$$\Psi(x,t) = N\int_{k-\frac{1}{2}\delta k}^{k+\frac{1}{2}\delta k} \exp\left\{ikx - ik^2\hbar t/2m\right\} dk$$

$$\approx N\int_{k-\frac{1}{2}\delta k}^{k+\frac{1}{2}\delta k}\left\{1 - ik^2\hbar t/2m\right\} e^{ikx}\,dk \qquad [e^x = 1 + x + \ldots]$$

$$\approx \Psi(x,0) - i(\hbar t/2m)N\int_{k-\frac{1}{2}\delta k}^{k+\frac{1}{2}\delta k} k^2 e^{ikx}\,dk$$

$$\approx \Psi(x,0) - i(\hbar t/m)Nf(x)(e^{ikx}/x)$$

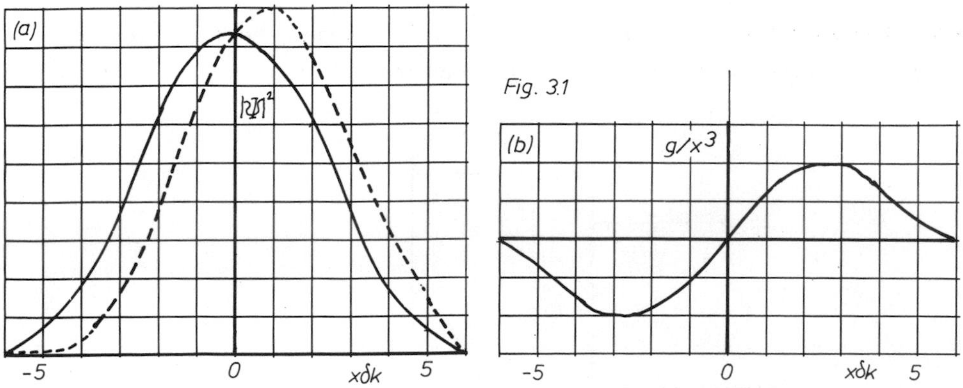

Fig. 3.1

with

$$f(x) = \tfrac{1}{2}\,x\,e^{-ikx}\int_{k-\frac{1}{2}\delta k}^{k+\frac{1}{2}\delta k} k^2\,e^{ikx}\,dk = f'(x) + i f''(x) \quad [f' = \mathrm{re}\,f\,,\ f'' = \mathrm{im}\,f]$$

and

$$f'(x) = \left(k^2 + \tfrac{1}{4}\,\delta k^2 - \frac{2}{x^2}\right)\sin\tfrac{1}{2}\delta kx + (\delta k/x)\cos\tfrac{1}{2}\,\delta kx$$

$$f''(x) = 2(k/x)\,\sin\tfrac{1}{2}\,\delta kx - k\delta k\,\cos\tfrac{1}{2}\,\delta kx\ .$$

Then, to first order in t:

$$|\Psi(x,t)|^2 = \left\{\Psi(x,0) - i(\hbar t/m) f(x)(e^{ikx}/x)\right\}$$

$$\times\left\{\Psi(x,0)^* + i(\hbar t/m)\,f^*(x)(e^{-ikx}/x)\right\}$$

$$\approx |\Psi(x,0)|^2 + (i\hbar t/m)\left\{\Psi(x,0)f^*(x)\,e^{-ikx} - \Psi(x,0)^* f(x)\,e^{ikx}\right\}/x$$

$$\approx |\Psi(x,0)|^2 + (4\hbar t/m)\,Nf''(x)\,\sin\tfrac{1}{2}\,\delta kx/x^2\ ,$$

$$|\Psi(x,t)|^2 - |\Psi(x,0)|^2 \approx k(4\hbar t/m)\,Ng(x)/x^3\ ,$$

$$g(x) = 1 - \cos\delta kx - \tfrac{1}{2}\,x\delta k\,\sin\delta kx\ .$$

The function $g(x)/(\delta kx)^3$ is plotted in Fig. 3.1(b) and its influence on $|\Psi(x,0)|^2$ is indicated by the broken line in Fig. 3.1(a), corresponding to a time t for which $4k\hbar tN/m \approx 1/\delta k^3$, so that

$$|\Psi(x,t)|^2 = |\Psi(x,0)|^2 + g(x)/(\delta kx)^3\ .$$

EXERCISE: Continue the development of the Gaussian wavepacket in the same way.

3.5 Consider the zones set out in Fig. 3.2; impose the condition of continuity of ψ and ψ' at each interface.

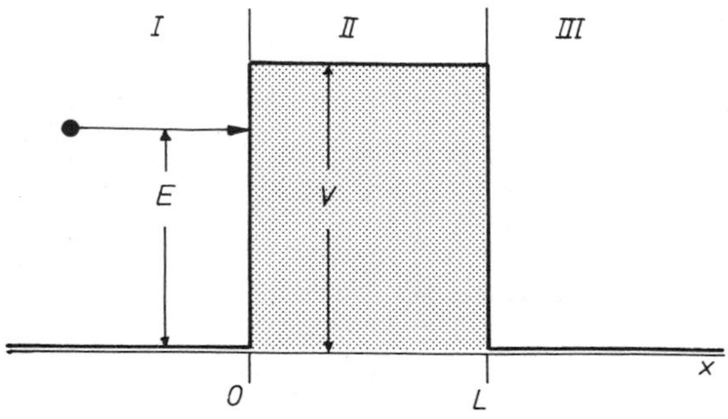

Fig. 3.2

$$\left.\begin{aligned}\psi_{\mathrm{I}} &= A\,\mathrm{e}^{\mathrm{i}kx} + B\mathrm{e}^{-\mathrm{i}kx}, \quad k^2 = 2mE/\hbar^2 \\ \psi_{\mathrm{II}} &= A'\mathrm{e}^{\mathrm{i}k'x} + B'\mathrm{e}^{-\mathrm{i}k'x}, \quad k'^2 = 2m(E-V)/\hbar^2\end{aligned}\right\} \quad \gamma = k/k'$$

$$\psi_{\mathrm{III}} = A''\mathrm{e}^{\mathrm{i}kx} \quad \text{[no particles incident from right]}$$

(1) $A + B = A' + B'$, $\quad$ [from $\psi_{\mathrm{I}}(0) = \psi_{\mathrm{II}}(0)$].

(2) $A'\,\mathrm{e}^{\mathrm{i}k'L} + B'\,\mathrm{e}^{-\mathrm{i}k'L} = A''\mathrm{e}^{\mathrm{i}kL}$, $\quad$ [from $\psi_{\mathrm{II}}(L) = \psi_{\mathrm{III}}(L)$].

(3) $kA - kB = k'A' - k'B'$, $\quad$ [from $\psi'_{\mathrm{I}}(0) = \psi'_{\mathrm{II}}(0)$].

(4) $k'A'\mathrm{e}^{\mathrm{i}k'L} - k'B'\mathrm{e}^{-\mathrm{i}k'L} = kA''\mathrm{e}^{\mathrm{i}kL}$, $\quad$ [from $\psi'_{\mathrm{II}}(L) = \psi'_{\mathrm{III}}(L)$].

From (1) and (3) :

$$A' = \tfrac{1}{2}(1+\gamma)A + \tfrac{1}{2}(1-\gamma)B; \qquad B' = \tfrac{1}{2}(1-\gamma)A + \tfrac{1}{2}(1+\gamma)B.$$

From (2) and (4)

$$A'' = A'\mathrm{e}^{(k'-k)L} + B'\mathrm{e}^{-\mathrm{i}(k'+k)L}$$

$$\gamma A'' = A'\mathrm{e}^{(k'-k)L} - B'\mathrm{e}^{-\mathrm{i}(k'+k)L},$$

so that

$$\tfrac{1}{2}(1+\gamma)A'' = A'\mathrm{e}^{\mathrm{i}(k'-k)L}, \quad \tfrac{1}{2}(1-\gamma)A'' = B'\,\mathrm{e}^{-\mathrm{i}(k'+k)L}.$$

Then

$$A''\mathrm{e}^{\mathrm{i}kL}\left\{(1+\gamma)^2\mathrm{e}^{-\mathrm{i}k'L} - (1-\gamma)^2\,\mathrm{e}^{\mathrm{i}k'L}\right\} = 4\gamma A$$

$$A''/A = 2\gamma\mathrm{e}^{-\mathrm{i}kL}\Big/\left\{2\gamma\cos k'L - \mathrm{i}(1+\gamma^2)\sin k'L\right\}.$$

The transmission coefficient (or tunnelling probability) is

$$P = |A''|^2/|A|^2 = |A''/A|$$

$$= \underline{4\gamma/\{4\gamma^2 + (1-\gamma^2)\sin^2 k'L\}}, \quad \gamma^2 = E/(E-V).$$

EXERCISE: Find the transmission coefficient for a particle incident on a rectangular dip in the potential energy.

3.6 Consider the zones set out in Fig. 3.3(a). Impose the conditions of continuity of ψ and ψ' at the single interface.

$$\psi_{\mathrm{I}} = A\mathrm{e}^{\mathrm{i}kx} + B\,\mathrm{e}^{-\mathrm{i}kx}, \qquad k^2 = 2mE/\hbar^2.$$

$$\psi_{\mathrm{II}} = A'\,\mathrm{e}^{\mathrm{i}k'x}, \qquad k'^2 = 2m(E-V)/\hbar^2.$$

(1) $A + B = A'$. $\quad$ [from $\psi_{\mathrm{I}}(0) = \psi_{\mathrm{II}}(0)$]

(2) $\mathrm{i}k(A-B) = \mathrm{i}k'A'$ $\quad$ [from $\psi'_{\mathrm{I}}(0) = \psi'_{\mathrm{II}}(0)$]

Note that

(a) $E < V$, $k' = \mathrm{i}(2m/\hbar^2)^{\frac{1}{2}} \surd(V-E) = \mathrm{i}\kappa$, κ real

(b) $E \geqslant V$, $k' = (2m/\hbar^2)^{\frac{1}{2}} \surd(E-V)$, k' real .

Fig. 3.3

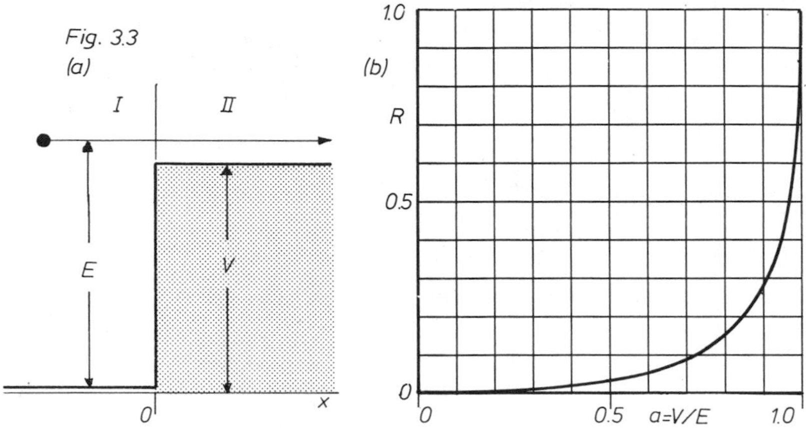

Therefore the second condition becomes :

(2a) $\mathrm{i}k(A-B) = -\kappa A'$, or $-\mathrm{i}\lambda(A-B) = A'$ with $\lambda = k/\kappa$.

Hence

$$B/A = -(1+\mathrm{i}\lambda)/(1-\mathrm{i}\lambda), \text{ so that } \underline{R = |B/A|^2 = 1} \text{ for } E < V.$$

(2b) $k(A-B) = k'A'$, or $\gamma(A-B) = A'$ with $\gamma = k/k'$.

Hence

$$B/A = (1-\gamma)/(1+\gamma),$$

$$R = (1-\gamma)^2/(1+\gamma)^2 = (k'-k)^2/(k'+k)^2$$

$$= \left\{\frac{(E-V)^{\frac{1}{2}} - E^{\frac{1}{2}}}{(E-V)^{\frac{1}{2}} + E^{\frac{1}{2}}}\right\}^2 = \underline{\left\{\frac{1-\surd(1-a)}{1+\surd(1-a)}\right\}^2}, \text{ with } a = V/E \text{ and } E > V.$$

Note that R is independent of m. If $E \approx 10.1$ eV and $V \approx 10.0$ eV, we have $\underline{R \approx 0.90}$. The function R is plotted in Fig. 3.3(b).

EXERCISE: Consider particles incident from the other side of the step with $E > V$. What is the reflection coefficient ?

3.7 Use the normalized wavefunctions in eqn (3.3.5) :

$$\psi_n = (2/L)^{\frac{1}{2}} \sin(n\pi x/L); \text{ also use}$$

$$\int \sin^2 ax \, \mathrm{d}x = \tfrac{1}{2}x - (1/4a)\sin 2ax .$$

(a) $P_n = \int_0^{\frac{1}{2}L} \psi_n^2 \, dx = (2/L) \int_0^{\frac{1}{2}L} \sin^2 (n\pi x/L) \, dx = \frac{1}{2}$ for all n.

(b) $P_n = \int_0^{\frac{1}{4}L} \psi_n^2 \, dx = (2/L) \int_0^{\frac{1}{4}L} \sin^2(n\pi x/L) \, dx = \frac{1}{4}\left\{1-(2/\pi n)\sin(\tfrac{1}{2}n\pi)\right\}$.

$P_1 = \frac{1}{4}\{1-(2/\pi)\} = 0.090\,85$.

(c) $P_n = \int_{\frac{1}{2}L-\delta}^{\frac{1}{2}L+\delta} \psi_n^2 \, dx = (2/L) \int_{\frac{1}{2}L-\delta}^{\frac{1}{2}L+\delta} \sin^2(n\pi x/L) \, dx$

$= (2/L)\{\delta-(L/2\pi n)\cos(n\pi)\sin(2n\pi\delta/L)\}$

$= (2/L)\{\delta-(-1)^n(L/2\pi n)\sin(2n\pi\delta/L)\}$.

$P_1 = (2/L)\{\delta+(L/2\pi)\sin(2\pi\delta/L)\} \approx 4\delta/L$ when $\delta/L \ll 1$.

Note that

$$\lim_{n\to\infty} P_n = \text{(a) } \tfrac{1}{2}, \quad \text{(b) } \tfrac{1}{4}, \quad \text{(c) } 2\delta/L,$$

the last corresponding to a uniform distribution (the classical limit).

EXERCISE: Find P_n (and P_1) for the particle being in a short region of length δ centred on the general point x.

3.8 The zero-point energy is $E_1 = h^2/8m_e L^2$ [eqn (3.3.5)]; hence, for $E_1 = m_e c^2$ we require $L = (h/m_e c)/\surd 8 = \lambda_C/\surd 8$, where λ_C is the Compton wavelength ($\lambda_C = 2.43$ pm). Hence $L = 0.86$ pm.

EXERCISE: At what length does the first excitation energy of a proton in a box equal its rest mass ?

3.9 Since $E_n = n^2 h^2/8mL^2$ [eqn (3.3.5)],

$F = -(dE_n/dL) = n^2h^2/4mL^3$.

For an electron ($m=m_e$) with $n=1$, $F = h^2/4m_e L^3$, hence

$L = (h^2/4m_e F)^{\frac{1}{3}} = 0.4939 \text{ pm}/(F/\text{N})^{\frac{1}{3}}$;

hence when $F = 1.0$ N, L = 0.49 pm.

EXERCISE: Consider the case of N particles in a cubic box. Find an expression for the product pV (p : pressure, V : volume). Discuss the relation of this result with the perfect gas equation of state.

3.10 Use the wavefunction $\psi_n = (2/L)^{\frac{1}{2}}\sin(n\pi x/L)$ and the integral

$$\int x\sin^2 ax \, dx = (1/4a^2)\left\{a^2x^2 - ax\sin(2ax) - \tfrac{1}{2}\cos(2ax)\right\}.$$

$$\langle x\rangle_n = \int_0^L x\,\psi_n^2\,\mathrm{d}x = (2/L)\int_0^L x\sin^2(n\pi x/L)\,\mathrm{d}x$$

$$= (L/2n^2\pi^2)\left\{n^2\pi^2 - n\pi\sin(2n\pi) - \tfrac{1}{2}[\cos(2n\pi)-1]\right\} = \underline{\tfrac{1}{2}L}.$$

The result is also obvious, by symmetry.

EXERCISE: Evaluate $\langle x\rangle$ when the particle is in the normalized mixed state $\psi_1\cos\beta + \psi_2\sin\beta$. Account for its dependence on the parameter β.

3.11 Use the integral

$$\int x^2\sin^2 ax\,\mathrm{d}x = \tfrac{1}{6}x^3 + (1/4a^3)\left\{\tfrac{1}{2}\sin(2ax) - ax\cos(2ax) - a^2x^2\sin(2ax)\right\}.$$

$$\langle x^2\rangle_n = (2/L)\int_0^L x^2\sin^2(n\pi x/L)\,\mathrm{d}x = \underline{\tfrac{1}{3}L^2\left\{1-(3/2n^2\pi^2)\right\}}.$$

$$\langle x^2\rangle_1 = \tfrac{1}{3}L^2\left\{1-(3/2\pi^2)\right\} = 0.283\,L^2.$$

$$\langle x^2\rangle_\infty = \lim_{n\to\infty}\langle x^2\rangle_n = \tfrac{1}{3}L^2.$$

Classically,

$$\langle x^2\rangle = (1/L)\int_0^L x^2\,\mathrm{d}x = \tfrac{1}{3}L^2 = \langle x^2\rangle_\infty$$

$$\delta x_n = \surd\left\{\langle x^2\rangle_n - \langle x\rangle_n^2\right\} = \surd\left\{\tfrac{1}{3}L^2 - (1/2n^2\pi^2)L^2 - \tfrac{1}{4}L^2\right\}$$

$$= \underline{(L/2\surd 3)\left\{1-(6/n^2\pi^2)\right\}^{\frac{1}{2}}}.$$

$$\delta x_\infty = \lim_{n\to\infty}\delta x_n = L/2\surd 3.$$

Classically,

$$\delta x = \surd\left\{\tfrac{1}{3}L^2 - \tfrac{1}{4}L^2\right\} = L/2\surd 3.$$

EXERCISE: Evaluate $\langle x^2\rangle$ and δx for the mixed state $\psi_1\cos\beta + \psi_2\sin\beta$.

3.12 Intuitive solution: $\langle p\rangle_n = 0$ because the wavefunction is a standing wave.

Elegant solution [draw on material in Chapter 5] :

$\langle p\rangle = \langle n|p|n\rangle = \langle n|p|n\rangle^*$ [hermiticity] $= \langle n|p^*|n\rangle = -\langle n|p|n\rangle$ [$p^* = -p$].

Therefore since $\langle p\rangle = -\langle p\rangle$, $\langle p\rangle = 0$.

Straightforward solution:

$$\langle p\rangle_n = (\hbar/i)(2/L)\int_0^L \sin(n\pi x/L)(\mathrm{d}/\mathrm{d}x)\sin(n\pi x/L)\,\mathrm{d}x$$

$$= (2\hbar/iL)(n\pi/L)\int_0^L \sin(n\pi x/L)\cos(n\pi x/L)\,\mathrm{d}x = \underline{0}.$$

$$\langle p^2\rangle_n = 2mE_n = \underline{n^2h^2/4L^2}\ .$$

$$\delta p_n = \sqrt{\left\{\langle p^2\rangle_n - \langle p\rangle_n^2\right\}} = \sqrt{\langle p^2\rangle_n} = \underline{nh/2L}\ .$$

Using the value of δx_n from Problem 3.11:

$$\delta x_n\,\delta p_n = (L/2\sqrt{3})\left\{1-(6/n^2\pi^2)\right\}^{\frac{1}{2}}\ (nh/2L)$$

$$= (n/4\sqrt{3})\left\{1-(6/n^2\pi^2)\right\}^{\frac{1}{2}}h = \underline{(n\pi/\sqrt{3})\left\{1-(6/n^2\pi^2)\right\}^{\frac{1}{2}}(\hbar/2)}\ .$$

$$\delta x_1\,\delta p_1 = (\pi/\sqrt{3})\left\{1-(6/\pi^2)\right\}^{\frac{1}{2}}(\hbar/2) = \underline{1.1357\,(\hbar/2)} > \hbar/2\ ,$$

as required.

EXERCISE: Repeat the calculation for the mixed state $\psi_1\cos\beta+\psi_2\sin\beta$. What value of β minimizes the uncertainty product ?

3.13 Refer to Fig. 3.4(a). Consider the case $E < V$.

Fig. 3.4

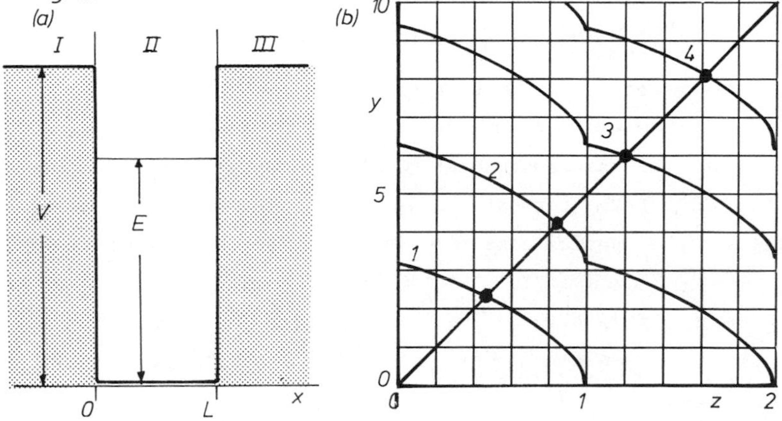

$$\psi_{\mathrm{I}} = A\,e^{-\kappa x} + Be^{\kappa x}\ , \qquad \kappa^2 = 2m(V-E)/\hbar^2$$

$$\psi_{\mathrm{II}} = A'\,e^{ikx} + B'\,e^{-ikx}\ , \qquad k^2 = 2mE/\hbar^2$$

$$\psi_{\mathrm{III}} = A''e^{-\kappa x} + B''e^{\kappa x}\ , \qquad \kappa^2 = 2m(V-E)/\hbar^2\ .$$

Since $\psi < \infty$ everywhere, $A=0$, $B''=0$ [consider $x\to-\infty$ and $x\to\infty$ respectively]. At the interfaces of the zones :

$$\psi'_{\mathrm{I}}(0)\,/\,\psi_{\mathrm{I}}(0) = -\kappa(A-B)/(A+B) = \kappa \qquad [A=0]$$

$$\psi'_{\mathrm{II}}(0)/\psi_{\mathrm{II}}(0) = ik(A'-B')/(A'+B')$$

$$\psi'_{\mathrm{II}}(L)/\psi_{\mathrm{II}}(L) = ik(A'e^{ikL} - B'e^{-ikL})/(A'e^{ikL} + B'e^{-ikL})$$

$$\psi'_{\mathrm{III}}(L)/\psi_{\mathrm{III}}(L) = -\kappa(A''e^{-\kappa L} - B''e^{\kappa L})/(A''e^{-\kappa L} + B''e^{\kappa L}) = -\kappa \quad [B''=0].$$

Since ψ'/ψ is continuous at each boundary,

$$(A' - B')/(A' + B') = \kappa/ik = -i\kappa/k = -i\gamma \qquad [\gamma = \kappa/k]$$

$$\left(A' e^{ikL} - B' e^{-ikL}\right)\Big/\left(A' e^{ikL} + B' e^{-ikL}\right) = -\kappa/ik = i\kappa/k = i\gamma$$

This pair of equations solves to

$$(1 + i\gamma)A' = (1 - i\gamma)B' , \qquad (1 - i\gamma)A' e^{ikL} = (1 + i\gamma)B' e^{-ikL} .$$

It follows that

$$(1 - \gamma^2)\sin kL - 2\gamma\cos kL = 0 , \quad \text{or} \quad \tan kL = 2\gamma/(1 - \gamma^2) .$$

Then, since

$$\tan kL = 2\tan(\tfrac{1}{2}kL)/[1 - \tan^2(\tfrac{1}{2}kL)], \quad \tan(\tfrac{1}{2}kL) = \gamma .$$

Consequently, $\cos(\frac{1}{2}kL) = 1/(1 + \gamma^2)^{\frac{1}{2}} = \hbar k/(2mV)^{\frac{1}{2}}$.

Therefore, $\quad kL = 2\arccos\left\{\hbar k/(2mV)^{\frac{1}{2}}\right\} + n\pi , \quad n = 0, 1, \ldots$

But $\arccos z = \frac{1}{2}\pi - \arcsin z$, and so

$$kL + 2\arcsin\left\{\hbar k/(2mV)^{\frac{1}{2}}\right\} = n\pi , \quad n = 1, 2, \ldots$$

Solve this equation for k by plotting $y = kL$ and

$$y = n\pi - 2\arcsin(\hbar^2 k^2/2mV)^{\frac{1}{2}} \quad \text{for} \quad n = 1, 2, \ldots$$

and finding the values of k at which the two graphs coincide, and then form $E_n = \hbar^2 k^2/2m$ for each value of n. This is illustrated in Fig. 3.4(b) for the special case $V = 4\hbar^2/2mL^2$, so that, with $kL = z$, $y = z$ and $y = n\pi - 2\arcsin z/2$, $E_n = z_n^2(\hbar^2/2mL^2)$ with z_n the intersection value of n. (Since $E < V$, $z < 2$.) We find $z = 0.48, 0.85, 1.20, 1.61$ for $n = 1, 2, 3, 4$; hence $E/(\hbar^2/2mL^2) = 0.23, 0.72, 1.44, 2.59$, for $n = 1, 2, 3, 4$.

When V is large in the sense $2mV \gg \hbar^2 k^2$, $\arcsin(\hbar^2 k^2/2mV)^{\frac{1}{2}} \approx 0$. Hence the equation to solve is $kL \approx n\pi$. Consequently $E_n \approx n^2 h^2/8mL^2$, in accord with the infinity deep square-wall solutions.

EXERCISE: First consider the special case $V = 6\hbar^2/2mL^2$, and find the allowed solutions. Then repeat the calculation for an unsymmetrical well in which the potential energy rises to V on the left and to $4V$ on the right.

3.14 $E_n = n^2(h^2/8m_e L^2)$ [eqn (3.3.5), $m = m_e$].

$$E_{n+1} - E_n = (2n+1)(h^2/8m_e L^2); \qquad hc/\lambda = E_{n+1} - E_n$$

Hence, $\lambda = 8m_e cL^2/(2n+1)h$.

Since $n = \frac{1}{2}N$ and $L = (N-1)R_{CC}$ [N is even],

$$\lambda = (8m_e c R_{CC}^2 / h)(N-1)^2/(N+1)$$

$$\lambda/\mathrm{nm} = 3.297 \times 10^{-3}\,(R_{CC}/\mathrm{pm})^2\,(N-1)^2/(N+1)$$

Taking $R_{CC} = 140$ pm and $N = 22$ gives $\lambda \approx 1240$ nm. In fact the rings at each end of the carotene are twisted so as to eliminate their conjugation with the chain joining them; hence take $N \approx 20$ for a better representation. Hence λ (carrots) ≈ 1166 nm. This corresponds to an absorption in the infrared, which is absurd (vegetables do absorb there, but not for that reason).

EXERCISE: The colour of tomatoes is due to a molecule resembling β-carotene, but in which the rings have undergone scission. This enables the two terminal double bands to conjugate. What colour are tomatoes if carrots are orange ?

3.15 $H = -(\hbar^2/2m)\{(\partial^2/\partial x^2) + (\partial^2/\partial y^2) + (\partial^2/\partial z^2)\}$; $H\psi = E\psi$. Write $\psi = XYZ$, then

$$H\psi = -(\hbar^2/2m)\{X''YZ + XY''Z + XYZ''\} = EXYZ,$$

$$-(\hbar^2/2m)\{(X''/X) + (Y''/Y) + (Z''/Z)\} = E.$$

Each term must be equal to a constant (by the same argument as in Section 3.4); therefore

$$-(\hbar^2/2m)\,Q''/Q = E^Q, \quad \text{or} \quad -(\hbar^2/2m)\,Q'' = E^Q Q$$

with $\quad Q = X, Y,$ and Z, and $E = E^X + E^Y + E^Z$.

The solutions, with $L_X = L_Y = L_Z = L$, are therefore

$$\psi_{n_x n_y n_z} = (2/L)^{\frac{3}{2}} \sin(n_x \pi x/L)\sin(n_y \pi y/L)\sin(n_z \pi z/L)$$

$$E_{n_x n_y n_z} = (n_x^2 + n_y^2 + n_z^2)(\hbar^2/8mL^2).$$

For othonormality,

$$\int \psi_{n'_x n'_y n'_z}\psi_{n_x n_y n_z}\,d\tau = (2/L)^3 \int_0^L \sin(n'_x \pi x/L)\sin(n_x \pi x/L)\,dx$$
$$\times \int_0^L \sin(n'_y \pi y/L)\sin(n_y \pi y/L)\,dy$$
$$\times \int_0^L \sin(n'_z \pi z/L)\sin(n_z \pi z/L)\,dz$$
$$= \begin{cases} 0 \text{ if } n'_x \neq n_x,\ n'_y \neq n_y, \text{ and } n'_z \neq n_z \\ 1 \text{ if } n'_x = n_x,\ n'_y = n_y, \text{ and } n'_z = n_z. \end{cases}$$

The degeneracy of $E = 14(\hbar^2/8mL^2)$ is equal to the number of ways of satisfying $n_x^2 + n_y^2 + n_z^2 = 14$ with n_x, n_y, n_z positive integers greater than zero. In this case the states (1,2,3), (1,3,2), (3,1,2), (3,2,1), (2,1,3), and (2,3,1) are all degenerate; hence the degeneracy is six.

EXERCISE: Show that a leaky rectangular box, in which the potential energy rises to V outside the box, is also separable. Find the energy levels when V has the value employed in Problem 3.13.

3.16 The Schrödinger equation is

$$-(\hbar^2/2m)(\mathrm{d}^2\psi/\mathrm{d}x^2) + \tfrac{1}{2}kx^2\psi = E\psi$$

Substitute $y = (m\omega/\hbar)^{\frac{1}{2}}x$ with $\omega^2 = k/m$; then $\psi'' - y^2\psi = -\lambda\psi$, with $\lambda = E/\tfrac{1}{2}\hbar\omega$ and $\psi'' = \mathrm{d}^2\psi/\mathrm{d}y^2$.

Substitute eqn. (3.5.5): $\psi = N_v H_v \mathrm{e}^{-y^2/2}$:

$$(\mathrm{d}^2/\mathrm{d}y^2)(H_v \mathrm{e}^{-y^2/2}) - y^2 H_v \mathrm{e}^{-y^2/2} = -\lambda H_v \mathrm{e}^{-y^2/2}.$$

Use

$$(\mathrm{d}^2/\mathrm{d}y^2)(H_v \mathrm{e}^{-y^2/2}) = (H'' - 2yH_v' - H_v + y^2 H_v)\,\mathrm{e}^{-y^2/2}$$

$$= (2yH_v' - 2vH_v' - H_v + y^2 H_v)\,\mathrm{e}^{-y^2/2} \qquad \text{[given]}$$

$$= \{y^2 H_v - (2v+1)H_v\}\,\mathrm{e}^{-y^2/2}.$$

Then $\{y^2 H_v - (2v+1)H_V - y^2 H_v\}\,\mathrm{e}^{-y^2/2} = -\lambda H_v \mathrm{e}^{-y^2/2}$,

so that $\lambda = 2v+1$, or $E = \tfrac{1}{2}(2v+1)\hbar\omega = (v+\tfrac{1}{2})\hbar\omega$, as required.

3.17 $\omega = (k/m)^{\frac{1}{2}}$, $E_v = (v+\tfrac{1}{2})\hbar\omega$ [eqn (3.5.4)].

$$m_H = M_r u = 1.008\,u = 1.674 \times 10^{-27}\ \mathrm{kg}; \quad \nu = \omega/2\pi.$$

$$\omega = \{(313.8\ \mathrm{N\,m^{-1}})/(1.674 \times 10^{-27}\ \mathrm{kg})\}^{\frac{1}{2}}$$

$$= \underline{4.33 \times 10^{14}\ \mathrm{s^{-1}} \mathrel{\hat{=}} 6.89 \times 10^{13}\ \mathrm{Hz}}.$$

The separation of energy levels is

$$E_{v+1} - E_v = \hbar\omega = 4.57 \times 10^{-20}\ \mathrm{J} \mathrel{\hat{=}} 27.5\ \mathrm{kJ\,mol^{-1}}.$$

Use the Boltzmann distribution, eqn (1.1.8):

$$p_v = (1/q)\,\mathrm{e}^{-\varepsilon_v/kT}, \quad q = \sum_v \mathrm{e}^{-\varepsilon_v/kT}.$$

In the present case, $\varepsilon_v = E_v = (v+\tfrac{1}{2})\hbar\omega$

$$q = e^{-\hbar\omega/2kT} \sum_{v=0}^{\infty} e^{-v\hbar\omega/kT} = e^{-\hbar\omega/2kT} \sum_{v} \left(e^{-\hbar\omega/kT}\right)^{v}$$

$$= e^{-\hbar\omega/2kT}/(1-e^{-\hbar\omega/kT}) \quad [1+x+x^2+\ldots = 1/(1-x)].$$

Therefore

$$p_v = (1/q)\, e^{-(v+\frac{1}{2})\hbar\omega/kT} = e^{-v\hbar\omega/kT}\,(1-e^{-\hbar\omega/kT}).$$

EXERCISE: Consider the effect of deuteration on ν, p_0 and p_1.

3.18 $P_v = \psi_v^2\,(x = 16.1\text{ pm})/\psi_v^2(0)$ [x is the displacement from equilibrium].

When $x = 16.1$ pm, $y = (m\omega/\hbar)^{\frac{1}{2}}x = 1.33$ [ω, m in Problem 3.17]. Use the wavefunctions

$$\psi_v = N_v H_v(y)\, e^{-y^2/2}, \quad \text{so that} \quad \psi_v^2 = N_v^2 H_v^2(y)\, e^{-y^2}$$

$$P_v = H_v^2(y)\, e^{-y^2}/H_v^2(0) \qquad \text{with} \quad y = 1.33.$$

Use Table 3.1 for the Hermite polynomials:

(a) $P_0 = \left\{H_0^2(y)/H_0^2(0)\right\} e^{-y^2} = e^{-y^2} = e^{-1.77} = \underline{0.171}$.

(b) $P_4 = \left\{H_4^2(y)/H_4^2(0)\right\} e^{-y^2} = (1/144)\left\{16y^4 - 48y^2 + 12\right\}^2 e^{-y^2} = \underline{0.620}$.

EXERCISE: Repeat the calculation for the deuterated species and $v = 2, 6$.

3.19 $\langle x\rangle = 0$ by symmetry, $\langle p\rangle = 0$ by reality [Problem 3.12]. Alternatively

$$\langle p\rangle \propto \int_{-\infty}^{\infty} e^{-y^2/2}\,(d/dy)\, e^{-y^2/2}\, dy \propto \int_{-\infty}^{\infty} y\, e^{-y^2}\, dy = 0 \quad \text{by symmetry.}$$

$$\langle x^2\rangle = N_0^2 \int_{-\infty}^{\infty} x^2 e^{-y^2}\, dy$$

$\left[\text{the wavefunctions are normalized in the sense that } \int_{-\infty}^{\infty} \psi^2\, dy = 1\right]$.

$$\langle x^2\rangle = (1/\pi^{\frac{1}{2}})(\hbar/m\omega) \int_{-\infty}^{\infty} y^2 e^{-y^2}\, dy = \tfrac{1}{2}(\hbar/m\omega) = \tfrac{1}{2}(\hbar\omega/k).$$

$$\langle p^2\rangle = \int \psi_0^*(\hbar/i)^2 (d^2/dx^2)\,\psi_0\, d\tau = -\hbar^2 N_0^2 \int_{-\infty}^{\infty} e^{-y^2/2}\,(d/dx)^2\, e^{-y^2/2}\, dy$$

$$= -\hbar^2 N_0^2\,(m\omega/\hbar) \int_{-\infty}^{\infty} e^{-y^2/2}(d/dy)^2\, e^{-y^2/2}\, dy$$

$$= -\hbar^2 N_0^2\,(m\omega/\hbar) \int_{-\infty}^{\infty} (y^2-1)\, e^{-y^2}\, dy$$

$$= -\hbar^2 N_0^2\,(m\omega/\hbar)\,\{\tfrac{1}{2}\sqrt{\pi}-\sqrt{\pi}\} = \tfrac{1}{2}\,\hbar m\omega = \tfrac{1}{2}\,\hbar k/\omega\,.$$

(Note that these results can be obtained more simply by using the virial theorem : $\langle p^2\rangle/2m = \frac{1}{2}\,k\,\langle x^2\rangle = \frac{1}{2}E = \frac{1}{4}\,\hbar\omega$.) It follows that

$$\left.\begin{aligned}\delta x &= \sqrt{\{\langle x^2\rangle - \langle x\rangle^2\}} = \sqrt{\langle x^2\rangle} = (\hbar\omega/k)^{\frac{1}{2}}/\sqrt{2}\\ \delta p &= \sqrt{\{\langle p^2\rangle - \langle p\rangle^2\}} = \sqrt{\langle p^2\rangle} = (\hbar k/\omega)^{\frac{1}{2}}/\sqrt{2}\end{aligned}\right\}\ \delta x\,\delta p = \tfrac{1}{2}\hbar\,,$$

which is in accord with $\delta x\delta p \geqslant \frac{1}{2}\hbar$. A Gaussian function is a *minimum uncertainty* function.

EXERCISE: Deduce that $\delta x_v\,\delta p_v = (v+\frac{1}{2})\,\hbar$.

4 Exact solutions: rotational motion

4.1 $E = m_l^2(\hbar^2/2I)$ [eqn (4.1.5)]; $I = m_H R^2$.

$E = m_l^2(\hbar^2/2m_H R^2)$.

$m_H = 1.674 \times 10^{-27}$ kg, $R = 160$ pm; $\hbar^2/2m_H R^2 = 1.30 \times 10^{-22}$ J.

Hence,

$\underline{E = (1.30 \times 10^{-22} \text{ J})\, m_l^2}$ and $\Delta E = (1.30 \times 10^{-22} \text{ J})(1-0) = 1.30 \times 10^{-22}$ J.

$\lambda = hc/\Delta E = 1.53 \times 10^{-3}$ m $= \underline{1.53 \text{ mm}}$.

EXERCISE: Calculate the effect of deuteration on E and λ $(1 \leftarrow 0)$.

4.2 $\int_0^{2\pi} \Phi^*_{m_l'} \Phi_{m_l}\, d\phi = (1/2\pi) \int_0^{2\pi} e^{i(m_l - m_l')\phi}\, d\phi$

$$= (1/2\pi) \left\{ \frac{e^{2i(m_l - m_l')\pi} - 1}{(m_l - m_l')\, i} \right\} = 0 \text{ if } m_l' \neq m_l$$

$[e^{2in\pi} = 1$, n an integer$]$.

(Note that when $m_l' = m_l$ the integral has the value 2π.)

EXERCISE: Normalize the wavefunction $e^{i\phi}\cos\beta + e^{-i\phi}\sin\beta$, and find an orthogonal linear combination of $e^{+i\phi}$ and $e^{-i\phi}$.

4.3 The moment of inertia of a solid uniform disc of mass M and radius R is

$$I = \tfrac{1}{2} MR^2; \text{ hence } I = 7.5 \times 10^{-4} \text{ kg m}^2.$$

Then

$$E = m_l^2(\hbar^2/2I) = \underline{(7.4 \times 10^{-66} \text{ J})\, m_l^2}.$$

33 rpm corresponds to a frequency $\nu = (33/60)$ Hz $= 0.55$ Hz. Hence $\omega = 2\pi\nu = 3.5$ s^{-1}. The angular momentum is $I\omega = 2.6 \times 10^{-3}$ kg m^2 s^{-1}. If this is set equal to $|m_l|\hbar$ we require $|m_l| = 2.5 \times 10^{31}$. Since the turntable rotates anticlockwise when seen from below, m_l is negative. Hence, $\underline{m_l = -2.5 \times 10^{31}}$.

EXERCISE: How much more energy is required to raise the disc into its next rotational state?

4.4 See Figs. 4.1(a) and 4.1(b). We have plotted

$$\text{re}\,\Phi_{m_l} = (1/2\pi)^{\frac{1}{2}} \cos m_l\phi = \begin{cases} (1/2\pi)^{\frac{1}{2}} \cos 3\phi & \text{for } m_l = 3 \\ (1/2\pi)^{\frac{1}{2}} \cos 4\phi & \text{for } m_l = 4. \end{cases}$$

Fig. 4.1

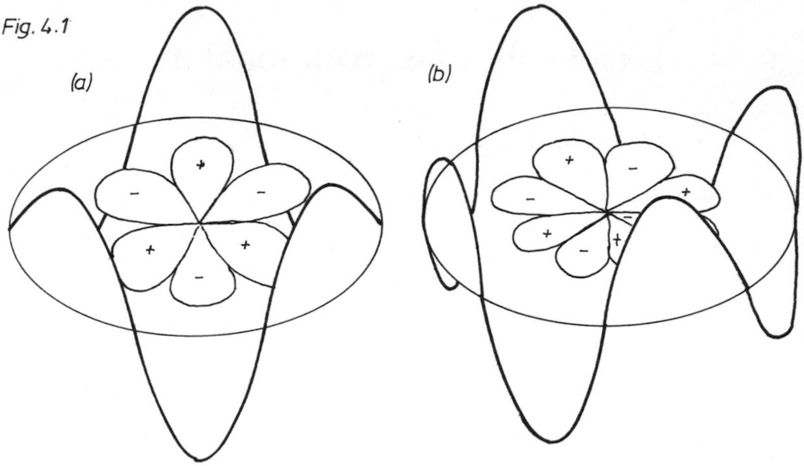

EXERCISE: Superimpose the imaginary parts of Φ_{m_l} on the diagrams. Draw re Φ for the superposition $e^{i\phi}\cos\beta + e^{-i\phi}\sin\beta$.

4.5 $\Psi = N \sum_{m_l=0}^{\infty} (1/m_l!)\, e^{im_l\phi} = N \sum_{m_l=0}^{\infty} (1/m_l)(e^{i\phi})^{m_l}$

$= N \exp(e^{i\phi}) = N \exp(\cos\phi + i\sin\phi)$.

For normalization, $\Psi^*\Psi = N^2 e^{\cos\phi - i\sin\phi} e^{\cos\phi + i\sin\phi} = N^2 e^{2\cos\phi}$;

$$\int_0^{2\pi} |\Psi|^2 d\phi = N^2 \int_0^{2\pi} e^{2\cos\phi} d\phi = 2\pi N^2 I_0(2) = 1 \quad \text{[required]}.$$

Therefore, $N = 1/\sqrt{\{2\pi I_0(2)\}} = 0.2642$ $[I_0(2) = 2.280$, given]. The form of $|\Psi|^2$ is

$$|\Psi|^2 = N^2 e^{2\cos\phi} = 0.069\,80\ e^{2\cos\phi},$$

which is plotted in Fig. 4.2.

Fig. 4.2

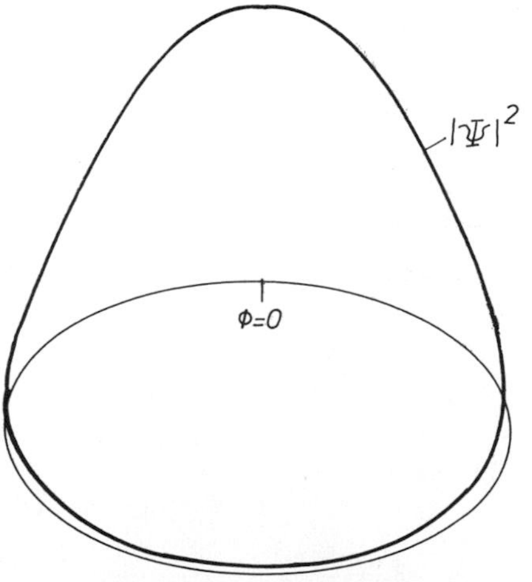

$$\langle\phi\rangle = N^2 \int_0^{2\pi} \phi\, e^{2\cos\phi}\, d\phi$$

$$= N^2 \int_{-\pi}^{\pi} \phi\, e^{2\cos\phi}\, d\phi \qquad \begin{bmatrix}\text{same integration range}\\ \text{but differently expressed}\end{bmatrix}$$

$$= \underline{0} \qquad [\phi \text{ odd under } \phi\to-\phi,\ \cos\phi \text{ even under } \phi\to-\phi].$$

$$\langle\sin\phi\rangle = N^2 \int_0^{2\pi} \sin\phi\, e^{2\cos\phi}\, d\phi = \underline{0} \qquad [\text{same argument as for } \phi].$$

$$\langle l_z\rangle = N^2 \int_0^{2\pi} e^{\cos\phi - i\sin\phi}\, l_z\, e^{\cos\phi + i\sin\phi}\, d\phi$$

$$= (\hbar/i)\, N^2 \int_0^{2\pi} e^{\cos\phi - i\sin\phi}\, (d/d\phi)\, e^{\cos\phi + i\sin\phi}\, d\phi$$

$$= (\hbar/i)\, N^2 \int_0^{2\pi} e^{\cos\phi - i\sin\phi}\, (-\sin\phi + i\cos\phi)\, e^{\cos\phi + i\sin\phi}\, d\phi$$

$$= \hbar N^2 \int_0^{2\pi} \cos\phi\, e^{2\cos\phi}\, d\phi = 2\pi\hbar N^2 I_1(2)$$

$$= \left\{I_1(2)/I_0(2)\right\}\hbar = \underline{0.698\,\hbar}.$$

$\langle l_z\rangle < \hbar$ because it is the weighted average value, and $m_l = 0$ contributes significantly to the sum.

EXERCISE: Repeat the calculation for a wavepacket found in the same way but with the omission of $m_l = 0$.

4.6 $$\Psi = N \sum_{m_l=0}^{\infty} (\alpha^{m_l}/m_l!)\, e^{im_l\phi} = N \sum_{m_l=0}^{\infty} (1/m_l!)(\alpha e^{i\phi})^{m_l}$$

$$= N\, e^{\alpha(\cos\phi + i\sin\phi)}.$$

Normalization: $$\int_0^{2\pi} |\Psi|^2\, d\phi = N^2 \int_0^{2\pi} e^{2\alpha\cos\phi}\, d\phi = 2\pi N^2 I_0(2\alpha) = 1;$$

$$N = 1/\surd\left\{2\pi I_0(2\alpha)\right\}.$$

Expectation value:

$$\langle l_z\rangle = (\hbar/i)\, N^2 \int_0^{2\pi} e^{\alpha(\cos\phi - i\sin\phi)}\, (d/d\phi)\, e^{\alpha(\cos\phi + i\sin\phi)}\, d\phi$$

$$= \alpha(\hbar/i)\, N^2 \int_0^{2\pi} e^{\alpha(\cos\phi - i\sin\phi)}\, (-\sin\phi + i\cos\phi)\, e^{\alpha(\cos\phi + i\sin\phi)}\, d\phi$$

$$= \alpha\hbar N^2 \int_0^{2\pi} \cos\phi\, e^{2\alpha\cos\phi}\, d\phi = 2\pi\alpha\hbar N^2 I_1(2\alpha) = \underline{\alpha\hbar\,\{I_1(2\alpha)/I_0(2\alpha)\}}.$$

Therefore, when α is large, $\lim_{\alpha\to\infty} I_1(2\alpha)/I_0(2\alpha) = 1$, and so $\langle l_z \rangle \sim \alpha\hbar$.

EXERCISE: The uncertainty relation for cyclic variables was quoted on p. 62 of the text: $(\delta l_z)^2\,(\delta \sin\phi)^2 \geqslant \frac{1}{4}\hbar^2 \langle \cos\phi \rangle^2$. Investigate it for the wavepacket of this Problem, and examine the limit $\alpha\to\infty$.

4.7 The Schrödinger equation is given in eqn (4.2.6):

$\Lambda^2\psi = -(2IE/\hbar^2)\,\psi$; $\Lambda^2 = (1/\sin^2\theta)(\partial^2/\partial\phi^2) + (1/\sin\theta)(\partial/\partial\theta)\sin\theta\,(\partial/\partial\theta)$.

Write $\psi = \Theta\Phi$; then with $\Theta' = d\Theta/d\theta$ and $\Phi' = d\Phi/d\phi$, etc.,

$$(1/\sin^2\theta)\,\Theta\Phi'' + (1/\sin\theta)\,\Phi(d/d\theta)\sin\theta\,\Theta' = -(2IE/\hbar^2)\,\Theta\Phi$$

$$\Phi''/\Phi + (1/\Theta)\sin\theta\,(d/d\theta)\sin\theta\Theta' = -(2IE/\hbar^2).$$

Write $\Phi''/\Phi = -m_l^2$, a constant; then

$$(1/\Theta)\sin\theta\,(d/d\theta)\sin\theta\Theta' = m_l^2 - (2IE/\hbar^2).$$

Since $(d/d\theta)\sin\theta\Theta' = \Theta'\cos\theta + \Theta''\sin\theta$, this rearranges into

$$\Theta''\sin^2\theta + \Theta'\sin\theta\cos\theta - \{m_l^2 - (2IE/\hbar^2)\}\,\Theta = 0.$$

EXERCISE: Identify this equation in A & S, and write down its solutions.

4.8 It is sufficient to show that the Y_{lm_l} satisfy $\Lambda^2 Y_{lm_l} = -l(l+1)\,Y_{lm_l}$ [eqn (4.2.8)].

(a) $Y_{11} = -\frac{1}{2}(3/2\pi)^{\frac{1}{2}}\sin\theta\, e^{i\phi}$

$$\begin{aligned}\Lambda^2\sin\theta\, e^{i\phi} &= (1/\sin^2\theta)(\partial^2/\partial\phi^2)\sin\theta\, e^{i\phi} \\ &\quad + (1/\sin\theta)(\partial/\partial\theta)\sin\theta\,(\partial/\partial\theta)\sin\theta\, e^{i\phi} \\ &= -(1/\sin\theta)\,e^{i\phi} + (1/\sin\theta)\,e^{i\phi}\,(d/d\theta)\sin\theta\cos\theta \\ &= -(1/\sin\theta)\,e^{i\phi} + (1/\sin\theta)\,e^{i\phi}\,(\cos^2\theta - \sin^2\theta) \\ &= -(1/\sin\theta)\,e^{i\phi} + (1/\sin\theta)\,e^{i\phi}(1 - 2\sin^2\theta) = -2\sin\theta\, e^{i\phi}.\end{aligned}$$

Hence, $\Lambda^2 Y_{11} = -2Y_{11}$, in accord with $l = 1$.

(b) $Y_{20} = \frac{1}{4}(5/\pi)^{\frac{1}{2}}\,(3\cos^2\theta - 1)$.

$$\begin{aligned}\Lambda^2(3\cos^2\theta - 1) &= (1/\sin\theta)\,(d/d\theta)\sin\theta\,(d/d\theta)\,(3\cos^2\theta - 1) \\ &= -6(1/\sin\theta)\,(d/d\theta)\sin^2\theta\cos\theta \\ &= -6(1/\sin\theta)\,\{2\sin\theta\cos^2\theta - \sin^3\theta\} \\ &= -6\,\{2\cos^2\theta - \sin^2\theta\} = -6(3\cos^2\theta - 1).\end{aligned}$$

Hence, $\Lambda^2 Y_{20} = -6\,Y_{20}$, in accord with $l = 2$.

(a) $\int |Y_{11}|^2 \, d\tau = \frac{1}{4}(3/2\pi) \int_0^{\pi} \sin^2\theta \sin\theta \, d\theta \int_0^{2\pi} d\phi$

$= \frac{3}{4} \int_{-1}^{1} (1-x^2) \, dx \quad [x = \cos\theta] = 1.$

(b) $\int |Y_{20}|^2 \, d\tau = \frac{1}{16}(5/\pi) \int_0^{\pi} (3\cos^2\theta - 1)^2 \sin\theta \, d\theta \int_0^{2\pi} d\phi$

$= (5/8) \int_{-1}^{1} (3x^2 - 1)^2 \, dx = 1.$

(c) $\int Y_{11}^* Y_{20} \, d\tau \propto \int_0^{2\pi} e^{-i\phi} \, d\phi = 0.$

EXERCISE: Repeat the calculation for Y_{21} and Y_{31}.

4.9 From eqn (4.2.10), $E = l(l+1)(\hbar^2/2I) = (1.3 \times 10^{-22}\ \mathrm{J})\ l(l+1)$ [Problem 4.1].

Draw up the following Table, using degeneracy $g_l = 2l + 1$:

l	$E/10^{-22}$ J	g_l
0	0	1
1	2.60	3
2	7.80	5

$\Delta E\,(1-0) = 2.60 \times 10^{-22}$ J ,

$\lambda\,(1-0) = hc/\Delta E\,(1-0) = 7.64 \times 10^{-4}$ m $=$ 0.764 mm.

EXERCISE: Calculate the same quantities for the deuterated species.

4.1 Refer to Fig. 4.3.

$\cos\theta = m_l / \surd\{l(l+1)\}$, hence

$\theta = \arccos\{m_l/\surd[l(l+1)]\}$.

Fig. 4.3

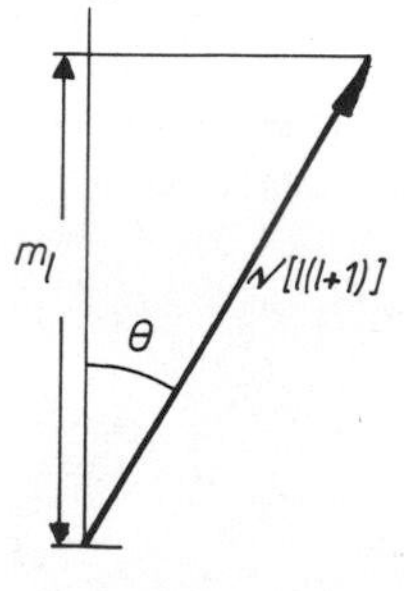

The minimum angle is obtained when m_l takes its maximum value ($m_l = l$).

$\theta_{min} = \arccos\{l/\surd[l(l+1)]\}$

$= \arccos\{\surd[l/(l+1)]\}$.

$\lim_{l\to\infty} \theta_{min} = \arccos\{\surd[l/l]\} = \arccos 1 = \underline{0}$.

Draw up the following Table (angles in degrees)

m_l	3	2	1	0	−1	−2	−3
$l=1,\ \theta=\arc\cos(m_l/\sqrt{2})$			45.0	90	135.0		
$l=2,\ \theta=\arc\cos(m_l/\sqrt{6})$		35.3	65.9	90	114.1	144.7	
$l=3,\ \theta=\arc\cos(m_l/\sqrt{12})$	30	54.7	73.2	90	106.8	125.3	150

EXERCISE: The chlorine nucleus has a spin $I=\frac{3}{2}$. What values of θ are allowed ?

4.11 See Fig. 4.4.

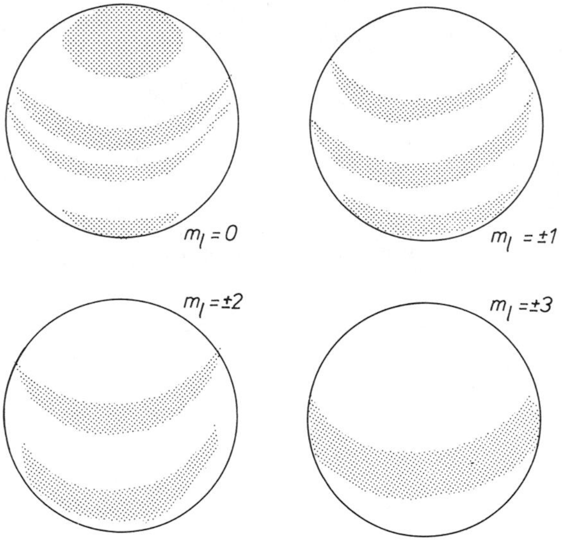

Fig. 4.4

EXERCISE: Draw the corresponding diagrams for $l=4$.

4.12 $H=-(\hbar^2/2\mu)\nabla^2-(Ze^2/4\pi\varepsilon_0)(1/r)$.

Since e^2 has been replaced by Ze^2, the e^4 in eqn (4.3.7) is replaced by Z^2e^4. Therefore,

$$\underline{E=-(\mu e^4Z^2/32\pi^2\varepsilon_0^2\hbar^2)(1/n^2)}\ ;\quad n=1,2,\ldots$$

EXERCISE: Evaluate the energy levels of positronium.

4.13 $\psi_{1s}=R_{10}Y_{00}=(1/a)^{\frac{3}{2}}\left\{2e^{-r/a}\right\}\left\{1/2\sqrt{\pi}\right\}=(1/\pi a^3)^{\frac{1}{2}}e^{-r/a}$ [Tables 4.1, 4.2].

$$\langle T\rangle=\int\psi_{1s}^*(-\hbar^2/2\mu)\nabla^2\psi_{1s}\,d\tau$$

$$\nabla^2\psi_{1s}=(1/\pi a^3)^{\frac{1}{2}}(1/r)(d^2/dr^2)\,r\,e^{-r/a}\qquad[\Lambda^2\psi_{1s}=0]$$

$$=(1/\pi a^3)^{\frac{1}{2}}\left\{-(2/ar)+(1/a^2)\right\}e^{-r/a}.$$

$$\langle T\rangle = -(\hbar^2/2\mu)(1/\pi a^3)\int_0^{2\pi} d\phi \int_0^{\pi} \sin\theta \, d\theta \int_0^{\infty}\left\{-(2/ar)+(1/a^2)\right\} e^{-2r/a} r^2 \, dr$$

$$= -(2\hbar^2/\mu a^3)\int_0^{\infty}\left\{-(2r/a)+(r^2/a^2)\right\} e^{-2r/a} dr \quad \left[\int_0^{\infty} x^n e^{-ax} dx = n!/a^{n+1}\right]$$

$$= -(2\hbar^2/\mu a^3)(-a/4) = \underline{\hbar^2/2\mu a^2 = -E_{1s}} \, .$$

$$\langle V\rangle = \int \psi^*_{1s} \, (-e^2/4\pi\varepsilon_0 r) \, \psi_{1s} \, d\tau$$

$$= (1/\pi a^3)(-e^2/4\pi\varepsilon_0)\int_0^{2\pi} d\phi \int_0^{\pi} \sin\theta \, d\theta \int_0^{\infty} (1/r) \, e^{-2r/a} r^2 \, dr$$

$$= -(e^2/4\pi^2\varepsilon_0 a^3)\{2\pi\}\{2\}\{\tfrac{1}{4}a^2\} = \underline{-e^2/4\pi\varepsilon_0 a = 2E_{1s}} \, .$$

Hence, $\langle T\rangle + \langle V\rangle = 2E_{1s} - E_{1s} = E_{1s}$.

According to the virial theorem, for $V \propto 1/r$, $\underline{\langle T\rangle = -\tfrac{1}{2}\langle V\rangle}$, as here.

EXERCISE: Repeat the calculation for a 2s-orbital.

4.14 Use eqn (4.3.4) in the form

$$(1/r)(d^2/dr^2)\, rR + \left\{(\mu e^2/2\pi\varepsilon_0 \hbar^2 r) - l(l+1)/r^2\right\} R = -(2\mu E/\hbar^2)\, R ,$$

with $E = -(\mu e^4/32\pi^2\varepsilon_0^2\hbar^2)(1/n^2)$ [eqn (4.3.7)].

(a) $R_{10}: n=1,\ l=0;\ E = -(\mu e^4/32\pi^2\varepsilon_0^2\hbar^2);\quad l=0$.

$$(1/r)(d^2/dr^2)\, rR_{10} + (\mu e^2/2\pi\varepsilon_0\hbar^2)(1/r)\, R_{10} = -(2\mu E/\hbar^2)\, R_{10} \, .$$

Then, since $R_{10} \propto e^{-r/a}$,

$$(d^2/dr^2)\, rR_{10} = 2R'_{10} + rR''_{10} = -(2/a)\, R_{10} + (r/a^2)\, R_{10}$$

$$-(2/ar) + (1/a^2) + (\mu e^2/2\pi\varepsilon_0\hbar^2)(1/r) = -(2\mu E/\hbar^2) \, .$$

But $2/a = \mu e^2/2\pi\varepsilon_0\hbar^2$; hence $1/a^2 = -2\mu E/\hbar^2$, so that $E = -\hbar^2/2\mu a^2$, as required.

(b) $R_{20} \propto (2-\rho)\, e^{-\rho/2} = (2-r/a)\, e^{-r/2a};\quad E_{2s} = -\tfrac{1}{4}(\hbar^2/2\mu a^2)$.

$$(d^2/dr^2) rR_{20} = 2R'_{20} + rR''_{20} \propto \left\{-(4/a) + (5r/2a^2) - (r^2/4a^3)\right\} e^{-r/2a} ,$$

$$-(4/ar) + (5/2a^2) - (r/4a^3) + \underbrace{(\mu e^2/2\pi\varepsilon_0\hbar^2)}_{2/a}(1/r)(2-r/a)$$

$$= -(2\mu E/\hbar^2)(2-r/a) ,$$

$$-(4/ar) + (5/2a^2) - (r/4a^3) + (4/ar) - (2/a^2) = (1/2a^2) - (r/4a^3)$$

$$= (1/4a^2)(2-r/a) \, .$$

Hence, $-2\mu E/\hbar^2 = 1/4a^2$, as required.

(c) $R_{31} \propto (4-\rho)\,e^{-\rho/2}$, $\rho = 2r/3a$, $l(l+1) = 2$; then proceed as above, obtaining $-2\mu E/\hbar^2 = 1/9a^2$.

EXERCISE: Confirm that R_{11} and R_{30} satisfy the wave equation.

4.15 The radial nodes are at the zeros of R_{nl}; denote them r_0.

(a) ψ_{2s} : $R_{2s} = 0$ when $2-\rho = 0$; $\rho = r/a$.

Hence, $r_0/a = 2$ or $\underline{r_0 = 2a = 105.8\text{ pm}}$.

(b) ψ_{3s} : $R_{3s} = 0$ when $6-6\rho+\rho^2 = 0$, $\rho = 2r/3a$. The solutions are $\rho_0 = 3\pm\sqrt{3}$, or $r_0 = (3\pm\sqrt{3})(3a/2) = \underline{1.90a,\ 7.10a \text{ or } 101\text{ pm},\ 376\text{ pm}}$.

EXERCISE: Find (a) the Z-dependence of these node locations, (b) the location of the radial nodes of (i) 2p-orbitals, (ii) 4s-orbitals. [A general point in this connection is that A & S lists the locations of zeros of many functions.]

4.16 (a) $\langle r\rangle = \int r\psi^2\,d\tau = \int |Y|^2 \sin\theta\,d\theta\,d\phi \int_0^\infty rR^2r^2dr = \int_0^\infty r^3R^2dr$ [Y normalized].

Use $\int_0^\infty x^n e^{-ax}\,dx = n!/a^{n+1}$.

$$\langle r\rangle_{1s} = 4(Z/a)^3\int_0^\infty r^3 e^{-2Zr/a}\,dr = 4(Z/a)^3\{3!/(2Z/a)^4\} = \underline{\tfrac{3}{2}(a/Z)}.$$

$$\langle r\rangle_{2s} = \tfrac{1}{8}(Z/a)^3\int_0^\infty (2-Zr/a)^2 r^3 e^{-Zr/a}\,dr$$

$$= \tfrac{1}{8}(Z/a)^3\int_0^\infty (4r^3 - 4Zr^4/a + Z^2r^5/a^2)\,e^{-Zr/a}\,dr$$

$$= \tfrac{1}{8}(Z/a)^3\left\{24a^4/Z^4 - 96a^4/Z^4 + 120a^4/Z^4\right\} = \underline{6(a/Z)}.$$

$$\langle r\rangle_{3s} = (1/243)(Z/a)^3\int_0^\infty (6 - 6Zr/a + 4Z^2r^2/9a^2)^2 r^3 e^{-2Zr/3a}\,dr$$

$$= \underline{(27/2)(a/Z)}.$$

The *general expression* is

$$\langle r\rangle_{nlm_l} = n^2\{1+\tfrac{1}{2}[1-l(l+1)/n^2]\}(a/Z).$$

(b) $\langle r^2\rangle = \int \psi^2 r^2\,d\tau = \int_0^\infty r^2R^2\,dr$.

$$\langle r^2\rangle_{1s} = 4(Z/a)^3\int_0^\infty r^4 e^{-2Zr/a}\,dr = \underline{3(a/Z)^2}.$$

$$\langle r^2\rangle_{2s} = \tfrac{1}{8}(Z/a)^3\int_0^\infty (2 - Zr/a)^2 r^4 e^{-Zr/a}\,dr = \underline{42\,(a/Z)^2}.$$

$$\langle r^2\rangle_{3s} = (1/243)(Z/a)^3\int_0^\infty (6 - 4Zr/a + 4Z^2r^2/9a^2)^2 r^4 e^{-2Zr/3a}\,dr$$
$$= \underline{207\,(a/Z)^2}.$$

The *general expression* is

$$\langle r^2\rangle_{nlm_l} = n^4\left\{1 + \tfrac{3}{2}[1 - l(l+1)/n^2 + 1/3n^2]\right\}(a/Z)^2.$$

(c) The most probable radius, r_*, is where $(d/dr)\,r^2R^2 = 0$, or, equivalently, where $(d/d\rho)\,\rho^2R^2 = 0$. Select the principal maximum (the outermost in each case.

(i) $\rho^2R^2_{1s} \propto \rho^2 e^{-\rho}$; $(d/d\rho)\,\rho^2R^2 \propto (2\rho - \rho^2)e^{-\rho} = 0$ at $\rho = \rho_*$; therefore, $\rho_* = 2$, implying $\underline{r_* = a/Z}$.

(ii) $\rho^2R^2_{2s} \propto \rho^2(2-\rho)^2 e^{-\rho}$;
$(d/d\rho)\,\rho^2R^2 \propto \left\{2\rho(2-\rho)^2 - 2\rho^2(2-\rho) - \rho^2(2-\rho)^2\right\}e^{-\rho} = 0$ at $\rho = \rho_*$;

therefore, solve $\rho_*(\rho_* - 2)(\rho_*^2 - 6\rho_* + 4) = 0$, finding $\rho_* = 0$ (at nucleus) $\rho_* = 3 - \sqrt{5}$ (first antinode), $\rho_* = 2$ (first node), $\rho_* = 3 + \sqrt{5}$ (second antinode, principal). Hence $\rho_* = 3 + \sqrt{5} = 5.236\,07$ and so $\underline{r_* = 5.24\;a/Z}$.

(iii) $\rho^2R^2_{3s} \propto \rho^2(6 - 6\rho + \rho^2)^2 e^{-\rho}$;
$(d/d\rho)\,\rho^2R^2_{3s} \propto \rho\,(\rho^5 - 18\rho^4 + 108\rho^3 - 264\rho^2 + 252\rho - 77)\,e^{-\rho} = 0$ at $\rho = \rho_*$;

therefore, solve $\rho_*(\rho_*^5 - 18\rho_*^4 + 108\rho_*^3 - 264\rho_*^2 + 252\rho_* - 72) = 0$, finding ρ_* (outermost) $= 8.7160$. Hence, $r_* = 3\rho_*a/2Z = \underline{13.07\;a/Z}$.

EXERCISE: Evaluate $\langle r^2\rangle_{4s}$ and $\langle r\rangle_{2p}$. Find the most probable radius for an electron in a 4s-orbital.

4.17 Evaluate $|\psi^2_{ns}(0)|^2 = |Y_{00}|^2 R^2_{n0} = R^2_{n0}/4\pi$.

$\psi^2_{1s}(0) = 4\,(Z/a)^3/4\pi = \underline{(1/\pi)(Z/a)^3} = 2.15\times10^{-6}\ \text{pm}^{-3}$ for hydrogen.

$\psi^2_{2s}(0) = \tfrac{1}{2}(Z/a)^3/4\pi = \underline{(1/8\pi)(Z/a)^3} = 2.68\times10^{-7}\ \text{pm}^{-3}$.

$\psi^2_{3s}(0) = (6/9\sqrt{3})^2(Z/a)^3/4\pi = \underline{(1/27\pi)(Z/a)^3} = 7.95\times10^{-8}\ \text{pm}^{-3}$.

EXERCISE: Evaluate the probability density for a 4s-orbital.

4.18 $\int \psi^*_{2s}\,\psi_{1s}\,d\tau \propto \int_0^\infty R_{20}\,R_{10}r^2\,dr \propto \int_0^\infty (2-Zr/a)\,e^{-Zr/2a}\,e^{-Zr/a}\,r^2\,dr$

$$\propto \int_0^\infty (2r^2 - Zr^3/a)\,e^{-3Zr/a}\,dr = (2^5a^3/3^3Z^3) - (2^5a^3/3^3Z^3) = 0.$$

EXERCISE: Confirm that ψ_{2s} and ψ_{3s} are orthogonal.

4.19 $\langle 1/r^3\rangle = \int_0^\infty (1/r^3)\,R_{21}^2\,r^2\,dr = \int_0^\infty (1/r)\,R_{21}^2\,dr$

$$= (Z/a)^3(1/2\sqrt{6})^2(Z/a)^2\int_0^\infty (1/r)\,r^2\,e^{-Zr/a}\,dr = \underline{(1/24)(Z/a)^3}.$$

The *general expression* is

$$\langle 1/r^3\rangle_{nlm_l} = (Z/a)^3/n^3l(l+\tfrac{1}{2})(l+1).$$

EXERCISE: Evaluate (a) $\langle 1/r^2\rangle$ for a $2p_z$-orbital, and (b) $\langle (1-3\cos^2\theta)/r^3\rangle$ for (i) a 2s-orbital, (ii) a $2p_z$-orbital.

4.20 $E = -(\mu e^4/32\pi^2\varepsilon_0^2\hbar^2)(1/n^2) = -(2.178\,72\times 10^{-19}\ \mathrm{J})/n^2.$

$\tilde{\nu} = E/hc = -R_{\mathrm{H}}/n^2 = -(109\,678\ \mathrm{cm}^{-1})/n^2.$

$E_{1s} = -2.178\,72\times 10^{-19}\ \mathrm{J} \mathrel{\hat{=}} \underline{-109\,678\ \mathrm{cm}^{-1}}.$

$E_{2s} = -0.544\,68\times 10^{-19}\ \mathrm{J} \mathrel{\hat{=}} \underline{-27\,420\ \mathrm{cm}^{-1}} = E_{2p}.$

$E_{3s} = -0.242\,08\times 10^{-19}\ \mathrm{J} \mathrel{\hat{=}} \underline{-12\,186\ \mathrm{cm}^{-1}} = E_{3p} = E_{3d}.$

$\Delta E(3d-2p) = 0.302\,60\times 10^{-19}\ \mathrm{J} \mathrel{\hat{=}} \underline{15\,234\ \mathrm{cm}^{-1}};\quad \lambda = \underline{656.43\ \mathrm{nm}}.$

$\Delta E(2p-2s) = \underline{0}.$

$\Delta E(2p-1s) = 1.634\,04\times 10^{-19}\ \mathrm{J} \mathrel{\hat{=}} \underline{82\,258\ \mathrm{cm}^{-1}};\quad \lambda = \underline{121.57\ \mathrm{nm}}.$

EXERCISE: Evaluate the same quantities for (a) positronium, (b) He^+.

4.21 $I = hcR$ (i.e. $I = -E_{1s}$).

$I(\mathrm{H}) - I(\mathrm{D}) = hc(R_{\mathrm{H}} - R_{\mathrm{D}}) = hc(\mu_{\mathrm{H}} - \mu_{\mathrm{D}})\,e^4/8h^2\varepsilon_0^2$

$$= (\mu_{\mathrm{H}} - \mu_{\mathrm{D}})\,hcR_\infty/m_{\mathrm{e}}.$$

$\mu_{\mathrm{H}} = m_{\mathrm{e}}m_{\mathrm{p}}/(m_{\mathrm{e}} - m_{\mathrm{p}}),\quad \mu_{\mathrm{D}} = m_{\mathrm{e}}m_{\mathrm{d}}/(m_{\mathrm{e}} + m_{\mathrm{d}}):$

$(\mu_{\mathrm{H}} - \mu_{\mathrm{D}})/m_{\mathrm{e}} = [m_{\mathrm{p}}/(m_{\mathrm{e}} + m_{\mathrm{p}})] - [m_{\mathrm{d}}/(m_{\mathrm{e}} + m_{\mathrm{d}})]$

$$= m_{\mathrm{e}}(m_{\mathrm{p}} - m_{\mathrm{d}})/(m_{\mathrm{e}} + m_{\mathrm{p}})(m_{\mathrm{e}} + m_{\mathrm{d}}) = m_{\mathrm{e}}(m_{\mathrm{H}} - m_{\mathrm{D}})m_{\mathrm{H}}\,m_{\mathrm{D}}.$$

$m_{\mathrm{e}} = 9.109\,53\times 10^{31}\ \mathrm{kg},\ m_{\mathrm{H}} = 1.6735\times 10^{-27}\ \mathrm{kg},\ m_{\mathrm{D}} = 3.3445\times 10^{-27}\ \mathrm{kg};$

$(\mu_{\mathrm{H}} - \mu_{\mathrm{D}})/m_{\mathrm{e}} = -2.7196\times 10^{-4}.$

Consequently,

$$I(\mathrm{H}) - I(\mathrm{D}) = -(2.7196 \times 10^{-4}) \times (2.1799 \times 10^{-18}\ \mathrm{J})$$

$$= -5.9285 \times 10^{-22}\ \mathrm{J} \mathrel{\hat{=}} \underline{-0.357\ \mathrm{kJ\ mol^{-1}}} \mathrel{\hat{=}} -3.70\ \mathrm{meV}\ .$$

The experimental values are 109 678.758 $\mathrm{cm^{-1}}$ and 109 708.596 $\mathrm{cm^{-1}}$, so that $\{I(\mathrm{H}) - I(\mathrm{D})\}/\mathrm{cm^{-1}} = -29.838\ \mathrm{cm^{-1}} \mathrel{\hat{=}} -0.357\ \mathrm{kJ\ mol^{-1}}$.

EXERCISE: Evaluate the ionization energy of positronium on the basis of the ionization energy of $^{1}\mathrm{H}$.

4.22 For a given value of l there are $2l+1$ values of m_l. For a given n there are n values of l. Hence, the degeneracy g is

$$g = \sum_{l=0}^{n-1} (2l+1) = n(n-1) + n = \underline{n^2}\ .$$

EXERCISE: Calculate the average value of m_l^2 for an atom in a state with principal quantum number equal to n but with l, m_l unspecified.

4.23 Refer to Fig. 4.5. The rotation (a) → (b) corresponds to 3p becoming 3d, the rotation (b) → (c) corresponds to 3d becoming 3s.

Fig. 4.5

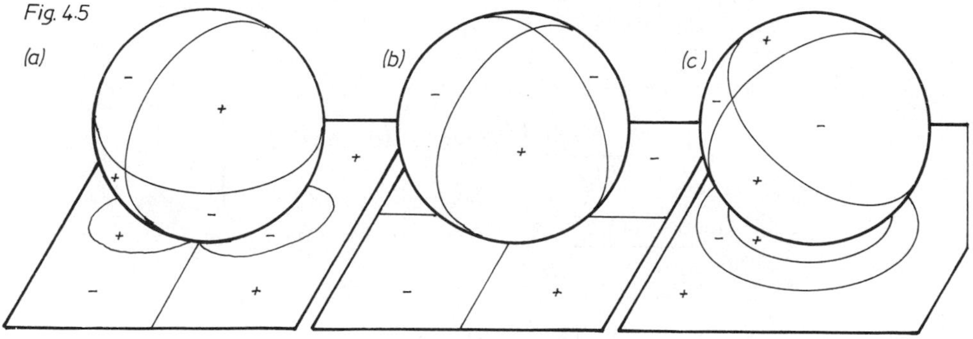

5 The properties of operators

5.1 (a) $(d/dx)\,e^{ax} = a\,e^{ax}$; e^{ax} is an eigenfunction, eigenvalue a.

$(d/dx)\,e^{ax^2} = 2ax\,e^{ax^2} = 2a\left\{x\,e^{ax^2}\right\}$; e^{ax^2} not an e.f.

$(d/dx)x = 1$; x not an e.f.

$(d/dx)\,x^2 = 2x$; x^2 not an e.f.

$(d/dx)(ax+b) = a$; $ax+b$ not an e.f.

$(d/dx)\sin x = \cos x$; $\sin x$ not an e.f.

(b) $(d^2/dx^2)\,e^{ax} = a^2\,e^{ax}$; e^{ax} is an eigenfunction, eigenvalue a^2.

$(d^2/dx^2)\,e^{ax^2} = 2a\,e^{ax^2} + 4a^2x^2\,e^{ax^2}$; e^{ax^2} not an e.f.

$(d^2/dx^2)\,x = 0 = 0x$; x is an e.f.; e.v. is 0.

$(d^2/dx^2)(ax+b) = 0 = 0\,(ax+b)$; $ax+b$ is an e.f.; e.v. is 0.

$(d^2/dx^2)\sin x = -\sin x$; $\sin x$ is an e.f.; e.v. is -1.

EXERCISE: Find the operator of which e^{ax^2} is an eigenfunction. Find the eigenfunction of the operator 'multiplication by x^2'.

5.2 Use the correspondence in eqn (5.1.10)

(a) $T = p^2/2m = \underline{-(\hbar^2/2m)(d^2/dx^2)}$ in one dimension.

$$T = p^2/2m = -(\hbar^2/2m)\left\{(\partial^2/\partial x^2) + (\partial^2/\partial y^2) + (\partial^2/\partial z^2)\right\}$$
$$= \underline{-(\hbar^2/2m)\nabla^2} \text{ in three dimensions.}$$

(b) $1/x \longrightarrow$ multiplication by $(1/x)$.

(c) $\boldsymbol{\mu} = \sum_i e_i\mathbf{r}_i \longrightarrow$ multiplication by $\sum_i e_i\mathbf{r}_i$.

(d) $l_z = xp_y - yp_x = \underline{(\hbar/i)\{x(\partial/\partial y) - y(\partial/\partial x)\}}$

$= \underline{(\hbar/i)(\partial/\partial\phi)}$ for $x = r\cos\phi$, $y = r\sin\phi$.

(e) $\delta x^2 = x^2 - \langle x\rangle^2 \longrightarrow$ multiplication by $\{x^2 - \langle x\rangle^2\}$.

$\delta p^2 = p^2 - \langle p\rangle^2 \longrightarrow \{-\hbar^2(\partial^2/\partial x^2) - \langle p\rangle^2\}$.

EXERCISE: Devise operators for $1/r$, xp_x, and e^{ax}.

5.3 Use the correspondence in eqn (5.1.11).

(a) $T = p^2/2m \longrightarrow$ multiplication by $p^2/2m$.

(b) $1/x = x^{-1}$; since $x = i\hbar(\partial/\partial p_x)$, $\underline{x^{-1} = (1/i\hbar)\int dp_x(\ldots)}$.

(c) $\mu = \sum_i e_i \boldsymbol{r}_i = \underline{i\hbar \sum_i e_i \nabla_i}$, $\nabla = \hat{i}(\partial/\partial p_x) + \hat{j}(\partial/\partial p_y) + \hat{k}(\partial/\partial p_z)$.

(d) $l_z = xp_y - yp_x = \underline{i\hbar\{p_y(\partial/\partial p_x) - p_x(\partial/\partial p_y)\}}$.

(e) $\delta x^2 = \underline{-\{\hbar^2(\partial^2/\partial p_x^2) + \langle x\rangle^2\}}$

$\delta p^2 = p^2 - \langle p\rangle^2 \longrightarrow$ $\underline{\text{multiplication by } \{p^2 - \langle p\rangle^2\}}$.

EXERCISE: Devise operators for $1/r$, xp_x, and $e^{\alpha x}$.

5.4 $E^2 = p^2c^2 + m^2c^4$

$E = (p^2c^2 + m^2c^4)^{\frac{1}{2}} = mc^2\ (1 + p^2/m^2c^2)^{\frac{1}{2}}$

$= mc^2(1 + p^2/2m^2c^2 + \ldots) \approx mc^2 + p^2/2m$.

Write $E \longrightarrow i\hbar(\partial/\partial t)$, $p = (\hbar/i)(\partial/\partial x)$, etc.; then

$$-\hbar^2(\partial^2/\partial t^2) = -\hbar^2c^2(\partial^2/\partial x^2) + m^2c^4.$$

Therefore, the relativistic analogue of the Schrödinger equation is

$\underline{-\hbar^2c^2(\partial^2\Psi/\partial x^2) + m^2c^4\Psi = -\hbar^2(\partial^2\Psi/\partial t^2)}$, the *Klein-Gordon equation*.

Since $\Psi \propto \cos\omega t$ is a solution, the probability is not conserved (recall Problem 2.6).

5.5 (a) $\int \psi_a^* T\psi_b\, d\tau = -(\hbar^2/2m)\int_{-\infty}^{\infty} \psi_a^*(d^2/dx^2)\psi_b\, dx$.

$$\int \psi_a^*(d^2/dx^2)\psi_b\, dx = \int \psi_a^*\ (d/dx)(d\psi_b/dx)\, dx$$

$$= \psi_a^*(d\psi_b/dx) - \int (d\psi_a^*/dx)(d\psi_b/dx)\, dx$$

$[\int u\,dv = uv - \int v\,du]$

$$= \psi_a^*(d\psi_b/dx) - (d\psi_a^*/dx)\psi_b + \int (d^2\psi_a^*/dx^2)\psi_b\, dx.$$

At each limit $\psi^*\psi'$ and $\psi^{*\prime}\psi$ disappear; therefore

$$\int \psi_a^* T\psi_b\, d\tau = -(\hbar^2/2m)\int_{-\infty}^{\infty}(d^2\psi_a^*/dx^2)\psi_b\, dx = \int (T\psi_a^*)\psi_b\, d\tau.$$

But $T^* = T$; hence $\int \psi_a^* T\psi_b\ d\tau = \int (T\psi_a)^*\psi_b\, d\tau$, and so T is hermitian [by eqn (5.2.2)].

(b) $$\int \psi_a^* l_z \psi_b \, d\tau = (\hbar/i) \int_0^{2\pi} \psi_a^* (\partial\psi_b/\partial\phi) \, d\phi$$
$$= (\hbar/i) \left\{ \psi_a^* \psi_b - \int (\partial\psi_a^*/\partial\phi) \psi_b \, d\phi \right\} \Big|_0^{2\pi}$$
$$= -(\hbar/i) \int_0^{2\pi} (\partial\psi_a^*/\partial\phi) \psi_b \, d\phi = \int_0^{2\pi} (l_z \psi_a)^* \psi_b \, d\phi \qquad [l_z^* = -l_z]$$
$$= \int (l_z \psi_a)^* \psi_b \, d\tau ,$$

and so l_z is hermitian.

EXERCISE: Confirm that p_x, l_z^2, and l_z^3 are hermitian.

5.6 $$\langle m|A + iB|n\rangle = \langle m|A|n\rangle + i\langle m|B|n\rangle$$
$$= \langle n|A|m\rangle^* + i\langle n|B|m\rangle \qquad [A, B \text{ hermitian, eqn (5.2.6)}]$$
$$= \{\langle n|A|m\rangle - i\langle n|B|m\rangle\}^* = \langle n|A - iB|m\rangle^* .$$

Hence, $A - iB$ is the hermitian conjugate of $A + iB$ (and $A + iB$ is not *self-conjugate*, another term for hermitian).

EXERCISE: Confirm that $x + (d/dx)$ and $x - (d/dx)$ are hermitian conjugates.

5.7 Use the relations $[x, p_x] = i\hbar$, $[y, p_y] = i\hbar$, $[z, p_z] = i\hbar$, all others zero.

(a) $[x, y] = \underline{0}$ [basic relation].

(b) $[p_x, p_y] = \underline{0}$ [basic relation].

(c) $[x, p] = \underline{i\hbar}$ [basic relation].

(d) $$[x^2, p] = x^2p - px^2 = x^2p - pxx = x^2p - xpx + [x, p]x$$
$$= x^2p - x^2p + x[x, p] + [x, p]x = xi\hbar + i\hbar x = \underline{2i\hbar x}.$$

(e) $$[x^n, p] = x^np - px^n = x^np - xpx^{n-1} + [x\ p]\,x^{n-1}$$
$$= x^np - x^2px^{n-2} + x[x, p]\,x^{n-2} + [x, p]\,x^{n-1}$$
$$= x^np - x^np + x^{n-1}[x, p] + x^{n-2}[x, p]\,x + \\ + \ldots + x[x, p]x^{n-2} + [x, p]x^{n-1}$$
$$= \underline{ni\hbar x^{n-1}} .$$

(f) $[1/x, p]$; use the position representation.

$$[1/x, p] = [x^{-1}, (\hbar/i)d/dx] = (\hbar/i)\left\{x^{-1}(d/dx) - (d/dx)\,x^{-1}\right\}$$
$$= (\hbar/i)\left\{x^{-1}(d/dx) - (dx^{-1}/dx) - x^{-1}(d/dx)\right\}$$
$$= -(\hbar/i)(dx^{-1}/dx) = \underline{(\hbar/i)\,x^{-2}} .$$

(g) $[1/x\,, p^2] = [x^{-1}, -\hbar^2(\mathrm{d}^2/\mathrm{d}x^2)]$

$$= -\hbar^2\left\{x^{-1}(\mathrm{d}^2/\mathrm{d}x^2) - (\mathrm{d}^2/\mathrm{d}x^2)x^{-1}\right\}$$
$$= -\hbar^2\left\{x^{-1}(\mathrm{d}^2/\mathrm{d}x^2) - (\mathrm{d}/\mathrm{d}x)[(\mathrm{d}x^{-1}/\mathrm{d}x) + x^{-1}(\mathrm{d}/\mathrm{d}x)]\right\}$$
$$= -\hbar^2\left\{x^{-1}(\mathrm{d}^2/\mathrm{d}x^2) - (\mathrm{d}/\mathrm{d}x)\,[-x^{-2} + x^{-1}(\mathrm{d}/\mathrm{d}x)]\right\}$$
$$= -\hbar^2\left\{x^{-1}(\mathrm{d}^2/\mathrm{d}x^2) + (\mathrm{d}x^{-2}/\mathrm{d}x) + x^{-2}(\mathrm{d}/\mathrm{d}x) - (\mathrm{d}x^{-1}/\mathrm{d}x)(\mathrm{d}/\mathrm{d}x) - x^{-1}(\mathrm{d}^2/\mathrm{d}x^2)\right\}$$
$$= -\hbar^2\left\{-2x^{-3} + 2x^{-2}(\mathrm{d}/\mathrm{d}x)\right\}$$
$$= 2\hbar^2/x^3 - 2\hbar^2 x^{-2}(\mathrm{i}/\hbar)p = \underline{(2\hbar/x^3)\{\hbar - \mathrm{i}xp\}}.$$

(h) $[xp_y - yp_x\,, yp_z - zp_y]$

$$= [xp_y\,, yp_z] - [xp_y\,, zp_y] - [yp_x\,, yp_z] + [yp_x\,, zp_y] \qquad \text{[eqn (5.4.2f)]}$$
$$= x[p_y\,, y]p_z - 0 - 0 + p_x[y\,, p_y]z \qquad \text{[eqn (5.4.2b)]}$$
$$= x(-\mathrm{i}\hbar)p_z + p_x(\mathrm{i}\hbar)z = \underline{\mathrm{i}\hbar(zp_x - xp_z)}.$$

(i) $[x^2(\partial^2/\partial y^2)\,, y(\partial/\partial x)]$

$$= x^2(\partial^2/\partial y^2)y(\partial/\partial x) - y(\partial/\partial x)x^2(\partial^2/\partial y^2)$$
$$= x^2(\partial/\partial x)(\partial^2/\partial y^2)y - (\partial/\partial x)\,x^2y\,(\partial^2/\partial y^2)$$
$$= x^2(\partial/\partial x)(\partial/\partial y)\left\{1 + y(\partial/\partial y)\right\} - \left\{2x + x^2(\partial/\partial x)\right\}y\,(\partial^2/\partial y^2)$$
$$= x^2(\partial/\partial x)\left\{2(\partial/\partial y) + y(\partial^2/\partial y^2)\right\} - 2xy\,(\partial^2/\partial y^2) - x^2(\partial/\partial x)y(\partial^2/\partial y^2)$$
$$= 2x^2(\partial/\partial x)(\partial/\partial y) - 2xy(\partial^2/\partial y^2)$$
$$= \underline{2x^2(\partial^2/\partial x\,\partial y) - 2xy(\partial^2/\partial y^2)}.$$

EXERCISE: Evaluate $[x\,, p^2]$, $[x\,, p^n]$, $[x^2\,, p^2]$, and $[xy(\partial^2/\partial x\,\partial y)\,, x^2(\partial^2/\partial y^2)]$.

5.8 (a) $[A, B] = AB - BA = -(BA - AB) = -[B, A]$.

(b) $[A^m, A^n] = A^mA^n - A^nA^m = A^{m+n} - A^{m+n} = 0$.

(c) $[A^2, B] = AAB - BAA = ABA + (AAB - ABA) - ABA + (ABA - BAA)$
$= A[A, B] + [A, B]A$.

(d) $[A,[B,C]] + [B,[C,A]] + [C,[A,B]]$
$= (ABC - ACB - BCA + CBA) + (BCA - BAC - CAB + ACB) + (CAB - CBA - ABC + BAC) = 0$.

$$[l_y\,,[l_y\,, l_z]] = [l_y\,, \mathrm{i}\hbar l_x] = \mathrm{i}\hbar(-\mathrm{i}\hbar l_z) = \underline{\hbar^2 l_z}.$$

EXCERCISE: Express $[A^2,B^2]$ $[A^3,B]$, and $[A,[B,[C,[D,E]]]]$ in terms of individual commutators. Evaluate $[l_x + l_y\,, [l_x + l_y\,, l_x + l_y + l_z]]$.

5.9 $e^{A}|a\rangle = \sum_{n}(1/n!)A^{n}|a\rangle$ [definition of e^{A}]

$= \sum_{n}(1/n!)a^{n}|a\rangle$ [$A^{n}|a\rangle = a^{n}|a\rangle$ if $A|a\rangle = a|a\rangle$]

$= e^{a}|a\rangle$ [definition of e^{a}]

Hence $|a\rangle$ is an eigenstate of e^{A} with eigenvalue e^{a}.

EXERCISE: Find expansions suitable for the definition of $(1+A)^{-1}$ and $(1+A)^{\frac{1}{2}}$, and find their eigenvalues corresponding to the states $|a\rangle$ for which $A|a\rangle = a|a\rangle$.

5.10 $e^{A}e^{B} = (1 + A + \frac{1}{2}A^{2} + \ldots)(1 + B + \frac{1}{2}B^{2} + \ldots)$

$= 1 + (A+B) + \frac{1}{2}(A^{2}+2AB+B^{2}) + \ldots$

$e^{A+B} = 1 + (A+B) + \frac{1}{2}(A+B)^{2} + \ldots$

$= 1 + (A+B) + \frac{1}{2}(A^{2} + AB + BA + B^{2}) + \ldots$

Therefore, $e^{A}e^{B} = e^{A+B}$ only if $AB = BA$, which is so if $[A,B] = 0$.

If $[A,[A,B]] = [B,[A,B]] = 0$, then

$$e^{A+B} = 1 + (A+B) + \tfrac{1}{2}(A^{2} + AB + BA + B^{2}) + (1/3!)(A^{3} + A^{2}B + ABA + BAA + BBA + BAB + ABB + B^{3}) + \ldots$$

$$= 1 + (A+B) + \tfrac{1}{2}(A^{2} + 2AB + B^{2}) - \tfrac{1}{2}[A,B] + (1/3!)(A^{3} + 3A^{2}B + 3AB^{2} + B^{3}) - \tfrac{1}{2}(A+B)[A,B] + \ldots$$

$$= \underline{e^{A}e^{B}e^{-\frac{1}{2}[A,B]}}.$$

EXERCISE: Find expressions for $\cos A \cos B$ and $\cos A \sin B$, where A and B operators such that

$$[A,[A,B]] = [B,[A,B]] = 0.$$

(Use $\cos A = \frac{1}{2}(e^{iA} + e^{-iA})$, etc.)

5.11 $[H, p_{x}] = [p^{2}/2m + V(x), p_{x}] = [V(x), p_{x}]$

$= (\hbar/i)\{V(x)(\partial/\partial x) - (\partial/\partial x)V(x)\}$

$= (\hbar/i)\{V(x)(\partial/\partial x) - (\partial V/\partial x) - V(x)(\partial/\partial x)\} = \underline{i\hbar(\partial V/\partial x)}.$

$[H, x] = [p^{2}/2m + V(x), x] = [p^{2}, x]/2m$

$= -(\hbar^{2}/2m)[d^{2}/dx^{2}, x]$

$= -(\hbar^{2}/2m)\{(d^{2}/dx^{2})x - x(d^{2}/dx^{2})\}$

$= -(\hbar^{2}/2m)\{(d/dx)[1 + x(d/dx)] - x(d^{2}/dx^{2})\}$

$= -(\hbar^{2}/2m)\{2(d/dx) + x(d^{2}/dx^{2}) - x(d^{2}/dx^{2})\}$

$= -(\hbar^{2}/m)(d/dx) = \underline{(\hbar/im)p_{x}}.$

(a) $V(x) = V$ entails $[H, p_x] = \underline{0}$, $[H, x] = \underline{(\hbar/im)\,p_x}$.

(b) $V(x) = \frac{1}{2}kx^2$ entails $[H, p_x] = \underline{i\hbar kx}$, $[H, x] = \underline{(\hbar/im)\,p_x}$.

(c) $V(x) \longrightarrow V(r) = e^2/4\pi\varepsilon_0 r$ entails

$$[H, p_x] = i\hbar(\partial V/\partial x) = (i\hbar e^2/4\pi\varepsilon_0)(\partial r^{-1}/\partial x).$$

Since $(\partial r^{-1}/\partial x) = (\partial/\partial x)(x^2+y^2+z^2)^{-\frac{1}{2}} = -x/r^3$ we have

$$[H, p_x] = \underline{-(i\hbar e^2/4\pi\varepsilon_0)(x/r^3)}.$$

$$[H, p_x] = \underline{(\hbar/im)\,p_x}.$$

EXERCISE: Evaluate $[H, T]$, where $T = p^2/2m$, and calculate it for the three cases (a) – (c).

5.12 $\delta A\,\delta B \geqslant \frac{1}{2}\left|\langle c\rangle\right|$ with $C = [A, B]/i$ [eqn (5.4.9)].

(a) $A = x$, $B = p_x$; $[A, B] \longrightarrow [x, p_x] = i\hbar \longrightarrow C = \hbar$.

Hence, $\underline{\delta x\,\delta p_x \geqslant \frac{1}{2}\hbar}$.

(b) $A = p_x$, $B = p_z$ (etc.); $[p_x, p_y] = 0$; hence $\underline{\delta p_x\,\delta p_y \geqslant 0}$ (unrestricted).

(c) $A = (1/2m)\,p_x^2$, $B = V(x)$;

$$\begin{aligned}[A, B] \longrightarrow [p_x^2, V]/2m &= -(\hbar^2/2m)\left\{(\partial^2/\partial x^2)V - V(\partial^2/\partial x^2)\right\}\\ &= -(\hbar^2/2m)\left\{V'' + 2V'(\partial/\partial x) + V(\partial^2/\partial x^2) - V(\partial^2/\partial x^2)\right\}\\ &= -(\hbar^2/2m)\left\{V'' + 2V'(\partial/\partial x)\right\}\\ &= -(\hbar^2/2m)\,V'' - (i\hbar/m)\,V'p_x.\end{aligned}$$

Therefore, $\underline{\delta T\,\delta V \geqslant \frac{1}{2}\left|(\hbar^2/2m)\langle V''\rangle + (i\hbar/m)\langle V'p_x\rangle\right|}$,

and in general the r.h.s. is non-zero.

(d) $A = -ex$, $B = p^2/2m + V$;

$$[A, B] \longrightarrow -e[x, p_x]/2m = (-ie\hbar/m)\,p_x \qquad \text{[Problem 5.11]}.$$

$$\underline{\delta\mu\,\delta E \geqslant \tfrac{1}{2}(e\hbar/m)\langle p\rangle}.$$

For systems described by real wavefunctions $\langle p\rangle = 0$; hence then $\delta\mu\,\delta E \geqslant 0$, and $\delta\mu\,\delta E$ is unrestricted.

(e) $A = x$, $B = p^2/2m$; $[A, B] \longrightarrow [x, p^2]/2m = (i\hbar/m)p$ as above.

$\delta x\,\delta T \geqslant \underline{(\hbar/2m)\langle p\rangle = 0}$ for real wavefunctions (non-degenerate states).

EXERCISE: Determine the restrictions on the simultaneous determination of the values of x^2 and (a) the momentum, (b) the kinetic energy.

5.13 $(d/dt)\langle\Omega\rangle = (i/\hbar)\langle[H,\Omega]\rangle$ [eqn (5.5.1)].

For a harmonic oscillator, $H = p^2/2m + \frac{1}{2}kx^2$, and

$$[H,x] = [p^2/2m,x] = -(i\hbar/m)p \qquad \text{[Problem 5.11b]}$$
$$[H,p] = [\tfrac{1}{2}kx^2,p] = i\hbar kx \qquad \text{[Problem 5.11b]}$$
$$(d/dt)\langle x\rangle = \underline{(1/m)\langle p\rangle}; \quad (d/dt)\langle p\rangle = -\underline{k\langle x\rangle}.$$

Therefore,

$$(d^2/dt^2)\langle x\rangle = (1/m)(d/dt)\langle p\rangle = -(k/m)\langle x\rangle.$$

The solution of $(d^2/dt^2)\langle x\rangle = -(k/m)\langle x\rangle$ is

$$\langle x\rangle = A\cos\omega t + B\sin\omega t, \quad \omega^2 = k/m,$$

$$\langle p\rangle = m(d/dt)\langle x\rangle = -Am\omega\sin\omega t + Bm\omega\cos\omega t$$

which is the classical trajectory.

EXERCISE: Find the equation of motion of the expectation values of x and p for a quartic oscillator ($V \propto x^4$).

5.14 Use eqn (5.5.2). Since $l_z = (\hbar/i)(\partial/\partial\phi)$, $V(\phi) = V$, a constant, and $H = (1/2mr^2)l_z^2 + V$:

$$[H,l_z] = (1/2mr^2)[l_z^2,l_z] + [V,l_z] = 0 \quad \Big[[V,l_z] \propto dV/d\phi = 0\Big].$$

Hence, $(d/dt)\langle l_z\rangle = 0$ [eqn (5.5.2)].

EXERCISE: Find the equation of motion for the expectation value of l_z for a particle on a vertical ring in a uniform gravitational field. Examine the equations for small displacements from the lowest point.

5.15 Construct the matrices $\|x_{v'v}\|$ and $\|p_{v'v}\|$:

$$\|x\| = \begin{bmatrix} x_{00} & x_{01} & x_{02} & \cdots \\ x_{10} & x_{11} & x_{12} & \cdots \\ x_{20} & x_{21} & x_{22} & \cdots \\ \vdots & \vdots & \vdots & \end{bmatrix} = (\hbar/2m\omega)^{\frac{1}{2}} \begin{bmatrix} 0 & 1 & 0 & 0 & 0 & \cdots \\ 1 & 0 & \sqrt{2} & 0 & 0 & \cdots \\ 0 & \sqrt{2} & 0 & \sqrt{3} & 0 & \cdots \\ 0 & 0 & \sqrt{3} & 0 & \sqrt{4} & \cdots \\ \vdots & \vdots & \vdots & \vdots & \vdots & \end{bmatrix}$$

$$\|p\| = \begin{bmatrix} p_{00} & p_{01} & p_{02} & \cdots \\ p_{10} & p_{11} & p_{12} & \cdots \\ p_{20} & p_{21} & p_{22} & \cdots \\ \vdots & \vdots & \vdots & \end{bmatrix} = -i(\hbar m\omega/2)^{\frac{1}{2}} \begin{bmatrix} 0 & 1 & 0 & 0 & 0 & \cdots \\ -1 & 0 & \sqrt{2} & 0 & 0 & \cdots \\ 0 & -\sqrt{2} & 0 & \sqrt{3} & 0 & \cdots \\ 0 & 0 & -\sqrt{3} & 0 & \sqrt{4} & \cdots \\ \vdots & \vdots & \vdots & \vdots & \vdots & \end{bmatrix}$$

$$xp - px = -i(\hbar/2)\left\{\begin{bmatrix}0 & 1 & 0 & \cdots\\ 1 & 0 & \surd 2 & \cdots\\ 0 & \surd 2 & 0 & \cdots\\ \vdots & \vdots & \vdots & \end{bmatrix}\begin{bmatrix}0 & 1 & 0 & \cdots\\ -1 & 0 & \surd 2 & \cdots\\ 0 & -\surd 2 & 0 & \cdots\\ \vdots & \vdots & \vdots & \end{bmatrix} - \begin{bmatrix}0 & 1 & 0 & \cdots\\ -1 & 0 & \surd 2 & \cdots\\ 0 & -\surd 2 & 0 & \cdots\\ \vdots & \vdots & \vdots & \end{bmatrix}\begin{bmatrix}0 & 1 & 0 & \cdots\\ 1 & 0 & \surd 2 & \cdots\\ 0 & \surd 2 & 0 & \cdots\\ \vdots & \vdots & \vdots & \end{bmatrix}\right\}$$

$$= -i(\hbar/2)\left\{\begin{bmatrix}-1 & 0 & \surd 2 & \cdots\\ 0 & -1 & 0 & \cdots\\ -\surd 2 & 0 & -1 & \cdots\\ \vdots & \vdots & \vdots & \end{bmatrix} - \begin{bmatrix}-1 & 0 & \surd 2 & \cdots\\ 0 & 1 & 0 & \cdots\\ -\surd 2 & 0 & 1 & \cdots\\ \vdots & \vdots & \vdots & \end{bmatrix}\right\} = i\hbar\begin{bmatrix}1 & 0 & 0 & \cdots\\ 0 & 1 & 0 & \cdots\\ 0 & 0 & 1 & \cdots\\ \vdots & \vdots & \vdots & \cdots\end{bmatrix} = \underline{i\hbar\mathbb{1}}.$$

$H = p^2/2m + \frac{1}{2}kx^2$

$$= -(\hbar\omega/4)\begin{bmatrix}0 & 1 & 0 & \cdots\\ -1 & 0 & \surd 2 & \cdots\\ 0 & -\surd 2 & 0 & \cdots\\ \vdots & \vdots & \vdots & \end{bmatrix}\begin{bmatrix}0 & 1 & 0 & \cdots\\ -1 & 0 & \surd 2 & \cdots\\ 0 & -\surd 2 & 0 & \cdots\\ \vdots & \vdots & \vdots & \end{bmatrix} + (\hbar k/4m\omega)\begin{bmatrix}0 & 1 & 0 & \cdots\\ 1 & 0 & \surd 2 & \cdots\\ 0 & \surd 2 & 0 & \cdots\\ \vdots & \vdots & \vdots & \end{bmatrix}\begin{bmatrix}0 & 1 & 0 & \cdots\\ 1 & 0 & \surd 2 & \cdots\\ 0 & \surd 2 & 0 & \cdots\\ \vdots & \vdots & \vdots & \end{bmatrix}$$

$$= -\tfrac{1}{4}\hbar\omega\begin{bmatrix}-1 & 0 & \surd 2 \cdots\\ 0 & -3 & 0 \cdots\\ \surd 2 & 0 & -5 \cdots\\ \vdots & \vdots & \vdots\end{bmatrix} + \tfrac{1}{4}\hbar\omega\begin{bmatrix}1 & 0 & \surd 2 & \cdots\\ 0 & 3 & 0 & \cdots\\ \surd 2 & 0 & 5 & \cdots\\ \vdots & \vdots & \vdots & \end{bmatrix} = \tfrac{1}{2}\hbar\omega\begin{bmatrix}1 & 0 & 0 & \cdots\\ 0 & 3 & 0 & \cdots\\ 0 & 0 & 5 & \cdots\\ \vdots & \vdots & \vdots & \end{bmatrix}.$$

The eigenvalues of H are therefore $\underline{(V + \frac{1}{2})\hbar\omega}$.

EXERCISE: Evaluate the commutators $[H,x]$ and $[H,p]$, and find the equations of motion. What matrix represents the rate of change of the product xp?

5.16 $$\langle v|xp^2x|v\rangle = \sum_{v'}\langle v|xp^2|v'\rangle\langle v'|x|v\rangle$$
$$= \langle v|xp^2|v+1\rangle\langle v+1|x|v\rangle + \langle v|xp^2|v-1\rangle\langle v-1|x|v\rangle$$
$$= \langle v|xp^2|v+1\rangle(\hbar/2m\omega)^{\frac{1}{2}}\surd(v+1) + \langle v|xp^2|v-1\rangle(\hbar/2m\omega)^{\frac{1}{2}}\surd v$$
$$= (\hbar/2m\omega)^{\frac{1}{2}}\left\{\langle v|xp^2|v+1\rangle + \langle v|xp^2|v-1\rangle\surd v\right\}.$$

$$\langle v|xp^2|v+1\rangle = \sum_{v'}\langle v|xp|v'\rangle\langle v'|p|v+1\rangle$$
$$= \langle v|xp|v+2\rangle\langle v+2|p|v+1\rangle + \langle v|xp|v\rangle\langle v|p|v+1\rangle$$
$$= i(\hbar m\omega/2)^{\frac{1}{2}}\left\{\langle v|xp|v+2\rangle\surd(v+2) - \langle v|xp|v\rangle\surd(v+1)\right\}$$

$$\langle v|xp|v+2\rangle = \sum_{v'}\langle v|x|v'\rangle\langle v'|p|v+2\rangle$$
$$= \langle v|x|v+1\rangle\langle v+1|p|v+2\rangle = -i(\hbar/2)\surd\{(v+1)(v+2)\}$$

$$\langle v|xp|v\rangle = \langle v|x|v+1\rangle\langle v+1|p|v\rangle + \langle v|x|v+1\rangle\langle v+1|p|v\rangle$$
$$= -i(\hbar/2)\{(v+1) - v\} = -\tfrac{1}{2}i\hbar$$

$$\langle v|xp^2|v-1\rangle = \langle v|xp|v\rangle\langle v|p|v-1\rangle + \langle v|xp|v-2\rangle\langle v-2|p|v-1\rangle$$
$$= \mathrm{i}(\hbar m\omega/2)^{\frac{1}{2}}\left\{\langle v|xp|v\rangle\surd v - \langle v|xp|v-2\rangle\surd(v-1)\right\}$$

$$\langle v|xp|v-2\rangle = \langle v|x|v-1\rangle\langle v-1|p|v-2\rangle = \tfrac{1}{2}\mathrm{i}\hbar\surd\left\{v(v-1)\right\}.$$

Then, combining all the elements:

$$\langle v|xp^2x|v\rangle = \underline{(\tfrac{1}{2}\hbar)^2(2v^2+2v+1)}.$$

EXERCISE: Evaluate $\langle v|x^4|v\rangle$ and $\langle v+2|xp^2x|v\rangle$.

6 Angular momentum

6.1 $[l_x, l_y] = (\hbar/i)^2 [y(\partial/\partial z) - z(\partial/\partial y), z(\partial/\partial x) - x(\partial/\partial z)]$ [eqn (6.1.4)]

$= (\hbar/i)^2 \{[y(\partial/\partial z), z(\partial/\partial x)] + [z(\partial/\partial y), x(\partial/\partial z)]\}$

[eqn (5.4.2f)]

$= (\hbar/i)^2 \{y[(\partial/\partial z), z](\partial/\partial x) + x[z, (\partial/\partial z)](\partial/\partial y)\}$

[eqn (5.4.2b)]

$[(\partial/\partial z), z] = (\partial/\partial z)z - z(\partial/\partial z) = 1 + z(\partial/\partial z) - z(\partial/\partial z) = 1.$

Therefore,

$[l_x, l_y] = (\hbar/i)^2 \{y(\partial/\partial x) - x(\partial/\partial y)\} = -(\hbar/i)\, l_z$ [eqn (6.1.14c]

$= \underline{i\hbar l_z}.$

EXERCISE: Evaluate $[l_y, l_x]$ in the position representation.

6.2 $[l_y^2, l_x] = l_y^2 l_x - l_x l_y^2 = l_y l_y l_x - l_x l_y^2$

$= l_y l_x l_y + l_y(l_y l_x - l_x l_y) - l_x l_y^2$

$= l_x l_y^2 + [l_y, l_x]l_y + l_y[l_y, l_x] - l_x l_y^2$

$= -i\hbar l_z l_y - i\hbar l_y l_z = \underline{-i\hbar(l_z l_y + l_y l_z)}.$

$[l_y^2, l_z^2] = l_y^2 l_x^2 - l_x^2 l_y^2 = l_y l_y l_x l_x - l_x^2 l_y^2$

$= l_y l_x l_y l_x + l_y[l_y, l_x]l_x - l_x^2 l_y^2$

$= l_x l_y l_y l_x + [l_y, l_x]l_y l_x + l_y[l_y, l_x]l_x - l_x^2 l_y^2$

$= l_x l_y l_x l_y + l_x l_y[l_y, l_x] + [l_y, l_x]l_y l_x + l_y[l_y, l_x]l_x - l_x^2 l_y^2$

$= l_x l_x l_y l_y + l_x[l_y, l_x]l_y + l_x l_y[l_y, l_x]$
$\quad + [l_y, l_x]l_y l_x + l_y[l_y, l_x]l_x - l_x^2 l_y^2$

$= \underline{-i\hbar\{l_x l_z l_y + l_x l_y l_z + l_z l_y l_x + l_y l_z l_x\}}.$

$[l_x, [l_x, l_y]] = i\hbar[l_x, l_z] = i\hbar(-i\hbar)l_y = \underline{\hbar^2 l_y}.$

EXERCISE: Evaluate $[l_z^2, l_x]$ and $[l_y^2, l_z^2]$.

6.3 $$\boldsymbol{l} \wedge \boldsymbol{l} = \begin{vmatrix} \hat{i} & \hat{j} & \hat{k} \\ l_x & l_y & l_z \\ l_x & l_y & l_z \end{vmatrix}$$

$$= \hat{i}(l_y l_z - l_z l_y) - \hat{j}(l_x l_z - l_z l_x) + \hat{k}(l_x l_y - l_y l_x)\,.$$

$$\mathrm{i}\hbar\boldsymbol{l} = \hat{i}(\mathrm{i}\hbar l_x) + \hat{j}(\mathrm{i}\hbar l_y) + \hat{k}(\mathrm{i}\hbar l_z).$$

Hence, equating both sides term by term reproduces the commutation rules, eqn (6.1.6).

EXERCISE: Show that if $\boldsymbol{l}_1 \wedge \boldsymbol{l}_1 = \mathrm{i}\hbar\boldsymbol{l}_1$ and $\boldsymbol{l}_2 \wedge \boldsymbol{l}_2 = \mathrm{i}\hbar\boldsymbol{l}_2$, then $\boldsymbol{l} \wedge \boldsymbol{l} = \mathrm{i}\hbar\boldsymbol{l}$, where $\boldsymbol{l} = \boldsymbol{l}_1 + \boldsymbol{l}_2$, but only if $[l_{1q}, l_{2q'}] = 0$ for all q, q'.

6.4 $$[s_x, s_y] = (\tfrac{1}{2}\hbar)^2 [\sigma_x, \sigma_y]$$

$$= (\tfrac{1}{2}\hbar)^2 \left\{\begin{pmatrix} 0 & 1 \\ 1 & 0 \end{pmatrix}\begin{pmatrix} 0 & -\mathrm{i} \\ \mathrm{i} & 0 \end{pmatrix} - \begin{pmatrix} 0 & -\mathrm{i} \\ \mathrm{i} & 0 \end{pmatrix}\begin{pmatrix} 0 & 1 \\ 1 & 0 \end{pmatrix}\right\}$$

$$= (\tfrac{1}{2}\hbar)^2 \left\{\begin{pmatrix} \mathrm{i} & 0 \\ 0 & -1 \end{pmatrix} - \begin{pmatrix} -\mathrm{i} & 0 \\ 0 & \mathrm{i} \end{pmatrix}\right\} = \tfrac{1}{2}\mathrm{i}\hbar^2 \begin{pmatrix} 1 & 0 \\ 0 & -1 \end{pmatrix} = \underline{\mathrm{i}\hbar s_z}.$$

$$s^2 = s_x^2 + s_y^2 + s_z^2 = (\tfrac{1}{2}\hbar)^2 \left\{\begin{pmatrix} 0 & 1 \\ 1 & 0 \end{pmatrix}^2 + \begin{pmatrix} 0 & -\mathrm{i} \\ \mathrm{i} & 0 \end{pmatrix}^2 + \begin{pmatrix} 1 & 0 \\ 0 & -1 \end{pmatrix}^2\right\}$$

$$= (\tfrac{1}{2}\hbar)^2 \left\{\begin{pmatrix} 1 & 0 \\ 0 & 1 \end{pmatrix} + \begin{pmatrix} 1 & 0 \\ 0 & 1 \end{pmatrix} + \begin{pmatrix} 1 & 0 \\ 0 & 1 \end{pmatrix}\right\} = \tfrac{3}{4}\hbar^2 \begin{pmatrix} 1 & 0 \\ 0 & 1 \end{pmatrix}.$$

The eigenvalues of s^2 are therefore $\frac{3}{4}\hbar^2$; hence, identifying them with $s(s+1)\hbar^2$ identifies $s = \frac{1}{2}$.

EXERCISE: Confirm that the following matrices constitutes a representation of an angular momentum with $l = 1$.

$$l_x = (1/\surd 2)\begin{bmatrix} 0 & 1 & 0 \\ 1 & 0 & 1 \\ 0 & 1 & 0 \end{bmatrix}, \quad l_y = (\mathrm{i}/\surd 2)\begin{bmatrix} 0 & -1 & 0 \\ 1 & 0 & -1 \\ 0 & 1 & 0 \end{bmatrix}, \quad l_z = \begin{bmatrix} 1 & 0 & 0 \\ 0 & 0 & 0 \\ 0 & 0 & -1 \end{bmatrix}.$$

6.5 $$s_x s_y = (\tfrac{1}{2}\hbar)^2 \begin{pmatrix} 0 & 1 \\ 1 & 0 \end{pmatrix}\begin{pmatrix} 0 & -\mathrm{i} \\ \mathrm{i} & 0 \end{pmatrix} = (\tfrac{1}{2}\hbar)^2 \begin{pmatrix} \mathrm{i} & 0 \\ 0 & -\mathrm{i} \end{pmatrix}$$

$$= \mathrm{i}(\tfrac{1}{2}\hbar)^2 \sigma_z = \underline{\tfrac{1}{2}\mathrm{i}\hbar s_z}.$$

$$s_x s_y^2 s_z^2 = (\tfrac{1}{2}\hbar)^5 \begin{pmatrix} 0 & 1 \\ 1 & 0 \end{pmatrix}\begin{pmatrix} 0 & -\mathrm{i} \\ \mathrm{i} & 0 \end{pmatrix}^2 \begin{pmatrix} 1 & 0 \\ 0 & -1 \end{pmatrix}^2 = (\tfrac{1}{2}\hbar)^5 \begin{pmatrix} 0 & 1 \\ 1 & 0 \end{pmatrix}$$

$$= (\tfrac{1}{2}\hbar)^5 \sigma_x = \underline{(\tfrac{1}{2}\hbar)^4 s_x}.$$

$$s_x^2 s_y^2 s_z^2 = \underline{(\tfrac{1}{2}\hbar)^6} \qquad [\sigma_q^2 = \mathbb{1} \text{ for each } q].$$

EXERCISE: Express $s_x s_y s_z$ as a single spin operator and $\mathrm{e}^{\mathrm{i}\alpha s_x}$ in the form $a + bs_x$. Evaluate $[\mathrm{e}^{\mathrm{i}\alpha s_x}, \mathrm{e}^{\mathrm{i}\beta s_y}]$.

6.6 Suppose $[l_x, l_y] = -i\hbar l_z$, $l^{\pm} = l_x \pm i l_y$; then

$$[l^+, l_z] = \hbar l^+,\ [l^-, l_z] = -\hbar l^-, \text{ and } [l^+, l^-] = -2\hbar l_z .$$

Then, following the development that led to eqn (6.3.5),

$$l_z l^+ | l, m\rangle = \{l^+ l_z + [l_z, l^+]\} | l, m\rangle = \{l^+ l_z - \hbar l^+\} | l, m\rangle$$
$$= \{l^+ m\hbar - \hbar l^+\} | l, m\rangle = (m-1)\hbar l^+ | l, m\rangle .$$

Consequently, $l^+ | l, m\rangle \propto | l, m-1\rangle$ and $l^- | l, m\rangle \propto | l, m+1\rangle$; therefore l^+ is a *lowering* operator and l^- is a *raising* operator.

EXERCISE: Find a matrix representation of these l_x and l_y 'angular momenta' corresponding to $l = 1$ (draw on the matrices in the *Exercise* to Problem 6.4).

6.7 $\langle 0,0 | l_z | 0,0\rangle = \underline{0}$ [eqn (6.3.18)].

$\langle 1,1 | l^+ | 0,0\rangle = \underline{0}$ [l^+ is diagonal in l].

$\langle 2,1 | l^+ | 2,0\rangle = \hbar\sqrt{6}$ [eqn (6.3.19)].

$$\langle 2,0 | l^- l^+ | 2,0\rangle = \langle 2,0 | l^+ | 2,-1\rangle\langle 2,-1 | l^- | 2,0\rangle \quad \text{[insert complete set].}$$
$$= (\hbar\sqrt{6})(\hbar\sqrt{6}) = 6\hbar^2 \quad \text{[eqn (6.3.19)]}$$

$$\langle 2,0 | l^- l^+ | 2,0\rangle = \langle 2,0 | l^+ l^- + [l^-, l^+] | 2,0\rangle$$
$$= \langle 2,0 | l^+ l^- | 2,0\rangle - 2\hbar\langle 2,0 | l_z | 2,0\rangle$$
$$= \langle 2,0 | l^+ l^- | 2,0\rangle = \underline{6\hbar^2} .$$

$$\langle 2,0 | l^{-2} l_z l^{+2} | 2,0\rangle = \langle 2,0 | l^- | 2,1\rangle\langle 2,1 | l^- | 2,2\rangle\langle 2,2 | l_z | 2,2\rangle$$
$$\times \langle 2,2 | l^+ | 2,1\rangle\langle 2,1 | l^+ | 2,0\rangle$$
$$= (\hbar\sqrt{6})(2\hbar)(2\hbar)(2\hbar)(\hbar\sqrt{6}) = \underline{48\hbar^5} .$$

EXERCISE: Evaluate $\langle 3,1 | l^{-2} l_z l^{+2} | 3,1\rangle$ and $\langle 3,3 | l_x l_y l_z | 3,1\rangle$.

6.8 In each case $l = 1$ and $p_x = (p_- - p_+)/\sqrt{2}$, $p_y = (p_- + p_+)(i/\sqrt{2})$ [the phases are taken from Table 4.1]. Then $p_+ \triangleq |1,1\rangle$, $p_- \triangleq |1,-1\rangle$ and $p_z \triangleq |1,0\rangle$ in the notation $|l, m_l\rangle$. The l label will be omitted henceforth.

$$\langle p_x | l_z | p_y\rangle = (i/2)\{\langle -1 | - \langle 1 |\}\, l_y\, \{| -1\rangle + | 1\rangle\}$$
$$= (i/2)\{\langle -1 | l_z | -1\rangle - \langle 1 | l_y | 1\rangle\} = (i/2)\{-\hbar - \hbar\}$$
$$= \underline{-i\hbar} .$$

$$\langle p_x | l^+ | p_y\rangle = (i/2)\{\langle -1 | - \langle 1 |\}\, l^+ \{| -1\rangle + | 1\rangle\} = \underline{0} .$$

$$\langle p_z|l_y|p_x\rangle = (1/2i)(1/\sqrt{2})\langle 0|(l^+ - l^-)\{|-1\rangle - |1\rangle\}$$
$$= (1/2i\sqrt{2})\{\langle 0|l^+|-1\rangle + \langle 0|l^-|1\rangle\}$$
$$= (1/2i\sqrt{2})\{\hbar\sqrt{2} + \hbar\sqrt{2}\} = \underline{-i\hbar}.$$

$$\langle p_z|l_x|p_y\rangle = (1/2)(i/\sqrt{2})\langle 0|(l^+ + l^-)\{|-1\rangle + |1\rangle\}$$
$$= (i/2\sqrt{2})\{\langle 0|l^+|-1\rangle + \langle 0|l^-|1\rangle\}$$
$$= (i/2\sqrt{2})\{\hbar\sqrt{2} + \hbar\sqrt{2}\} = \underline{i\hbar}.$$

$$\langle p_z|l_x|p_x\rangle = (1/2)(1/\sqrt{2})\langle 0|(l^+ + l^-)\{|-1\rangle - |1\rangle\}$$
$$= (1/2\sqrt{2})\{\langle 0|l^+|-1\rangle - \langle 0|l^-|1\rangle\} = \underline{0}.$$

EXERCISE. Evaluate $\langle p_y|l^-|p_z\rangle$, $\langle p_x|l_y|p_z\rangle$, $\langle p_x|l^+l^-|p_y\rangle$, and $\langle d_{xy}|l_x|d_{xz}\rangle$.

6.9 $[l_x, l_y] = \hbar^2[\sin\phi(\partial/\partial\theta) + \cot\theta\cos\phi(\partial/\partial\phi),$
$\cos\phi(\partial/\partial\theta) - \cot\theta\sin\phi(\partial/\partial\phi)]$

$$= \hbar^2\Big\{-[\sin\theta(\partial/\partial\theta), \cot\theta\sin\phi(\partial/\partial\phi)]$$
$$+ [\cot\theta\cos\phi(\partial/\partial\phi), \cos\phi(\partial/\partial\theta)]$$
$$- [\cot\theta\cos\phi(\partial/\partial\phi), \cot\theta\sin\phi(\partial/\partial\phi)]\Big\}.$$

$[\sin\phi(\partial/\partial\theta), \cot\theta\sin\phi(\partial/\partial\phi)]$

$$= \sin\phi(\partial/\partial\theta)\cot\theta\sin\phi(\partial/\partial\phi) - \cot\theta\sin\phi(\partial/\partial\phi)\sin\phi(\partial/\partial\theta)$$
$$= \sin^2\phi(\partial\cot\theta/\partial\theta)(\partial/\partial\phi) + \sin^2\phi\cot\theta(\partial^2/\partial\theta\,\partial\phi)$$
$$- \cot\theta\sin\phi(\partial\sin\phi/\partial\phi)(\partial/\partial\theta) - \cot\theta\sin^2\phi(\partial^2/\partial\theta\,\partial\phi)$$
$$= -\sin^2\phi\ \mathrm{cosec}^2\theta(\partial/\partial\phi) - \cot\theta\sin\phi\cos\phi(\partial/\partial\theta).$$

$[\cot\theta\cos\phi(\partial/\partial\phi), \cos\phi(\partial/\partial\theta)]$

$$= \cot\theta\cos\phi(\partial/\partial\phi)\cos\phi(\partial/\partial\theta) - \cos\phi(\partial/\partial\theta)\cot\theta\cos\phi(\partial/\partial\phi)$$
$$= \cot\theta\cos\phi(\partial\cos\phi/\partial\phi)(\partial/\partial\theta) + \cot\theta\cos^2\phi(\partial^2/\partial\phi\,\partial\theta)$$
$$- \cos^2\phi(\partial\cot\theta/\partial\theta)(\partial/\partial\phi) - \cos^2\phi\cot\theta(\partial^2/\partial\theta\,\partial\phi)$$
$$= -\cot\theta\cos\phi\sin\phi(\partial/\partial\theta) + \cos^2\phi\ \mathrm{cosec}^2\theta(\partial/\partial\phi).$$

$[\cot\theta\cos\phi(\partial/\partial\phi), \cot\theta\sin\phi(\partial/\partial\phi)]$

$$= \cot^2\theta[\cos\phi(\partial/\partial\phi), \sin\phi(\partial/\partial\phi)]$$
$$= \cot^2\theta\{\cos\phi(\partial/\partial\phi)\sin\phi(\partial/\partial\phi) - \sin\phi(\partial/\partial\phi)\cos\phi(\partial/\partial\phi)\}$$
$$= \cot^2\theta\{\cos^2\phi(\partial/\partial\phi) + \sin^2\phi(\partial/\partial\phi)\} = \cot^2\theta(\partial/\partial\phi).$$

$$[l_x , l_y] = \hbar^2 \Big\{\sin^2\phi \operatorname{cosec}^2\theta\,(\partial/\partial\phi) + \cot\theta \sin\phi \cos\phi\,(\partial/\partial\theta)$$
$$+ \cos^2\phi \operatorname{cosec}^2\theta(\partial/\partial\phi) - \cot\theta \cos\phi \sin\phi(\partial/\partial\theta) - \cot^2\theta(\partial/\partial\phi)\Big\}$$
$$= \hbar^2 \Big\{(\sin^2\phi + \cos^2\phi) \operatorname{cosec}^2\theta\,(\partial/\partial\phi) - \cot^2\theta(\partial/\partial\phi)\Big\}$$
$$= \hbar^2 \Big\{\operatorname{cosec}^2\theta(\partial/\partial\phi) - \cot^2\theta\Big\}(\partial/\partial\phi) = \hbar^2(\partial/\partial\phi) = \underline{i\hbar l_z},$$

as required.

EXERCISE: Evaluate $[l_z , l_x]$ in this representation.

6.10 $l^-\psi_{2,m_l} = \hbar\{6 - m_l(m_l - 1)\}^{\frac{1}{2}}\psi_{2,m_l-1}$; $\psi_{l,l} = N\sin^l\theta\, e^{il\phi}$.

$\psi_{2,1} = (1/2\hbar)\, l^-\psi_{2,2}$; $\psi_{2,0} = (1/\hbar\sqrt{6})\, l^-\psi_{2,1}$;

$\psi_{2,-1} = (1/\hbar\sqrt{6})\, l^-\psi_{2,0}$; $\psi_{2,-2} = (1/2\hbar)\, l^-\psi_{2,-1}$.

$$l^- = -\hbar e^{-i\phi}\{(\partial/\partial\theta) - i\cot\theta(\partial/\partial\phi)\} \qquad \text{[eqn (6.4.2b)]}$$

$$(1/\hbar)\, l^-\psi_{2,2} = -e^{-i\phi}\{(\partial/\partial\theta) - i\cot\theta(\partial/\partial\phi)\}N\sin^2\theta\, e^{2i\phi}$$
$$= -N e^{-i\phi}\{2\sin\theta\cos\theta\, e^{2i\phi} - i(2i)\sin\theta\cos\theta\, e^{2i\phi}\}.$$
$$= -4N\sin\theta\cos\theta\, e^{i\phi} \;;\; \text{hence} \;\; \psi_{2,1} = -2N\sin\theta\cos\theta\, e^{i\phi}.$$

$$(1/\hbar)\, l^-\psi_{2,1} = -e^{-i\phi}\{(\partial/\partial\theta) - i\cot\theta(\partial/\partial\phi)\}(-2N\sin\theta\cos\theta)e^{i\phi}$$
$$= 2N\{\cos^2\theta - \sin^2\theta + \cos^2\theta\}$$
$$= 2N(3\cos^2\theta - 1) \;;\; \text{hence} \;\; \psi_{2,0} = (2/3)^{\frac{1}{2}} N(3\cos^2\theta - 1).$$

$$(1/\hbar)\, l^-\psi_{2,0} = -(2/3)^{\frac{1}{2}} N e^{-i\phi}\{(\partial/\partial\theta) - i\cot\theta(\partial/\partial\phi)\}(3\cos^2\theta - 1)$$
$$= (\sqrt{6})N e^{-i\phi}\, 2\cos\theta\sin\theta \;;\; \text{hence} \;\; \psi_{2,-1} = 2N\sin\theta\cos\theta\, e^{-i\phi}.$$

$$(1/\hbar)\, l^-\psi_{2,-1} = -2N e^{-i\phi}\{(\partial/\partial\theta) - i\cot\theta(\partial/\partial\phi)\}\sin\theta\cos\theta\, e^{-i\phi}$$
$$= -2N e^{-i\phi}\{\cos^2\theta - \sin^2\theta - \cos^2\theta\}\, e^{-i\phi}$$
$$= 2N\sin^2\theta\, e^{-2i\phi} \;;\; \text{hence} \;\; \psi_{2,-2} = N\sin^2\theta\, e^{-2i\phi}.$$

Since $N = (1/2^3)\{5!/4\pi\}^{\frac{1}{2}} = (15/32\pi)^{\frac{1}{2}}$, we have

$$\underline{\psi_{2,2} = (15/32\pi)^{\frac{1}{2}}\sin^2\theta\, e^{2i\phi}} \qquad \underline{\psi_{2,-2} = (15/32\pi)^{\frac{1}{2}}\sin^2\theta\, e^{-2i\phi}}$$
$$\underline{\psi_{2,1} = -(15/8\pi)^{\frac{1}{2}}\sin\theta\cos\theta\, e^{i\phi}} \qquad \underline{\psi_{2,-1} = (15/8\pi)^{\frac{1}{2}}\sin\theta\cos\theta\, e^{-i\phi}}$$
$$\underline{\psi_{2,0} = (5/16\pi)^{\frac{1}{2}}(3\cos^2\theta - 1)}.$$

These accord (including the phases) with the entries in Table 4.1.

EXERCISE: Find expressions for the f-orbital wavefunctions.

6.11 $$\boldsymbol{j}_1 \wedge \boldsymbol{j}_2 = \begin{vmatrix} \hat{i} & \hat{j} & \hat{k} \\ j_{1x} & j_{1y} & j_{1z} \\ j_{2x} & j_{2y} & j_{2z} \end{vmatrix}$$

$$= \hat{i}(j_{1y}j_{2z} - j_{1z}j_{2y}) - \hat{j}(j_{1x}j_{2z} - j_{1z}j_{2x}) + \hat{k}(j_{1x}j_{2y} - j_{1y}j_{2x})$$

$$= \hat{i}(j_{2z}j_{1y} - j_{2y}j_{1z}) - \hat{j}(j_{2z}j_{1x} - j_{2x}j_{1z}) + \hat{k}(j_{2y}j_{1x} - j_{2x}j_{1y})$$

$$= \begin{vmatrix} \hat{i} & \hat{j} & \hat{k} \\ -j_{2x} & -j_{2y} & -j_{2z} \\ j_{1x} & j_{1y} & j_{1z} \end{vmatrix} = -\boldsymbol{j}_2 \wedge \boldsymbol{j}_1 .$$

$$\boldsymbol{j} \wedge \boldsymbol{j} = (\boldsymbol{j}_1 + \boldsymbol{j}_2) \wedge (\boldsymbol{j}_1 + \boldsymbol{j}_2)$$

$$= \boldsymbol{j}_1 \wedge \boldsymbol{j}_1 + \boldsymbol{j}_2 \wedge \boldsymbol{j}_2 + \boldsymbol{j}_1 \wedge \boldsymbol{j}_2 + \boldsymbol{j}_2 \wedge \boldsymbol{j}_1 = \boldsymbol{j}_1 \wedge \boldsymbol{j}_1 + \boldsymbol{j}_2 \wedge \boldsymbol{j}_2$$

$$= i\hbar \boldsymbol{j}_1 + \hbar \boldsymbol{j}_2 = i\hbar \boldsymbol{j}.$$

$$(\boldsymbol{j}_1 - \boldsymbol{j}_2) \wedge (j_1 - j_1) = j_1 \wedge j_1 + j_2 \wedge j_2 - j_1 \wedge j_2 - j_2 \wedge j_1,$$

$$= \boldsymbol{j}_1 \wedge \boldsymbol{j}_1 + \boldsymbol{j}_2 \wedge \boldsymbol{j}_2 = i\hbar \boldsymbol{j} \neq i\hbar\, (j_1 - \boldsymbol{j}_2).$$

As $\boldsymbol{j}_1 - \boldsymbol{j}_2$ does not satisfy the commutation relation

$$(\boldsymbol{j}_1 - \boldsymbol{j}_2) \wedge (\boldsymbol{j}_1 - \boldsymbol{j}_2) = i\hbar(j_1 - \boldsymbol{j}_2),$$

it is not an angular momentum.

EXERCISE: Under what circumstances do j_1 and j_2 satisfy the vector relations set out in eqn (A20.3) ?

6.12 $$[j^2, j_{1z}] = 2i\hbar(j_{1x}j_{2y} - j_{1y}j_{2x}) \qquad \text{[eqn (6.6.4)]}$$

$$= \hbar(j_1^- j_2^+ - j_1^+ j_2^-) \qquad [\text{using } j^{\pm} = j_x \pm i j_y].$$

Therefore,

$$[j^2, j_{1z}]|j_1 j_1 ; j_2 j_2\rangle = \hbar\left\{j_1^- j_2^+ |j_1 j_1 ; j_2 j_2\rangle - j_1^+ j_2^- |j_1 j_1 ; j_2 j_2\rangle\right\}$$

$$= \hbar\left\{j_i^- |j_1 j_1\rangle\, j_2^+ |j_2 j_2\rangle - j_1^+ |j_1 j_1\rangle\, j_2^- |j_2 j_2\rangle\right\} = 0 ,$$

because $j_i^+|j_i j_i\rangle = 0$. Similarly,

$$[j^2, j_{1z}]|j_1, -j_1; j_2, -j_2\rangle = \hbar\left\{j_1^- |j_1, -j_1\rangle j_2^+ |j_2, -j_2\rangle - j_1^+ |j_1, -j_1\rangle j_2^- |j_2, -j_2\rangle\right\}$$
$$= 0$$

because $j_i^- |j_i j_i\rangle = 0$.

EXERCISE: Evaluate $[j^2, j_{1x}]|j_1 j_1 ; j_2 j_2\rangle$. What is the significance of the result in the light of the conclusion for $[j^2, j_{1z}]$?

6.13 Use the Clebsch-Gordan series, eqn (6.6.7).

(a) $j = 3 + 4,\ 3 + 4 - 1, \ldots |3 - 4| = \underline{7, 6, \ldots 1}$.

(bi) $L = 1 + 1,\ 1 + 1 - 1, \ldots |1 - 1| = \underline{2, 1, 0}$.

(bii) $L = 2 + 2,\ 2 + 2 - 1, \ldots |2 - 2| = \underline{4, 3, \ldots 0}$.

(biii) $L = 2 + 1,\ 2 + 1 - 1, \ldots |2 - 1| = \underline{3, 2, 1}$.

(c) $s_{12} = \frac{1}{2} + \frac{1}{2},\ \frac{1}{2} + \frac{1}{2} - 1, \ldots |\frac{1}{2} - \frac{1}{2}| = 1, 0$

$$s_{123} = \begin{cases} 1 + \frac{1}{2},\ 1 + \frac{1}{2} - 1, \ldots |1 - \frac{1}{2}| = \frac{3}{2}, \frac{1}{2} \\ 0 + \frac{1}{2},\ 0 + \frac{1}{2} - 1, \ldots |0 - \frac{1}{2}| = \frac{1}{2} \end{cases}$$

$$S = \begin{cases} \frac{3}{2} + \frac{1}{2},\ \frac{3}{2} + \frac{1}{2} - 1, \ldots |\frac{3}{2} - \frac{1}{2}| = 2, 1 \\ \frac{1}{2} + \frac{1}{2},\ \frac{1}{2} + \frac{1}{2} - 1, \ldots |\frac{1}{2} - \frac{1}{2}| = 1, 0 \\ \frac{1}{2} + \frac{1}{2},\ \frac{1}{2} + \frac{1}{2} - 1, \ldots |\frac{1}{2} - \frac{1}{2}| = 1, 0 . \end{cases}$$

That is, $S = \underline{2, 1, 1, 1, 0, 0}$.

EXERCISE: Determine the total angular momenta for the configurations fd and f^3, and for a system of six electrons.

6.14 Construct the vector coupling coefficients for $j = \frac{3}{2}, \frac{1}{2}$. The starting is $|j = \frac{3}{2}, m = \frac{3}{2}\rangle = |1, m_{j1} = 1; \frac{1}{2}, m_{j2} = \frac{1}{2}\rangle \equiv |1, \alpha\rangle$. Generate $|\frac{3}{2}\ \frac{1}{2}\rangle$ using $j^- |\frac{3}{2}, \frac{3}{2}\rangle = \hbar\sqrt{3}\, |\frac{3}{2}, \frac{1}{2}\rangle$ and

$$j^- |\tfrac{3}{2}, \tfrac{3}{2}\rangle = (j_1^- + j_2^-)|1, \alpha\rangle = \hbar\sqrt{2}\, |0, \alpha\rangle + \hbar\sqrt{}\, |1, \beta\rangle .$$

Therefore,

$$|\tfrac{3}{2}, \tfrac{1}{2}\rangle = (2/3)^{\frac{1}{2}} |0, \alpha\rangle + (1/3)^{\frac{1}{2}} |1, \beta\rangle .$$

Next, generate $|\frac{3}{2}, -\frac{1}{2}\rangle$ using $j^- |\frac{3}{2}, \frac{1}{2}\rangle = 2\hbar\, |\frac{3}{2}, -\frac{1}{2}\rangle$ and

$$j^- |\tfrac{3}{2}, \tfrac{1}{2}\rangle = (j_1^- + j_2^-) \left\{(2/3)^{\frac{1}{2}} |0, \alpha\rangle + (1/3)^{\frac{1}{2}} |1, \beta\rangle\right\}$$
$$= \hbar (2/3)^{\frac{1}{2}} \sqrt{2}\, |-1, \alpha\rangle + \hbar (2/3)^{\frac{1}{2}} |0, \beta\rangle + (2/3)^{\frac{1}{2}} \hbar\, |0, \beta\rangle$$
$$= \hbar (2/3)^{\frac{1}{2}} \sqrt{2}\, |-1, \alpha\rangle + 2\hbar (2/3)^{\frac{1}{2}} |0, \beta\rangle .$$

Therefore,

$$|\tfrac{3}{2}, -\tfrac{1}{2}\rangle = (1/3)^{\frac{1}{2}} |-1, \alpha\rangle + (2/3)^{\frac{1}{2}} |0, \beta\rangle .$$

There is only one possibility for $|\frac{3}{2}, -\frac{3}{2}\rangle = |-1, \beta\rangle$.

The $j = \frac{1}{2}$ states are orthogonal to the $j = \frac{3}{2}$. Hence, from $\langle \frac{3}{2}, \frac{1}{2} | \frac{1}{2}, \frac{1}{2}\rangle = 0$,

$$|\tfrac{1}{2}, \tfrac{1}{2}\rangle = -(1/3)^{\frac{1}{2}} |0, \alpha\rangle + (2/3)^{\frac{1}{2}} |1, \beta\rangle ,$$

and from

$$\langle \tfrac{3}{2}, -\tfrac{1}{2} | \tfrac{1}{2}, -\tfrac{1}{2}\rangle = 0 ,$$
$$|\tfrac{1}{2}, -\tfrac{1}{2}\rangle = -(2/3)^{\frac{1}{2}} |-1, \alpha\rangle + (1/3)^{\frac{1}{2}} |0, \beta\rangle .$$

These results lead to the following table of coefficients:

m_{j1}	m_{j2}	$\lvert\frac{3}{2},\frac{3}{2}\rangle$	$\lvert\frac{3}{2},\frac{1}{2}\rangle$	$\lvert\frac{1}{2},\frac{1}{2}\rangle$	$\lvert\frac{1}{2},-\frac{1}{2}\rangle$	$\lvert\frac{1}{2},-\frac{1}{2}\rangle$	$\lvert\frac{3}{2},-\frac{3}{2}\rangle$
1	$\frac{1}{2}$ (α)	1	0	0	0	0	0
1	$-\frac{1}{2}$ (β)	0	$\sqrt{(1/3)}$	$\sqrt{(2/3)}$	0	0	0
0	$\frac{1}{2}$ (α)	0	$\sqrt{(2/3)}$	$-\sqrt{(1/3)}$	0	0	0
0	$-\frac{1}{2}$ (β)	0	0	0	$\sqrt{(2/3)}$	$\sqrt{(1/3)}$	0
−1	$\frac{1}{2}$ (α)	0	0	0	$\sqrt{(1/3)}$	$-\sqrt{(2/3)}$	0
−1	$-\frac{1}{2}$ (β)	0	0	0	0	0	1

This is in accord with Appendix 9.

For the matrix elements

$$\langle \tfrac{3}{2},\tfrac{3}{2}|j_{1z}|\tfrac{3}{2},\tfrac{3}{2}\rangle = \langle 1,\alpha|j_{1z}|1,\alpha\rangle = \hbar .$$

$$\langle \tfrac{3}{2},\tfrac{3}{2}|j_{1z}|\tfrac{3}{2},\tfrac{1}{2}\rangle = \langle 1,\alpha|j_{1z}\left\{(2/3)^{\frac{1}{2}}|0,\alpha\rangle + (1/3)^{\frac{1}{2}}|1,\beta\rangle\right\} = 0 .$$

Likewise, all other $\langle \tfrac{3}{2},\tfrac{3}{2}|j_{1z}|jm\rangle = 0$.

$$\langle \tfrac{3}{2},-\tfrac{3}{2}|j_{1z}|\tfrac{3}{2},-\tfrac{3}{2}\rangle = \langle -1,\beta|j_{1z}|-1,\beta\rangle = -\hbar ,$$

and $\langle \tfrac{3}{2},-\tfrac{3}{2}|j_{1z}|jm\rangle = 0$ for all other j, m.

$$\begin{aligned}
\langle \tfrac{3}{2},\tfrac{1}{2}|j_{1z}|\tfrac{3}{2},\tfrac{1}{2}\rangle &= \left\{(2/3)^{\frac{1}{2}}\langle 0,\alpha| + (1/3)^{\frac{1}{2}}\langle 1,\beta|\right\} j_{1z} \\
&\quad \left\{(2/3)^{\frac{1}{2}}|0,\alpha\rangle + (1/3)^{\frac{1}{2}}|1,\beta\rangle\right\} \\
&= \left\{(2/3)^{\frac{1}{2}}\langle 0,\alpha| + (1/3)^{\frac{1}{2}}\langle 1,\beta|\right\}(1/3)^{\frac{1}{2}}\hbar|1,\beta\rangle = \tfrac{1}{3}\hbar .
\end{aligned}$$

$$\langle \tfrac{3}{2},-\tfrac{1}{2}|j_{1z}|\tfrac{3}{2},\tfrac{1}{2}\rangle = \left\{(1/3)^{\frac{1}{2}}\langle -1,\alpha| + (2/3)^{\frac{1}{2}}\langle 0,\beta|\right\}(1/3)^{\frac{1}{2}}\hbar|1,\beta\rangle = 0.$$

$$\begin{aligned}
\langle \tfrac{3}{2},-\tfrac{1}{2}|j_{1z}|\tfrac{3}{2},-\tfrac{1}{2}\rangle &= \left\{(1/3)^{\frac{1}{2}}\langle -1,\alpha| + (2/3)^{\frac{1}{2}}\langle 0,\beta|\right\} j_{1z} \\
&\quad \left\{(1/3)^{\frac{1}{2}}|-1,\alpha\rangle + (2/3)^{\frac{1}{2}}|0,\beta\rangle\right\} \\
&= \left\{(1/3)^{\frac{1}{2}}\langle -1,\alpha| + (2/3)^{\frac{1}{2}}\langle 0,\beta|\right\}(1/3)^{\frac{1}{2}}(-\hbar)|-1,\alpha\rangle = -\tfrac{1}{3}\hbar .
\end{aligned}$$

$$\langle \tfrac{1}{2},\tfrac{1}{2}|j_{1z}|\tfrac{1}{2},\tfrac{1}{2}\rangle = \left\{-(1/3)^{\frac{1}{2}}\langle 0,\alpha| + (2/3)^{\frac{1}{2}}\langle 1,\beta|\right\}(2/3)^{\frac{1}{2}}\hbar|1,\beta\rangle = \tfrac{2}{3}\hbar .$$

$$\langle \tfrac{1}{2},-\tfrac{1}{2}|j_{1z}|\tfrac{1}{2},\tfrac{1}{2}\rangle = \left\{-(2/3)^{\frac{1}{2}}\langle -1,\alpha| + (1/3)^{\frac{1}{2}}\langle 0,\beta|\right\}(2/3)^{\frac{1}{2}}\hbar|1,\beta\rangle = 0.$$

$$\langle \tfrac{3}{2},-\tfrac{1}{2}|j_{1z}|\tfrac{1}{2},\tfrac{1}{2}\rangle = \left\{(1/3)^{\frac{1}{2}}\langle -1,\alpha| + (2/3)^{\frac{1}{2}}\langle 0,\beta|\right\}(2/3)^{\frac{1}{2}}\hbar|1,\beta\rangle = 0 .$$

$$\langle \tfrac{3}{2},\tfrac{1}{2}|j_{1z}|\tfrac{1}{2},\tfrac{1}{2}\rangle = \left\{(2/3)^{\frac{1}{2}}\langle 0,\alpha| + (1/3)^{\frac{1}{2}}\langle 1,\beta|\right\}(2/3)^{\frac{1}{2}}\hbar|1,\beta\rangle = \tfrac{1}{3}\hbar\sqrt{2} .$$

$$\langle \tfrac{3}{2},-\tfrac{1}{2}|j_{1z}|\tfrac{1}{2},-\tfrac{1}{2}\rangle = \left\{(1/3)^{\frac{1}{2}}\langle -1,\alpha| + (2/3)^{\frac{1}{2}}\langle 0,\beta|\right\}(2/3)^{\frac{1}{2}}\hbar|-1,\alpha\rangle = \tfrac{1}{3}\hbar\sqrt{2} .$$

$$\langle \tfrac{1}{2},-\tfrac{1}{2}|j_{1z}|\tfrac{1}{2},-\tfrac{1}{2}\rangle = \left\{-(2/3)^{\frac{1}{2}}\langle -1,\alpha| + (1/3)^{\frac{1}{2}}\langle 0,\beta|\right\}(2/3)^{\frac{1}{2}}\hbar|-1,\alpha\rangle = -\tfrac{2}{3}\hbar .$$

$$\langle \tfrac{3}{2},\tfrac{1}{2}|j_{1z}|\tfrac{1}{2},-\tfrac{1}{2}\rangle = \left\{(2/3)^{\frac{1}{2}}\langle 0,\alpha| + (1/3)^{\frac{1}{2}}\langle 1,\beta|\right\}(2/3)^{\frac{1}{2}}\hbar|-1,\alpha\rangle = 0 .$$

The remaining elements can be obtained from these and the hermiticity of j_{1z} . This lets us construct the following table of values of $(1/\hbar)\langle j'm'|j_{1z}|jm\rangle$:

$\langle j'm'\|j_{1z}\|jm\rangle/\hbar$	$\|\frac{3}{2},\frac{3}{2}\rangle$	$\|\frac{3}{2},\frac{1}{2}\rangle$	$\|\frac{1}{2},\frac{1}{2}\rangle$	$\|\frac{3}{2},-\frac{1}{2}\rangle$	$\|\frac{1}{2},-\frac{1}{2}\rangle$	$\|\frac{3}{2},-\frac{3}{2}\rangle$
$\langle\frac{3}{2},\frac{3}{2}\|$	1	0	0	0	0	0
$\langle\frac{3}{2},\frac{1}{2}\|$	0	$\frac{1}{3}$	$\frac{1}{3}\sqrt{2}$	0	0	0
$\langle\frac{1}{2},\frac{1}{2}\|$	0	$\frac{1}{3}\sqrt{2}$	$\frac{2}{3}$	0	0	0
$\langle\frac{3}{2},-\frac{1}{2}\|$	0	0	0	$-\frac{1}{3}$	$\frac{1}{3}\sqrt{2}$	0
$\langle\frac{1}{2},-\frac{1}{2}\|$	0	0	0	$\frac{1}{3}\sqrt{2}$	$-\frac{1}{3}$	0
$\langle\frac{3}{2},-\frac{3}{2}\|$	0	0	0	0	0	1

EXERCISE: Repeat the procedure for a system with $j_1 = 2$, $j_2 = \frac{1}{2}$.

6.15 $2\boldsymbol{s}_1\cdot\boldsymbol{s}_2 = (\boldsymbol{s}_1+\boldsymbol{s}_2)^2 - s_1^2 - s_2^2 = S^2 - s_1^2 - s_2^2$.

$$(J/\hbar)\boldsymbol{s}_1\cdot\boldsymbol{s}_2|\tfrac{1}{2},\tfrac{1}{2};\,SM_S\rangle = (J/2\hbar)\left\{S^2 - s_1^2 - s_2^2\right\}\left|\tfrac{1}{2},\tfrac{1}{2};\,S,M_S\right\rangle$$
$$= (J/2\hbar)\,\hbar^2\left\{S(S+1) - \tfrac{3}{2} - \tfrac{3}{4}\right\}\left|\tfrac{1}{2},\tfrac{1}{2};S,M_S\right\rangle$$
$$= \tfrac{1}{2}\hbar J\left\{S(S+1) - \tfrac{3}{2}\right\}\left|\tfrac{1}{2},\tfrac{1}{2};\,S,M_S\right\rangle .$$

Therefore, for $S = 0$ the eigenvalue is $-\frac{3}{4}\hbar J$, and for $S = 1$ the eigenvalue is $+\frac{1}{4}\hbar J$ for all three values of M_S.

$(J/\hbar)\boldsymbol{s}_1\cdot\boldsymbol{s}_2$ is diagonal in the coupled representation; the magnetic interaction has off-diagonal components. Use the vector coupling coefficients in Appendix 9:

$$H_{\mathbf{mag}}|\tfrac{1}{2},\tfrac{1}{2};\,1,1\rangle \equiv H_{\mathbf{mag}}|1,1\rangle = (\mu_{\mathrm{B}}/\hbar)\,B\,(g_1 s_{1z} + g_2 s_{2z})|\alpha\alpha\rangle$$
$$= (\mu_{\mathrm{B}}/\hbar)\,B\,(g_1\tfrac{1}{2}\hbar + g_2\tfrac{1}{2}\hbar)|\alpha\alpha\rangle = \tfrac{1}{2}\mu_{\mathrm{B}}B\,(g_1+g_2)|1,1\rangle .$$

$$H_{\mathbf{mag}}|1,0\rangle = (\mu_{\mathrm{B}}/\hbar)\,B\,(g_1 s_{1z} + g_2 s_{2z})(1/\sqrt{2})\left\{|\alpha\beta\rangle + |\beta\alpha\rangle\right\}$$
$$= (\mu_{\mathrm{B}}/\hbar)(B/\sqrt{2})(\tfrac{1}{2}\hbar)\left\{(g_1-g_2)|\alpha\beta\rangle + (-g_1+g_2)|\beta\alpha\rangle\right\}$$
$$= (1/2\sqrt{2})\,\mu_{\mathrm{B}}B\,(g_1-g_2)\left\{|\alpha\beta\rangle - |\beta\alpha\rangle\right\} = \tfrac{1}{2}\mu_{\mathrm{B}}B\,(g_1-g_2)|0,0\rangle .$$

$$H_{\mathbf{mag}}|1,-1\rangle = (\mu_{\mathrm{B}}/\hbar)\,B\,(g_1 s_{1z} + g_2 s_{2z})|\beta\beta\rangle = -\tfrac{1}{2}\mu_{\mathrm{B}}B(g_1+g_2)|1,-1\rangle .$$

$$H_{\mathbf{mag}}|0,0\rangle = (\mu_{\mathrm{B}}/\hbar)\,B\,(g_1 s_{1z} + g_2 s_{2z})(1/\sqrt{2})\left\{|\alpha\beta\rangle - |\beta\alpha\rangle\right\}$$
$$= \tfrac{1}{2}\mu_{\mathrm{B}}B(g_1-g_2)|1,0\rangle .$$

Therefore, on writing $g=\frac{1}{2}(g_1+g_2)$ and $\Delta g = \frac{1}{2}(g_1-g_2)$, the matrix is

$\langle S', M_S' \vert H \vert S, M_S \rangle$	$\vert 1,1\rangle$	$\vert 1,0\rangle$	$\vert 0,0\rangle$	$\vert 1,-1\rangle$
$\langle 1,1 \vert$	$\frac{1}{4}\hbar J + g\mu_{\mathbf{B}} B$	0	0	0
$\langle 1,0 \vert$	0	$\frac{1}{4}\hbar J$	$\Delta g \mu_{\mathbf{B}} B$	0
$\langle 0,0 \vert$	0	$\Delta g \mu_{\mathbf{B}} B$	$-\frac{3}{4}\hbar J$	0
$\langle 1,-1 \vert$	0	0	0	$\frac{1}{4}\hbar J - g\mu_{\mathbf{B}} B$

In the uncoupled representation, $H_{\mathbf{mag}}$ has only diagonal elements:

$$H_{\mathbf{mag}}|\alpha\alpha\rangle = \tfrac{1}{2}\mu_{\mathbf{B}} B(g_1+g_2)|\alpha\alpha\rangle = g\mu_{\mathbf{B}}B|\alpha\alpha\rangle$$
$$H_{\mathbf{mag}}|\alpha\beta\rangle = \tfrac{1}{2}\mu_{\mathbf{B}} B(g_1-g_2)|\alpha\beta\rangle = \Delta g\mu_{\mathbf{B}}B|\alpha\beta\rangle$$
$$H_{\mathbf{mag}}|\beta\alpha\rangle = \tfrac{1}{2}\mu_{\mathbf{B}} B(-g_1+g_2)|\beta\alpha\rangle = -\Delta g\mu_{\mathbf{B}}B|\beta\alpha\rangle$$
$$H_{\mathbf{mag}}|\beta\beta\rangle = -\tfrac{1}{2}\mu_{\mathbf{B}} B(g_1+g_2)|\beta\beta\rangle = -g\mu_{\mathbf{B}}B|\beta\beta\rangle .$$

On the other hand, $J\boldsymbol{s}_1\cdot\boldsymbol{s}_2$ has off-diagonal elements. In order to calculate these, use $\boldsymbol{s}_1\cdot\boldsymbol{s}_2 = s_{1z}s_{2z} + \frac{1}{2}(s_1^+ s_2^- + s_1^- s_2^+)$.

$$(J/\hbar)\boldsymbol{s}_1\cdot\boldsymbol{s}_2|\alpha\alpha\rangle = (J/\hbar)s_{1z}s_{2z}|\alpha\alpha\rangle = \tfrac{1}{4}\hbar J|\alpha\alpha\rangle$$
$$(J/\hbar)\boldsymbol{s}_1\cdot\boldsymbol{s}_2|\alpha\beta\rangle = (J/\hbar)\left\{s_{1z}s_{2z}|\alpha\beta\rangle + \tfrac{1}{2}s_1^+ s_2^-|\alpha\beta\rangle + \tfrac{1}{2}s_1^- s_2^+|\alpha\beta\rangle\right\}$$
$$= -\tfrac{1}{4}\hbar J|\alpha\beta\rangle + \tfrac{1}{2}\hbar J|\beta\alpha\rangle$$
$$(J/\hbar)\boldsymbol{s}_1\cdot\boldsymbol{s}_2|\beta\alpha\rangle = -\tfrac{1}{4}\hbar J|\beta\alpha\rangle + \tfrac{1}{2}\hbar J|\alpha\beta\rangle$$
$$(J/\hbar)\boldsymbol{s}_1\cdot\boldsymbol{s}_2|\beta\beta\rangle = \tfrac{1}{4}\hbar J|\beta\beta\rangle .$$

The overall matrix is therefore

$\langle m_{s1}' m_{s2}' \vert H \vert m_{s1} m_{s2}\rangle$	$\vert\alpha\alpha\rangle$	$\vert\alpha\beta\rangle$	$\vert\beta\alpha\rangle$	$\vert\beta\beta\rangle$
$\langle\alpha\alpha\vert$	$\frac{1}{4}\hbar J + g\mu_{\mathbf{B}}B$	0	0	0
$\langle\alpha\beta\vert$	0	$-\frac{1}{4}\hbar J + \Delta g\mu_{\mathbf{B}}B$	$\frac{1}{2}\hbar J$	0
$\langle\beta\alpha\vert$	0	$\frac{1}{2}\hbar J$	$-\frac{1}{4}\hbar J - \Delta g\mu_{\mathbf{B}}B$	0
$\langle\beta\beta\vert$	0	0	0	$\frac{1}{4}\hbar J - g\mu_{\mathbf{B}}B$

When $\hbar J \gg \mu_{\mathbf{B}}B$ the off-diagonal elements are negligible in the coupled representation, and so that is then a 'good' description. When $\mu_{\mathbf{B}}B \gg \hbar J$ the off-diagonal elements are negligible in the uncoupled representation, and so that is then a 'good' description.

EXERCISE: Consider a system of two spin-$\frac{1}{2}$ nuclei, one in a local field $(1-\sigma_1)B$, the other in a local field $(1-\sigma_2)B$, and coupled together by the spin-spin term $(J/\hbar)\boldsymbol{I}_1\cdot\boldsymbol{I}_2$. Set up the matrices of the hamiltonian and find the eigenvalues in the cases J large and J small.

6.16 $|G,4\rangle = |2,2\rangle$ [eqn (6.8.11)].

$|G,3\rangle = (1/\sqrt{2})\{|1,2\rangle + |2,1\rangle\}$ [eqn (6.8.12)].

$$L^-|G,3\rangle = \hbar\sqrt{(14)}\,|G,2\rangle$$
$$=(l_1^- + l_2^-)(1/\sqrt{2})\{|1,2\rangle + |2,1\rangle\}$$
$$=(\hbar/\sqrt{2})\{\sqrt{6}|0,2\rangle + 2|1,1\rangle + 2|1,1\rangle + \sqrt{6}|2,0\rangle\}$$
$$=(\hbar/\sqrt{2})\{\sqrt{6}|0,2\rangle + 4|1,1\rangle + \sqrt{6}|2,0\rangle\};$$

therefore

$$|G,2\rangle = (1/2\sqrt{7})\{\sqrt{6}|0,2\rangle + 4|1,1\rangle + \sqrt{6}|2,0\rangle\}$$

$$L^-|G,2\rangle = (3\sqrt{2})\hbar|G,1\rangle$$
$$=(l_1^- + l_2^-)(1/2\sqrt{7})\{\sqrt{6}|0,2\rangle + 4|1,1\rangle + \sqrt{6}|2,0\rangle\}$$
$$=(\hbar/2\sqrt{7})\{6|-1,2\rangle + 2\sqrt{6}|0,1\rangle + 4\sqrt{6}|0,1\rangle + 4\sqrt{6}|1,0\rangle + 2\sqrt{6}|1,0\rangle + 6|2,-1\rangle\}$$
$$=(3/\sqrt{7})\hbar\{|-1,2\rangle + \sqrt{6}|0,1\rangle + \sqrt{6}|1,0\rangle + |2,-1\rangle\};$$

therefore

$$|G,1\rangle = (1/\sqrt{14})\{|-1,2\rangle + \sqrt{6}|0,1\rangle + \sqrt{6}|1,0\rangle + |2,-1\rangle\}.$$

$$L^-|G,1\rangle = (2\sqrt{5})\hbar|G,0\rangle$$
$$=(l_1^- + l_2^-)(1/\sqrt{14})\{|-1,2\rangle + \sqrt{6}|0,1\rangle + \sqrt{6}|1,0\rangle + |2,-1\rangle\}$$
$$=(\hbar/\sqrt{14})\{2|-2,2\rangle + 2|-1,1\rangle + 6|-1,1\rangle + 6|0,0\rangle + 6|0,0\rangle + 6|1,-1\rangle + 2|1,-1\rangle + 2|2,-2\rangle\}$$
$$=(2\hbar/\sqrt{14})\{|-2,2\rangle + 4|-1,1\rangle + 6|0,0\rangle + 4|1,-1\rangle + |2,-2\rangle\};$$

therefore

$$|G,0\rangle = (1/\sqrt{70})\{|-2,2\rangle + 4|-1,1\rangle + 6|0,0\rangle + 4|1,-1\rangle + |2,-2\rangle\}.$$

Now evaluate $\langle G,M_L|l_{1z}|G,M_L\rangle$ for each M_L:

$$\langle G,4|l_{1z}|G,4\rangle = \langle 2,2|l_{1z}|2,2\rangle = 2\hbar.$$

$$\langle G,3|l_{1z}|G,3\rangle = (1/2)\{\langle 1,2| + \langle 2,1|\}l_{1z}\{|1,2\rangle + |2,1\rangle\}$$
$$= (\hbar/2)\{\langle 1,2| + \langle 2,1|\}\{|1,2\rangle + 2|2,1\rangle\} = \tfrac{3}{2}\hbar.$$

$$\langle G,2|l_{1z}|G,2\rangle = (\hbar/28)\left\{\sqrt{6}\langle 0,2| + 4\langle 1,1| + \sqrt{6}\langle 2,0|\right\} \times \left\{0 + 4|1,1\rangle + 2\sqrt{6}|2,0\rangle\right\}$$

$$= (\hbar/28)\left\{0 + 16 + 12\right\} = \hbar .$$

$$\langle G,1|l_{1z}|G,1\rangle = (\hbar/14)\left\{\langle -1,2| + \sqrt{6}\langle 1,0| + \sqrt{6}\langle 1,0| + \langle 2,-1|\right\} \times \left\{-|-1,2\rangle + 0 + \sqrt{6}|1,0\rangle + 2|2,-1\rangle\right\}$$

$$= (\hbar/14)\left\{-1 + 0 + 6 + 2\right\} = \tfrac{1}{2}\hbar .$$

$$\langle G,0|l_{1z}|G,0\rangle = (\hbar/70)\left\{\langle -2,2| + 4\langle -1,1| + 6\langle 0,0| + 4\langle 1,-1| + \langle 2,-2|\right\} \times\left\{-2|-2,2\rangle - 4|-1,1\rangle + 0 + 4|1,-1\rangle + 2|2,-2\rangle\right\}$$

$$= (\hbar/70)\left\{-2 - 16 + 0 + 16 + 2\right\} = 0 .$$

By symmetry

$$\langle G,-1|l_{1z}|G,-1\rangle = -\tfrac{1}{2}\hbar , \qquad \langle G,-2|l_{1z}|G,-2\rangle = -\hbar$$

$$\langle G,-3|l_{1z}|G,-3\rangle = -\tfrac{3}{2}\hbar , \qquad \langle G,-4|l_{1z}|G,-4\rangle = -2\hbar .$$

That is, in general,

$$\underline{\langle G,M_L|l_{1z}|G,M_L\rangle = \tfrac{1}{2}M_L\hbar} .$$

Since $\langle G,M_L|l_{1z}+l_{2z}|G,M_L\rangle = M_L\hbar$, it then follows that

$$\underline{\langle G,M_L|l_{2z}|G,M_L\rangle = \tfrac{1}{2}M_L\hbar} .$$

EXERCISE: Calculate $\langle F,M_L|l_{1z}|F,M_L\rangle$ for the same configuration,

6.17 $\langle j_1 j_2;jm_j|j_1 j_2;jm_j\rangle$

$$= \sum_{m'_{j1}}\sum_{m'_{j2}}\sum_{m_{j1}}\sum_{m_{j2}} C^*_{m'_{j1}m'_{j2}} C_{m_{j1}m_{j2}} \langle j_1 m'_{j1} j_2 m'_{j2}|j_1 m_{j1} j_2 m_{j2}\rangle$$

$$= \sum_{m'_{j1}}\sum_{m'_{j2}}\sum_{m_{j1}}\sum_{m_{j2}} C^*_{m'_{j1}m'_{j2}} C_{m_{j1}m_{j2}} \delta_{m'_{j1}m_{j1}} \delta_{m'_{j2}m_{j2}}$$

$$= \sum_{m_{j1}}\sum_{m_{j2}} C^*_{m_{j1}m_{j2}} C_{m_{j1}m_{j2}} = \sum_{m_{j1}}\sum_{m_{j2}} \left|C_{m_{j1}m_{j2}}\right|^2 .$$

But $\langle j_1 j_2;jm_j|j_1 j_2;jm_j\rangle = 1$, which completes the proof.

EXERCISE: Find a general expression for $\langle j_1 j_2;jm_j|j_{1z}|j_1 j_2;jm_j\rangle$ and evaluate it for $\langle G,M_L|l_{1z}|G,M_L\rangle$; see Problem 6.16.

7 Group theory

7.1 (a) H_2O: $E, C_2, 2\sigma_v$; hence $\underline{C_{2v}}$.

(b) CO_2: $E, C_\infty, C_2 \perp C_\infty, \sigma_h$; hence $\underline{D_{\infty h}}$.

(c) C_2H_4: $E, C_2, 2C_2' \perp C_2, \sigma_h$; hence $\underline{D_{2h}}$.

(d) *cis*-ClHC = CHCl: $E, C_2, 2\sigma_v$; hence $\underline{C_{2v}}$.

(e) *trans*-ClHC = CHCl: E, C_2, σ_h; hence $\underline{C_{2h}}$.

(f) Benzene: $E, C_6, 6C_2', \sigma_h$; hence $\underline{D_{6h}}$.

(g) Naphthalene: E, C_2, C_2', σ_h; hence $\underline{D_{2h}}$.

(h) CHClFBr: E; hence $\underline{C_1}$.

(i) $B(OH)_3$: E, C_3, σ_h; hence $\underline{C_{3h}}$.

EXERCISE: Classify chlorobenzene, anthracene, H_2O_2, S_8.

7.2 $\mu = \int \psi^*\mu\psi d\tau = 0$ unless $\Gamma(\psi^*) \times \Gamma(\mu) \times \Gamma(\psi)$ contains A_1. But $\Gamma(\psi^*) \times \Gamma(\psi) = A_1$; hence $\Gamma(\mu) = A_1$ is a necessary requirement. Since $\Gamma(\mu) = \Gamma(r)$, inspect the group character tables (Appendix 10, and *Tables for Group Theory*, P.W. Atkins, M.S. Child, and C.S.G. Phillips, Oxford University Press, 1970) to see if any translation transforms as A_1. We have:

C_{2v} : z ; $D_{\infty h}$: none ; D_{2h}: none ; C_{2h}: none ; D_{6h}: none ,
C_1 : x, y, z ; C_{3h}: none.

Hence only H_2O, *cis*–ClHC = CHCl, and CHClFBr may have permanent dipole moments.

EXERCISE: Which of chlorobenzene, anthracene, H_2O_2, and S_8 may have permanent electric dipole moments ?

7.3 Write $\boldsymbol{f} = (\text{H1s}_A, \text{H1s}_B, \text{O2s}, \text{O2p}_x, \text{O2p}_y, \text{O2p}_z)$; then $E\boldsymbol{f} = \boldsymbol{f} = \boldsymbol{f}\mathbb{1}$; hence $\boldsymbol{D}(E) = \mathbb{1}$, the 6×6 unit matrix.

$$C_2\boldsymbol{f} = (\text{H1s}_B, \text{H1s}_A, \text{O2s}, -\text{O2p}_x, -\text{O2p}_y, \text{O2p}_z)$$

$$= \boldsymbol{f}\begin{bmatrix} 0 & 1 & 0 & 0 & 0 & 0 \\ 1 & 0 & 0 & 0 & 0 & 0 \\ 0 & 0 & 1 & 0 & 0 & 0 \\ 0 & 0 & 0 & -1 & 0 & 0 \\ 0 & 0 & 0 & 0 & -1 & 0 \\ 0 & 0 & 0 & 0 & 0 & 1 \end{bmatrix} = \boldsymbol{f}\boldsymbol{D}(C_2).$$

$$\sigma_v \boldsymbol{f} = (\mathrm{H1s_B}\ ,\ \mathrm{H1s_A}\ ,\ \mathrm{O2s}\ ,\ \mathrm{O2p}_x\ ,\ -\mathrm{O2p}_y\ ,\ \mathrm{O2p}_z)$$

$$= \boldsymbol{f}\begin{bmatrix} 0 & 1 & 0 & 0 & 0 & 0 \\ 1 & 0 & 0 & 0 & 0 & 0 \\ 0 & 0 & 1 & 0 & 0 & 0 \\ 0 & 0 & 0 & 1 & 0 & 0 \\ 0 & 0 & 0 & 0 & -1 & 0 \\ 0 & 0 & 0 & 0 & 0 & 1 \end{bmatrix} = \boldsymbol{f}\boldsymbol{D}(\sigma_v).$$

$$\sigma'_v \boldsymbol{f} = (\mathrm{H1s_A}\ ,\ \mathrm{H1s_B}\ ,\ \mathrm{O2s}\ ,\ -\mathrm{O2p}_x,\ \mathrm{O2p}_y\ ,\ \mathrm{O2p}_z)$$

$$= \boldsymbol{f}\begin{bmatrix} 1 & 0 & 0 & 0 & 0 & 0 \\ 0 & 1 & 0 & 0 & 0 & 0 \\ 0 & 0 & 1 & 0 & 0 & 0 \\ 0 & 0 & 0 & -1 & 0 & 0 \\ 0 & 0 & 0 & 0 & 1 & 0 \\ 0 & 0 & 0 & 0 & 0 & 1 \end{bmatrix} = \boldsymbol{f}\boldsymbol{D}(\sigma'_v).$$

EXERCISE: Replace the p-orbitals by d-orbitals, and find the matrix representation.

7.4

$$\boldsymbol{D}(C_2)\boldsymbol{D}(C_2) = \begin{bmatrix} 0 & 1 & 0 & 0 & 0 & 0 \\ 1 & 0 & 0 & 0 & 0 & 0 \\ 0 & 0 & 1 & 0 & 0 & 0 \\ 0 & 0 & 0 & -1 & 0 & 0 \\ 0 & 0 & 0 & 0 & -1 & 0 \\ 0 & 0 & 0 & 0 & 0 & 1 \end{bmatrix}\begin{bmatrix} 0 & 1 & 0 & 0 & 0 & 0 \\ 1 & 0 & 0 & 0 & 0 & 0 \\ 0 & 0 & 1 & 0 & 0 & 0 \\ 0 & 0 & 0 & -1 & 0 & 0 \\ 0 & 0 & 0 & 0 & -1 & 0 \\ 0 & 0 & 0 & 0 & 0 & 1 \end{bmatrix}$$

$$= \begin{bmatrix} 1 & 0 & 0 & 0 & 0 & 0 \\ 0 & 1 & 0 & 0 & 0 & 0 \\ 0 & 0 & 1 & 0 & 0 & 0 \\ 0 & 0 & 0 & 1 & 0 & 0 \\ 0 & 0 & 0 & 0 & 1 & 0 \\ 0 & 0 & 0 & 0 & 0 & 1 \end{bmatrix} = \boldsymbol{D}(E);\ \text{reproducing}\ C_2^2 = E.$$

$$\boldsymbol{D}(\sigma_v)\boldsymbol{D}(C_2) = \begin{bmatrix} 0 & 1 & 0 & 0 & 0 & 0 \\ 1 & 0 & 0 & 0 & 0 & 0 \\ 0 & 0 & 1 & 0 & 0 & 0 \\ 0 & 0 & 0 & 1 & 0 & 0 \\ 0 & 0 & 0 & 0 & -1 & 0 \\ 0 & 0 & 0 & 0 & 0 & 1 \end{bmatrix}\begin{bmatrix} 0 & 1 & 0 & 0 & 0 & 0 \\ 1 & 0 & 0 & 0 & 0 & 0 \\ 0 & 0 & 1 & 0 & 0 & 0 \\ 0 & 0 & 0 & -1 & 0 & 0 \\ 0 & 0 & 0 & 0 & -1 & 0 \\ 0 & 0 & 0 & 0 & 0 & 1 \end{bmatrix}$$

$$= \begin{bmatrix} 1 & 0 & 0 & 0 & 0 & 0 \\ 0 & 1 & 0 & 0 & 0 & 0 \\ 0 & 0 & 1 & 0 & 0 & 0 \\ 0 & 0 & 0 & -1 & 0 & 0 \\ 0 & 0 & 0 & 0 & 1 & 0 \\ 0 & 0 & 0 & 0 & 0 & 1 \end{bmatrix} = D(\sigma_v'); \text{ reproducing } \sigma_v C_2 = \sigma_v' .$$

EXERCISE: Confirm these multiplications for the representatives constructed using d-orbitals.

7.5 The sums of the diagonal elements in the matrices in Problem 7.3 are

$$\chi(E) = 6,\ \chi(C_2) = 0,\ \chi(\sigma_v) = 2,\ \chi(\sigma_v') = 4 .$$

Use eqn (7.8.5) in the form

$$a_l = (1/4)\left\{6\chi^{(l)}(E) + 0 + 2\chi^{(l)}(\sigma_v) + 4\chi^{(l)}(\sigma_v')\right\} .$$

Then

$$a(A_1) = \tfrac{1}{4}\{6+0+2+4\} = 3,\quad a(A_2) = \tfrac{1}{4}\{6+0-2-4\} = 0,$$

$$a(B_1) = \tfrac{1}{4}\{6-0+2-4\} = 1,\quad a(B_2) = \tfrac{1}{4}\{6-0-2+4\} = 2 .$$

Hence, the reduction is into $\underline{3A_1 + B_1 + 2B_2}$.

Draw up the following Table :

	$H1s_A$	$H1s_B$	$O2s$	$O2p_x$	$O2p_y$	$O2p_z$
E	$H1s_A$	$H1s_B$	$O2s$	$O2p_x$	$O2p_y$	$O2p_z$
C	$H1s_B$	$H1s_A$	$O2s$	$-O2p_x$	$-O2p_y$	$O2p_z$
σ_v	$H1s_B$	$H1s_A$	$O2s$	$O2p_x$	$-O2p_y$	$O2p_z$
σ_v'	$H1s_A$	$H1s_B$	$O2s$	$-O2p_x$	$O2p_y$	$O2p_z$

Form $f^{(A_1)}$ using $p^{(A_1)} = \frac{1}{4}\sum_R \chi^{(A_1)}(R)R$. From column 1,

$$f^{(A_1)} = \tfrac{1}{4}\left\{H1s_A + H1s_B + H1s_B + H1s_A\right\} = \tfrac{1}{2}\left\{H1s_A + H1s_B\right\} .$$

From column 2, find the same. From column 3, $f^{(A_1)} = O2s$, from columns 4 and 5 obtain 0. From column 6, $f^{(A_1)} = O2p_z$. Hence

$$\underline{f^{(A_1)} = \left\{\tfrac{1}{2}(H1s_A + H1s_B),\ O2s,\ O2p_z\right\}} .$$

Form $f^{(B_1)}$: only column 4 gives a non-zero quantity.

$$\underline{f^{(B_1)} = O2p_x} .$$

Form $f^{(B_2)}$: columns 1,3,4,6 give zero; columns 2 and 5 give

$$\boldsymbol{f}^{(B_2)} = \left\{\tfrac{1}{2}(\mathrm{H1s_A} - \mathrm{H1s_B}),\ \mathrm{O2p}_y\right\}.$$

Only $f_1^{(A_1)}$ and $f_1^{(B_2)}$ involve linear combinations; the matrix of coefficients in eqn (7.5.1) is therefore given by

$$\left\{\tfrac{1}{2}(\mathrm{H1s_A}+\mathrm{H1s_B}),\ \tfrac{1}{2}(\mathrm{H1s_B}-\mathrm{H1s_A}),\ \mathrm{O2s},\ \mathrm{O2p}_x,\ \mathrm{O2p}_y,\ \mathrm{O2p}_z\right\}$$

$$= \left\{\mathrm{H1s_A},\ \mathrm{H1s_B},\ \mathrm{O2s},\ \mathrm{O2p}_x,\ \mathrm{O2p}_y,\ \mathrm{O2p}_z\right\}\begin{bmatrix} \frac{1}{2} & -\frac{1}{2} & 0 & 0 & 0 & 0 \\ \frac{1}{2} & \frac{1}{2} & 0 & 0 & 0 & 0 \\ 0 & 0 & 1 & 0 & 0 & 0 \\ 0 & 0 & 0 & 1 & 0 & 0 \\ 0 & 0 & 0 & 0 & 1 & 0 \\ 0 & 0 & 0 & 0 & 0 & 1 \end{bmatrix}.$$

Consequently,

$$\boldsymbol{C} = \begin{bmatrix} \frac{1}{2} & -\frac{1}{2} & 0 & 0 & 0 & 0 \\ \frac{1}{2} & \frac{1}{2} & 0 & 0 & 0 & 0 \\ 0 & 0 & 1 & 0 & 0 & 0 \\ 0 & 0 & 0 & 1 & 0 & 0 \\ 0 & 0 & 0 & 0 & 1 & 0 \\ 0 & 0 & 0 & 0 & 0 & 1 \end{bmatrix}, \qquad \boldsymbol{C}^{-1} = \begin{bmatrix} 1 & 1 & 0 & 0 & 0 & 0 \\ -1 & 1 & 0 & 0 & 0 & 0 \\ 0 & 0 & 1 & 0 & 0 & 0 \\ 0 & 0 & 0 & 1 & 0 & 0 \\ 0 & 0 & 0 & 0 & 1 & 0 \\ 0 & 0 & 0 & 0 & 0 & 1 \end{bmatrix}.$$

Then, from eqn (7.5.6), showing only the H1s—combinations:

$$\boldsymbol{D}'(E) = \begin{pmatrix} 1 & 1 \\ -1 & 1 \end{pmatrix}\begin{pmatrix} 1 & 0 \\ 0 & 1 \end{pmatrix}\begin{pmatrix} \frac{1}{2} & -\frac{1}{2} \\ \frac{1}{2} & \frac{1}{2} \end{pmatrix} = \begin{pmatrix} 1 & 0 \\ 0 & 1 \end{pmatrix}$$

$$\boldsymbol{D}'(C_2) = \begin{pmatrix} 1 & 1 \\ -1 & 1 \end{pmatrix}\begin{pmatrix} 0 & 1 \\ 1 & 0 \end{pmatrix}\begin{pmatrix} \frac{1}{2} & -\frac{1}{2} \\ \frac{1}{2} & \frac{1}{2} \end{pmatrix} = \begin{pmatrix} 1 & 0 \\ 0 & -1 \end{pmatrix}$$

$$\boldsymbol{D}'(\sigma_v) = \begin{pmatrix} 1 & 1 \\ -1 & 1 \end{pmatrix}\begin{pmatrix} 0 & 1 \\ 1 & 0 \end{pmatrix}\begin{pmatrix} \frac{1}{2} & -\frac{1}{2} \\ \frac{1}{2} & \frac{1}{2} \end{pmatrix} = \begin{pmatrix} 1 & 0 \\ 0 & -1 \end{pmatrix}$$

$$\boldsymbol{D}'(\sigma_v') = \begin{pmatrix} 1 & 1 \\ -1 & 1 \end{pmatrix}\begin{pmatrix} 1 & 0 \\ 0 & 1 \end{pmatrix}\begin{pmatrix} \frac{1}{2} & -\frac{1}{2} \\ \frac{1}{2} & \frac{1}{2} \end{pmatrix} = \begin{pmatrix} 1 & 0 \\ 0 & 1 \end{pmatrix}.$$

Since these are diagonal (and therefore also block-diagonal), and the remainder of $\boldsymbol{D}(R)$ are already diagonal, the entire representation is (block-) diagonal.

EXERCISE: Consider a representation using the basis (p_x, p_y, p_z) on each atom in a C_{2v} AB_2 molecule. Find the representatives, the symmetry-adapted combinations, and the block-diagonal representations.

7.6   Refer to Fig. 7.1. $C_3(A)$ refers to a rotation about the axis through
A, $S(AB)$ to an improper rotation around the bisector of AB, $\sigma_d(AB)$
to a reflection in the plane passing through AB, and $C_2(AB)$ to a
rotation around the bisector of AB.   Write $f = (A,B,C,D)$.

Fig. 7.1

$f = ($ ⬜ , ◆ , ▲ , ⬜ $)$

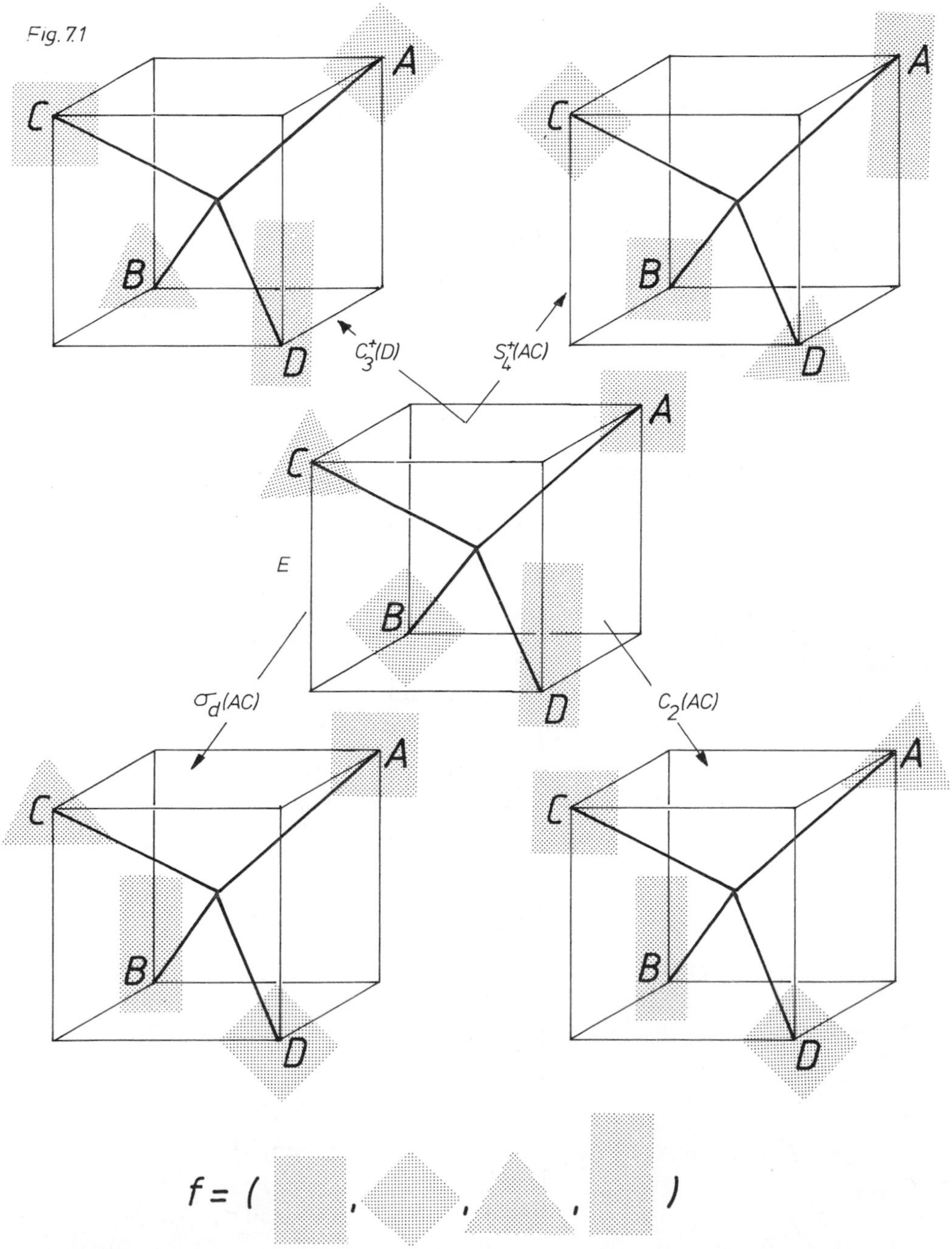

$Ef = f = f\mathbb{1}$; hence $D(E) = \mathbb{1}$ (the 4×4 unit matrix); $\chi = 4$.

$$C_3^+(\mathrm{D})f = (\mathrm{C},\mathrm{A},\mathrm{B},\mathrm{D}) = f\begin{bmatrix}0&1&0&0\\0&0&1&0\\1&0&0&0\\0&0&0&1\end{bmatrix} = fD(C_3^+(\mathrm{D})); \quad \chi = 1.$$

$$C_3^-(\mathrm{D})f = (\mathrm{B},\mathrm{C},\mathrm{A},\mathrm{D}) = f\begin{bmatrix}0&0&1&0\\1&0&0&0\\0&1&0&0\\0&0&0&1\end{bmatrix} = fD(C_3^-(\mathrm{D})); \quad \chi = 1.$$

$$C_3^+(\mathrm{C})f = (\mathrm{B},\mathrm{D},\mathrm{C},\mathrm{A}) = f\begin{bmatrix}0&0&0&1\\1&0&0&0\\0&0&1&0\\0&1&0&0\end{bmatrix} = fD(C_3^+(\mathrm{C})); \quad \chi = 1.$$

$$C_3^-(\mathrm{C})f = (\mathrm{D},\mathrm{A},\mathrm{C},\mathrm{B}) = f\begin{bmatrix}0&1&0&0\\0&0&0&1\\0&0&1&0\\1&0&0&0\end{bmatrix} = fD(C_3^-(\mathrm{C})); \quad \chi = 1.$$

$$C_3^+(\mathrm{B})f = (\mathrm{D},\mathrm{B},\mathrm{A},\mathrm{C}) = f\begin{bmatrix}0&0&1&0\\0&1&0&0\\0&0&0&1\\1&0&0&0\end{bmatrix} = fD(C_3^+(\mathrm{B})); \quad \chi = 1.$$

$$C_3^-(\mathrm{B})f = (\mathrm{C},\mathrm{B},\mathrm{D},\mathrm{A}) = f\begin{bmatrix}0&0&0&1\\0&1&0&0\\1&0&0&0\\0&0&1&0\end{bmatrix} = fD(C_3^-(\mathrm{B})); \quad \chi = 1.$$

$$C_3^+(\mathrm{A})f = (\mathrm{A},\mathrm{C},\mathrm{D},\mathrm{B}) = f\begin{bmatrix}1&0&0&0\\0&0&0&1\\0&1&0&0\\0&0&1&0\end{bmatrix} = fD(C_3^+(\mathrm{A})); \quad \chi = 1.$$

$$C_3^-(\mathrm{A})f = (\mathrm{A},\mathrm{D},\mathrm{B},\mathrm{C}) = f\begin{bmatrix}1&0&0&0\\0&0&1&0\\0&0&0&1\\0&1&0&0\end{bmatrix} = fD(C_3^-(\mathrm{A})); \quad \chi = 1.$$

$$S_4^+(AC)f = (B,C,D,A) = f\begin{bmatrix}0&0&0&1\\1&0&0&0\\0&1&0&0\\0&0&1&0\end{bmatrix} = fD(S_4^+(AC));\ \chi = 0.$$

$$S_4^-(AC)f = (D,A,B,C) = f\begin{bmatrix}0&1&0&0\\0&0&1&0\\0&0&0&1\\1&0&0&0\end{bmatrix} = fD(S_4^-(AC));\ \chi = 0.$$

$$S_4^+(CD)f = (C,D,B,A) = f\begin{bmatrix}0&0&0&1\\0&0&1&0\\1&0&0&0\\0&1&0&0\end{bmatrix} = fD(S_4^+(CD));\ \chi = 0.$$

$$S_4^-(CD)f = (D,C,A,B) = f\begin{bmatrix}0&0&1&0\\0&0&0&1\\0&1&0&0\\1&0&0&0\end{bmatrix} = fD(S_4^-(CD));\ \chi = 0.$$

$$S_4^+(AD)f = (C,A,D,B) = f\begin{bmatrix}0&1&0&0\\0&0&0&1\\1&0&0&0\\0&0&1&0\end{bmatrix} = fD(S_4^+(AD));\ \chi = 0.$$

$$S_4^-(AD)f = (B,D,A,C) = f\begin{bmatrix}0&0&1&0\\1&0&0&0\\0&0&0&1\\0&1&0&0\end{bmatrix} = fD(S_4^-(AD));\ \chi = 0.$$

$$\sigma_d(AC)f = (A,D,C,B) = f\begin{bmatrix}1&0&0&0\\0&0&0&1\\0&0&1&0\\0&1&0&0\end{bmatrix} = fD(\sigma_d(AC));\ \chi = 2.$$

$$\sigma_d(BD)f = (C,B,A,D) = f\begin{bmatrix}0&0&1&0\\0&1&0&0\\1&0&0&0\\0&0&0&1\end{bmatrix} = fD(\sigma_d(BD));\ \chi = 2.$$

$$\sigma_d(CD)f = (B,A,C,D) = f\begin{bmatrix}0&1&0&0\\1&0&0&0\\0&0&1&0\\0&0&0&1\end{bmatrix} = fD(\sigma_d(CD)); \quad \chi = 2 .$$

$$\sigma_d(AD)f = (A,C,B,D) = f\begin{bmatrix}1&0&0&0\\0&0&1&0\\0&1&0&0\\0&0&0&1\end{bmatrix} = fD(\sigma_d(AD)); \quad \chi = 2 .$$

$$\sigma_d(BC)f = (D,B,C,A) = f\begin{bmatrix}0&0&0&1\\0&1&0&0\\0&0&1&0\\1&0&0&0\end{bmatrix} = fD(\sigma_d(BC)); \quad \chi = 2.$$

$$\sigma_d(AB)f = (A,B,D,C) = f\begin{bmatrix}1&0&0&0\\0&1&0&0\\0&0&0&1\\0&0&1&0\end{bmatrix} = fD(\sigma_d(AB)); \quad \chi = 2.$$

$$C_2(AC)f = (C,D,A,B) = f\begin{bmatrix}0&0&1&0\\0&0&0&1\\1&0&0&0\\0&1&0&0\end{bmatrix} = fD(C_2(AC)); \quad \chi = 0 .$$

$$C_2(CD)f = (B,A,D,C) = f\begin{bmatrix}0&1&0&0\\1&0&0&0\\0&0&0&1\\0&0&1&0\end{bmatrix} = fD(C_2(CD)); \quad \chi = 0.$$

$$C_2(AD)f = (D,C,B,A) = f\begin{bmatrix}0&0&0&1\\0&0&1&0\\0&1&0&0\\1&0&0&0\end{bmatrix} = fD(C_2(AD)); \quad \chi = 0.$$

EXERCISE: Find the representation for the 1s-orbital basis at the corners of a regular trigonal bipyramid.

7.7 $D(C_3^+(\mathrm{A}))\,D(C_3^-(\mathrm{A}))$

$$= \begin{bmatrix} 1 & 0 & 0 & 0 \\ 0 & 0 & 0 & 1 \\ 0 & 1 & 0 & 0 \\ 0 & 0 & 1 & 0 \end{bmatrix} \begin{bmatrix} 1 & 0 & 0 & 0 \\ 0 & 0 & 1 & 0 \\ 0 & 0 & 0 & 1 \\ 0 & 1 & 0 & 0 \end{bmatrix} = \begin{bmatrix} 1 & 0 & 0 & 0 \\ 0 & 1 & 0 & 0 \\ 0 & 0 & 1 & 0 \\ 0 & 0 & 0 & 1 \end{bmatrix} = D(E)\,.$$

$D(S_4^+(\mathrm{AC}))\,D(C_3^-(\mathrm{B}))$

$$= \begin{bmatrix} 0 & 0 & 0 & 1 \\ 1 & 0 & 0 & 0 \\ 0 & 1 & 0 & 0 \\ 0 & 0 & 1 & 0 \end{bmatrix} \begin{bmatrix} 0 & 0 & 0 & 1 \\ 0 & 1 & 0 & 0 \\ 1 & 0 & 0 & 0 \\ 0 & 0 & 1 & 0 \end{bmatrix} = \begin{bmatrix} 0 & 0 & 1 & 0 \\ 0 & 0 & 0 & 1 \\ 0 & 1 & 0 & 0 \\ 1 & 0 & 0 & 0 \end{bmatrix} = D(S_4^-(\mathrm{CD}))\,.$$

$D(S_4^+(\mathrm{AC}))\,D(C_3^-(\mathrm{C}))$

$$= \begin{bmatrix} 0 & 0 & 0 & 1 \\ 1 & 0 & 0 & 0 \\ 0 & 1 & 0 & 0 \\ 0 & 0 & 1 & 0 \end{bmatrix} \begin{bmatrix} 0 & 1 & 0 & 0 \\ 0 & 0 & 0 & 1 \\ 0 & 0 & 1 & 0 \\ 1 & 0 & 0 & 0 \end{bmatrix} = \begin{bmatrix} 1 & 0 & 0 & 0 \\ 0 & 1 & 0 & 0 \\ 0 & 0 & 0 & 1 \\ 0 & 0 & 1 & 0 \end{bmatrix} = D(\sigma_d(\mathrm{AD}))\,.$$

EXERCISE: Check three of the group multiplications for the representation developed in the last *Exercise*.

7.8 Proceed *either* directly from the characters listed in Problem 7.6:

$$\chi = 4,1,0,2,0 \quad \text{for} \quad E, 8C_3, 3C_2, 6\sigma_d, 6S_4,$$

or, if Problem 7.6 has not been attempted (or completed), by the quick method set out in the *Example* on p.151 of the text. For the latter, note that under the operations E, C_3, C_2, σ_d, and S_4 there are respectively 4,1,0,2,0 basis orbitals left unchanged. Hence, the characters are as set out above. Then, applying eqn (7.8.5) with $h = 24$, we have

$$a_l = (1/24)\left\{4\chi^{(l)}(E) + 8\chi^{(l)}(C_3) + 12\chi^{(l)}(\sigma_d)\right\},$$

which gives

$$a(A_1) = (1/24)(4 + 8 + 12) = 1\,; \quad a(A_2) = (1/24)(4 + 8 - 12) = 0;$$

$$a(E) = (1/24)(8 - 8 + 0) = 0\,;$$

$$a(T_1) = (1/24)(12 + 0 - 12) = 0\,; \quad a(T_2) = (1/24)(12 + 0 + 12) = 0.$$

Hence, the basis spans $\underline{A_1 + T_2}$.

Now use the projection operator in eqn (7.9.6) :

$$p^{(l)} = (d_l/24)\sum_R \chi^{(l)}(R)\,R\,.$$

Proceed as in the *Example* on p.154 of the text :

	A	B	C	D
E	A	B	C	D
C_3^+(D)	C	A	B	D
C_3^-(D)	B	C	A	D
C_3^+(C)	B	D	C	A
C_3^-(C)	D	A	C	B
C_3^+(B)	D	B	A	C
C_3^-(B)	C	B	D	A
C_3^+(A)	A	C	D	B
C_3^-(A)	A	D	B	C
C_2 (AC)	C	D	A	B
C_2 (CD)	B	A	D	C
C_2 (AD)	D	C	B	A
S_4^+(AC)	B	C	D	A
S_4^-(AC)	D	A	B	C
S_4^+(CD)	C	D	B	A
S_4^-(CD)	D	C	A	B
S_4^+(AD)	C	A	D	B
S_4^-(AD)	B	D	A	C
σ_d (AC)	A	D	C	B
σ_d (BD)	C	B	A	D
σ_d (CD)	B	A	C	D
σ_d (AD)	A	C	B	D
σ_d (BC)	D	B	C	A
σ_d (AB)	A	B	D	C

$$p^{(A_1)}\text{A} = (1/24)\sum_R \chi^{(A_1)}(R)\,R\text{A} = \tfrac{1}{8}\,(\text{A}+\text{B}+\text{C}+\text{D}).$$

$$p^{(T_2)}\text{A} = (1/24)\sum_R \chi^{(T_2)}(R)\,R\text{A}$$

$$= \tfrac{1}{8}\,(3\text{A}-\text{C}-\text{B}-\text{D}-2\text{B}-2\text{C}-2\text{D}+3\text{A}+\text{B}+\text{C}+\text{D})$$

$$= \tfrac{1}{4}\,(3\text{A}-\text{B}-\text{C}-\text{D}).$$

$$p^{(T_2)}\text{B} = \tfrac{1}{8}\,(3\text{B}-\text{D}-\text{A}-\text{C}-2\text{A}-2\text{C}-2\text{D}+3\text{B}-\text{A}-\text{C}-\text{D})$$

$$= \tfrac{1}{4}\,(3\text{B}-\text{A}-\text{C}-\text{D}).$$

$$p^{(T_2)}C = \tfrac{1}{4}(3C - A - B - D).$$
$$p^{(T_2)}D = \tfrac{1}{4}(3D - A - B - C).$$

But there can be only three linearly independent combinations spanning T_2. The simplest procedure is to find linear combinations that have a definite symmetry under σ_d. For instance,

$$(3A - B - C - D) + (3D - A - B - C) = 2(A - B - C + D)$$
$$(3A - B - C - D) + (3B - A - C - D) = 2(A + B - C - D)$$
$$(3A - B - C - D) + (3C - A - B - D) = 2(A - B + C - D)$$

are symmetric under $\sigma_d(AD)$, $\sigma_d(AB)$, and $\sigma_d(AC)$ respectively. Hence,

$$\underline{f^{(A_1)} = A + B + C + D}$$
$$\underline{\boldsymbol{f}^{(T_2)} = (A - B - C + D\,,\; A + B - C - D\,,\; A - B + C - D)}$$

are the symmetry-adapted bases.

Construct the matrix of coefficients as follows:

$$\left(f^{(A_1)}, \boldsymbol{f}^{(T_2)}\right) = (A, B, C, D)\begin{bmatrix} 1 & 1 & 1 & 1 \\ 1 & -1 & 1 & -1 \\ 1 & -1 & -1 & 1 \\ 1 & 1 & -1 & -1 \end{bmatrix} = \boldsymbol{f}\boldsymbol{c}\,.$$

Using the rules of Appendix 8,

$$\boldsymbol{c}^{-1} = \frac{1}{4}\begin{bmatrix} 1 & 1 & 1 & 1 \\ 1 & -1 & -1 & 1 \\ 1 & 1 & -1 & -1 \\ 1 & -1 & 1 & -1 \end{bmatrix}.$$

Then, choosing two elements of the classes C_3 and S_4 to demonstrate the calculation,

$$D'(C_3^+(A)) = \boldsymbol{c}^{-1}D(C_3^+(A))\,\boldsymbol{c} \qquad \text{[eqn (7.5.6)]}$$

$$= \frac{1}{4}\begin{bmatrix} 1 & 1 & 1 & 1 \\ 1 & -1 & -1 & 1 \\ 1 & 1 & -1 & -1 \\ 1 & -1 & 1 & -1 \end{bmatrix}\begin{bmatrix} 1 & 0 & 0 & 0 \\ 0 & 0 & 0 & 1 \\ 0 & 1 & 0 & 0 \\ 0 & 0 & 1 & 0 \end{bmatrix}\begin{bmatrix} 1 & 1 & 1 & 1 \\ 1 & -1 & 1 & -1 \\ 1 & -1 & -1 & 1 \\ 1 & 1 & -1 & -1 \end{bmatrix}$$

$$= \frac{1}{4}\begin{bmatrix} 1 & 1 & 1 & 1 \\ 1 & -1 & -1 & 1 \\ 1 & 1 & -1 & -1 \\ 1 & -1 & 1 & -1 \end{bmatrix}\begin{bmatrix} 1 & 1 & 1 & 1 \\ 1 & 1 & -1 & -1 \\ 1 & -1 & 1 & -1 \\ 1 & -1 & -1 & 1 \end{bmatrix} = \left[\begin{array}{c|ccc} 1 & 0 & 0 & 0 \\ \hline 0 & 0 & 0 & 1 \\ 0 & 1 & 0 & 0 \\ 0 & 0 & 1 & 0 \end{array}\right].$$

$$D'(S_4^+(\mathrm{CD})) = \frac{1}{4}\begin{bmatrix}1 & 1 & 1 & 1\\ 1 & -1 & -1 & 1\\ 1 & 1 & -1 & -1\\ 1 & -1 & 1 & -1\end{bmatrix}\begin{bmatrix}0 & 0 & 0 & 1\\ 0 & 0 & 1 & 0\\ 1 & 0 & 0 & 0\\ 0 & 1 & 0 & 0\end{bmatrix}\begin{bmatrix}1 & 1 & 1 & 1\\ 1 & -1 & 1 & -1\\ 1 & -1 & -1 & 1\\ 1 & 1 & -1 & -1\end{bmatrix}$$

$$= \frac{1}{4}\begin{bmatrix}1 & 1 & 1 & 1\\ 1 & -1 & -1 & 1\\ 1 & 1 & -1 & -1\\ 1 & -1 & 1 & -1\end{bmatrix}\begin{bmatrix}1 & 1 & -1 & -1\\ 1 & -1 & -1 & 1\\ 1 & 1 & 1 & 1\\ 1 & -1 & 1 & -1\end{bmatrix} = \left[\begin{array}{c:ccc}1 & 0 & 0 & 0\\ \hdashline 0 & 0 & 0 & -1\\ 0 & 0 & -1 & 0\\ 0 & 1 & 0 & 0\end{array}\right].$$

EXERCISE: Repeat the procedure to find the symmetry-adapted linear combinations of the s-orbitals for the trigonal bipyramid (last *Exercise*), and show that the basis produces a block-diagonal representation.

7.9 (a) $\chi(A_2)\times\chi(B_1)\times\chi(B_2)$

$= (1,1,-1,-1)\times(1,-1,1,-1)\times(1,-1,-1,1) = (1,1,1,1) = \chi(A_1)$;

therefore, $\underline{A_2\times B_1\times B_2 = A_1}$ in C_{2v}.

(b) $\chi(A_1)\times\chi(A_2)\times\chi(E)$

$= (1,1,1)\times(1,1,-1)\times(2,-1,0) = (2,-1,0) = \chi(E)$;

therefore $\underline{A_1\times A_2\times E = E}$ in C_{3v}.

(c) $\chi(B_2)\times\chi(E_1) = (1,-1,1,-1,-1,1)\times(2,1,-1,-2,0,0)$

$= (2,-1,-1,2,0,0) = \chi(E_2)$;

therefore, $\underline{B_2\times E_1 = E_2}$ in C_{6v}.

(d) $\chi(E_1)\times\chi(E_1) = (2,-2,2\cos\phi,0)\times(2,-2,2\cos\phi,0)$

$= (4,4,4\cos^2\phi,0) = (4,4,2+2\cos\phi,0)$

$= \chi(A_1)+\chi(A_2)+\chi(E_2)$;

therefore, $\underline{E_1\times E_1 = A_1+A_2+E_2}$ in $C_{\infty v}$

(alternatively: $\Pi\times\Pi = \Sigma^+ + \Sigma^- + \Delta$).

(e) $\chi(T_1)\times\chi(T_2)\times\chi(E)$

$= (3,0,-1,1,-1)\times(3,0,-1,-1,1)\times(2,-1,2,0,0) = (18,0,2,0,0)$.

Decompose this using $a_l = (1/24)\{18\chi^{(l)}(E)+6\chi^{(l)}(C_2)\}$ [eqn (7.8.3)].

$a(A_1) = (1/24)\{18+6\} = 1$; $a(A_2) = (1/24)\{18+6\} = 1$;

$a(E) = (1/24)\{36+12\} = 2$;

$a(T_1) = (1/24)\{54-6\} = 2$; $a(T_2) = (1/24)\{54-6\} = 2$.

Therefore,

$\underline{T_1\times T_2\times E = A_1+A_2+2E+2T_1+2T_2}$ in O.

EXERCISE: Analyse the following direct products: $E \times E \times A_2$ in C_{3v}, $A_{2u} \times E_{1u}$ in D_{6h}, and $T_{1g}^2 \times T_{2g}^2 \times E_u$ in O_h.

7.10 $C_3^+(x,y,z) = (-\frac{1}{2}x - \frac{1}{2}\sqrt{3}y, \frac{1}{2}\sqrt{3}x - \frac{1}{2}y, z)$ [eqn (7.10.2)]

$C_3^-(x,y,z) = (-\frac{1}{2}x + \frac{1}{2}\sqrt{3}y, -\frac{1}{2}\sqrt{3}x - \frac{1}{2}y, z)$ [Table 7.6]

$\sigma_v(x,y,z) = (-x,y,z)$.

Consequently,

$$C_3^+(3x^2y - y^3) = C_3^+(3x^2 - y^2)y$$
$$= \left\{3(-\tfrac{1}{2}x - \tfrac{1}{2}\sqrt{3}y)^2 - (\tfrac{1}{2}\sqrt{3}x - \tfrac{1}{2}y)^2\right\}(\tfrac{1}{2}\sqrt{3}x - \tfrac{1}{2}y)$$
$$= \tfrac{1}{4}\left\{3(x^2 + 3y^2 + 2\sqrt{3}xy) - (3x^2 + y^2 - 2\sqrt{3}xy)\right\}(\tfrac{1}{2}\sqrt{3}x - \tfrac{1}{2}y)$$
$$= (y^2 + \sqrt{3}xy)(\sqrt{3}x - y) = 3x^2 - y^3.$$
$$\sigma_v(3x^2y - y^3) = (3x^2y - y^3),$$

and likewise for the other elements. Hence $R(3x^2y - y^3) = 3x^2y - y^3$ for all R, and so $3x^2y - y^3$ spans A_1.

EXERCISE: Show that $x(4z^2 - x^2 - y^2)$ and $y(4z^2 - x^2 - y^2)$ span E in C_{3v}.

7.11 $a^{(l)} = \frac{1}{4}\left\{4\chi^{(l)}(E)\right\} = \chi^{(l)}(E)$ [eqn (7.8.3)]. Hence,

$a(A_1) = 1$, $a(A_2) = 1$, $a(B_1) = 1$, $a(B_2) = 1$. Therefore f spans $\underline{A_1 + A_2 + B_1 + B_2}$.

EXERCISE: The same function spans a representation with characters 4, 0, 4, 0, 1 in C_{4v}. What symmetry species does it span?

7.12 $p^{(l)} = (1/4)\sum_R \chi^{(l)}(R)R$ [eqn (7.9.6); $h = 4$, $d_l = 1$ for each l].

$p^{(A_1)} = \frac{1}{4}(E + C_2 + \sigma_v + \sigma_v')$; $p^{(A_2)} = \frac{1}{4}(E + C_2 - \sigma_v - \sigma_v')$;

$p^{(B_1)} = \frac{1}{4}(E - C_2 + \sigma_v - \sigma_v')$; $p^{(B_2)} = \frac{1}{4}(E - C_2 - \sigma_v + \sigma_v')$.

$Ef(x,y,z) = f(x,y,z)$; $C_2f(x,y,z) = f(-x,-y,z)$;

$\sigma_vf(x,y,z) = f(x,-y,z)$; $\sigma_v'f(x,y,z) = f(-x,y,z)$.

Hence:

$$f^{(A_1)} = \underline{\tfrac{1}{4}\{f(x,y,z) + f(-x,-y,z) + f(x,-y,z) + f(-x,y,z)}$$
$$f^{(A_2)} = \underline{\tfrac{1}{4}\{f(x,y,z) + f(-x,-y,z) - f(x,-y,z) - f(-x,y,z)\}}$$
$$f^{(B_1)} = \underline{\tfrac{1}{4}\{f(x,y,z) - f(-x,-y,z) + f(x,-y,z) - f(-x,y,z)\}}$$
$$f^{(B_2)} = \underline{\tfrac{1}{4}\{f(x,y,z) - f(-x,-y,z) - f(x,-y,z) + f(-x,y,z)\}}.$$

EXERCISE: Find the bases for the irreps in the previous *Exercise*.

7.13 Refer to Fig. 7.2. Draw up the following Table:

Fig. 7.2

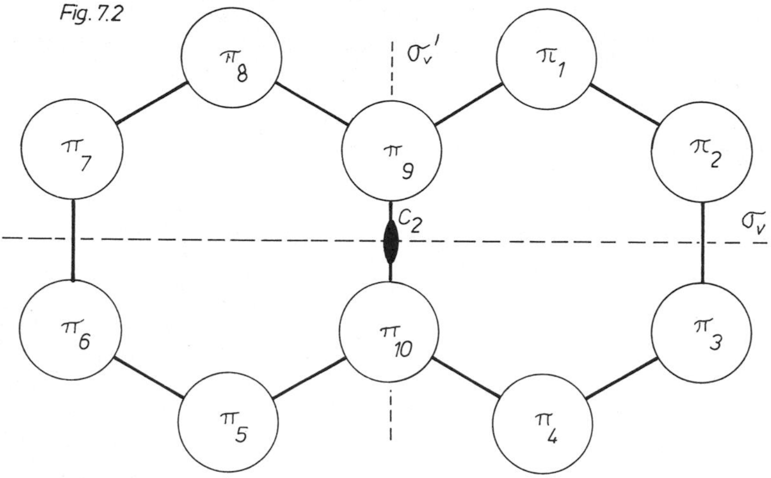

	π_1	π_2	π_3	π_4	π_5	π_6	π_7	π_8	π_9	π_{10}	χ
E	π_1	π_2	π_3	π_4	π_5	π_6	π_7	π_8	π_9	π_{10}	10
C_2	π_5	π_6	π_7	π_8	π_1	π_2	π_3	π_4	π_{10}	π_9	0
σ_v	π_4	π_3	π_2	π_1	π_8	π_7	π_6	π_5	π_{10}	π_9	0
σ_v'	π_8	π_7	π_6	π_5	π_4	π_3	π_2	π_1	π_9	π_{10}	2

The character (χ) of the operation is equal to the number of unchanged basis orbitals. The decomposition of $\chi = 10, 0, 0, 2$ is given by

$$a^{(l)} = \tfrac{1}{4}\left\{10\,\chi^{(l)}(E) + 2\chi^{(l)}(\sigma_v')\right\} \qquad [\text{eqn } (7.8.5)]\,.$$

$$a(A_1) = \tfrac{1}{4}(10+2) = 3\,, \qquad a(A_2) = \tfrac{1}{4}(10-2) = 2\,,$$

$$a(B_1) = \tfrac{1}{4}(10-2) = 2\,, \qquad a(B_2) = \tfrac{1}{4}(10+2) = 3\,.$$

Therefore, the orbitals span $3A_1 + 2A_2 + 2B_1 + 3B_2$. The projection operators are set out in Problem 7.12. Neglect of the normalizing factor $(\frac{1}{4})$ gives :

$$p(A_1)\pi_1 = \pi_1 + \pi_4 + \pi_5 + \pi_8\,, \qquad p(B_1)\pi_1 = \pi_1 - \pi_5 + \pi_4 - \pi_8$$

$$p(A_1)\pi_2 = \pi_2 + \pi_3 + \pi_6 + \pi_7\,, \qquad p(B_1)\pi_2 = \pi_2 - \pi_6 + \pi_3 - \pi_7$$

$$p(A_1)\pi_9 = \pi_9 + \pi_{10}\,, \qquad p(B_2)\pi_1 = \pi_1 - \pi_5 - \pi_4 + \pi_8$$

$$p(A_2)\pi_1 = \pi_1 + \pi_5 - \pi_4 - \pi_8\,, \qquad p(B_2)\pi_2 = \pi_2 - \pi_6 - \pi_3 + \pi_7$$

$$p(A_2)\pi_2 = \pi_2 + \pi_6 - \pi_3 - \pi_7\,, \qquad p(B_2)\pi_9 = \pi_9 - \pi_{10}\,.$$

In each case, we have listed only the linearly independent symmetry adapted combinations (e.g. $p(A_1)\pi_3$ repeats $p(A_1)\pi_2$).

EXERCISE: Find the π-orbital basis of anthracene.

7.14 Refer to Fig. 7.3. Draw up the following Table :

Fig. 7.3

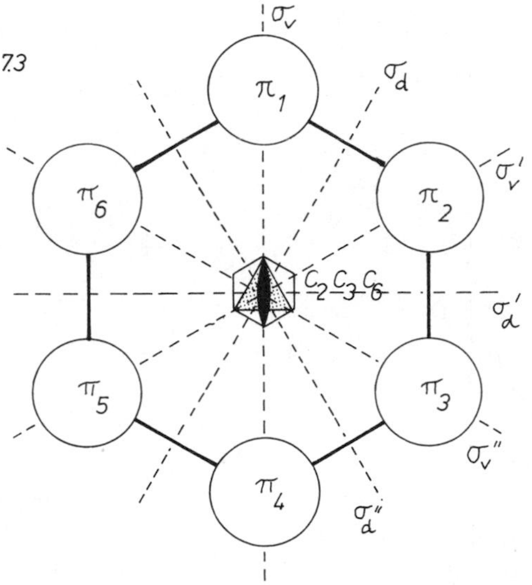

	π_1	π_2	π_3	π_4	π_5	π_6	χ
E	π_1	π_2	π_3	π_4	π_5	π_6	6
C_6^+	π_2	π_3	π_4	π_5	π_6	π_1	0
C_6^-	π_6	π_1	π_2	π_3	π_4	π_5	0
C_3^+	π_3	π_4	π_5	π_6	π_1	π_2	0
C_3^-	π_5	π_6	π_1	π_2	π_3	π_4	0
C_2	π_4	π_5	π_6	π_1	π_2	π_3	0
σ_v	π_1	π_6	π_5	π_4	π_3	π_2	2
σ_v'	π_3	π_2	π_1	π_6	π_5	π_4	2
σ_v''	π_5	π_4	π_3	π_2	π_1	π_6	2
σ_d	π_2	π_1	π_6	π_5	π_4	π_3	0
σ_d'	π_4	π_3	π_2	π_1	π_6	π_5	0
σ_d''	π_6	π_5	π_4	π_3	π_2	π_1	0

Therefore, in the order $(E, 2C_6, 2C_3, C_2, 3\sigma_v, 3\sigma_d)$ we have $\chi = (6,0,0,0,2,0)$. The decomposition is

$$a^{(l)} = (1/12)\left\{6\,\chi^{(l)}(E) + 6\chi^{(l)}(\sigma_v)\right\} = \tfrac{1}{2}\left\{\chi^{(l)}(E) + \chi^{(l)}(\sigma_v)\right\}.$$

Hence,

$$a(A_1)=1\ ,\ a(A_2)=0\ ,\ a(B_1)=1\ ,\ a(B_2)=0\ ,\ a(E_1)=1\ ,\ a(E_2)=1.$$

That is, the basis spans $\underline{A_1+B_1+E_1+E_2}$. The symmetry adapted linear combinations are then obtained from :

$$p(A_1) = (1/12)\left\{E + C_6^+ + C_6^- + C_3^+ + C_3^- + C_2 + \sigma_v + \sigma_v' + \sigma_v'' + \sigma_d + \sigma_d' + \sigma_d''\right\}$$

$$p(B_1) = (1/12)\left\{E - C_6^+ - C_6^- + C_3^+ + C_3^- - C_2 + \sigma_v + \sigma_v' + \sigma_v'' - \sigma_d - \sigma_d' - \sigma_d''\right\}$$

$$p(E_1) = (1/6)\left\{2E + C_6^+ + C_6^- - C_3^+ - C_3^- - 2C_2\right\}$$

$$p(E_2) = (1/6)\left\{2E - C_6^+ - C_6^- - C_3^+ - C_3^- + 2C_2\right\}.$$

Therefore,

$$p(A_1)\pi_1 = (1/6)\left\{\pi_1 + \pi_2 + \pi_3 + \pi_4 + \pi_5 + \pi_6\right\}$$

$$p(B_2)\pi_1 = (1/12)\left\{\pi_1 - \pi_2 - \pi_6 + \pi_3 + \pi_5 - \pi_4 + \pi_1 + \pi_3 + \pi_5 - \pi_2 - \pi_4 - \pi_6\right\}$$

$$= (1/6)\left\{\pi_1 - \pi_2 + \pi_3 - \pi_4 + \pi_5 - \pi_6\right\}.$$

$$p(E_1)\pi_1 = (1/6)\left\{2\pi_1 + \pi_2 + \pi_6 - \pi_3 - \pi_5 - 2\pi_4\right\}$$

$$\left.\begin{aligned} p(E_1)\pi_2 &= (1/6)\left\{2\pi_2 + \pi_3 + \pi_1 - \pi_4 - \pi_6 - 2\pi_5\right\} \\ p(E_1)\pi_3 &= (1/6)\left\{2\pi_3 + \pi_4 + \pi_2 - \pi_5 - \pi_1 - 2\pi_6\right\}\end{aligned}\right\}\ \tfrac{1}{4}(\pi_2 + \pi_3 - \pi_5 - \pi_6).$$

[The linear combination $\frac{1}{2}p(E_1)\pi_2 + \frac{1}{2}p(E_1)\pi_3$ is symmetric under σ_d' while $p(E_1)\pi_1$ is antisymmetric.]

$$p(E_2)\pi_1 = (1/6)\left\{2\pi_1 - \pi_2 - \pi_6 - \pi_3 - \pi_5 + 2\pi_4\right\}$$

$$\left.\begin{aligned} p(E_2)\pi_2 &= (1/6)\left\{2\pi_2 - \pi_3 - \pi_1 - \pi_4 - \pi_6 + 2\pi_5\right\} \\ p(E_2)\pi_3 &= (1/6)\left\{2\pi_3 - \pi_4 - \pi_2 - \pi_5 - \pi_1 + 2\pi_6\right\}\end{aligned}\right\}\ \tfrac{1}{4}(\pi_2 - \pi_3 + \pi_5 - \pi_6).$$

[The linear combination $\frac{1}{2}p(E_2)\pi_2 - \frac{1}{2}p(E_2)\pi_3$ is antisymmetric under σ_d' while $p(E_1)\pi_1$ is symmetric.]

In order to classify under D_{6h}, note that $\sigma_h\pi = -\pi$. The single A_1 basis function $\pi_1 + \pi_2 + \pi_3 + \pi_4 + \pi_5 + \pi_6$ is symmetric under E, C_2, C_3, C_6, and so is one of A_{1g}, A_{2g}, A_{1u}, A_{2u}. It is antisymmetric under σ_h, and so it is either A_{1g} or A_{2u}. It is symmetric under σ_v; therefore it is $\underline{A_{2u}\text{ in }D_{6h}}$.

The combination B_1, $\pi_1 - \pi_2 + \pi_3 - \pi_4 + \pi_5 - \pi_6$, has the characters $(1, -1, 1, -1)$ under $(E, 2C_6, 2C_3, C_2)$ in C_{6v}, and so is one of B_{1g}, B_{2g}, B_{1u}, B_{2u} in D_{6h}. It has character -1 under σ_h, and so is either B_{1g} or B_{2g}. Under $(3\sigma_v, 3\sigma_d)$ it is $(1, -1)$. Therefore it is $\underline{B_{2g}\text{ in }D_{6h}}$.

The pair spanning E_1 in C_{6v} has characters (2 , 1 , -1 , -2) under (E , $2C_6$, $2C_3$, C_2), and so spans either E_{1g} or E_{1u}. It is -2 under σ_h [both members of the set change sign], and so it is $\underline{E_{1g} \text{ in } D_{6h}}$.

The pair spanning E_2 in C_{6v} has characters (2 , -1 , -1 , 2) under (E , $2C_6$, $2C_3$, C_2), and so spans either E_{2g} or E_{2u}. It has character -2 under σ_h, and so it is $\underline{E_{2u} \text{ in } D_{6h}}$.

Therefore, overall :

C_{6v}	D_{6h}	Basis
A_1	A_{2u}	$\pi_1+\pi_2+\pi_3+\pi_4+\pi_5+\pi_6$
B_1	B_{2g}	$\pi_1-\pi_2+\pi_3-\pi_4+\pi_5-\pi_6$
E_1	E_{1g}	$\begin{cases} 2\pi_1+\pi_2-\pi_3-2\pi_4-\pi_5+\pi_6 \\ \pi_2+\pi_3-\pi_5-\pi_6 \end{cases}$
E_2	E_{2u}	$\begin{cases} 2\pi_1-\pi_2-\pi_3+2\pi_4-\pi_5-\pi_6 \\ \pi_2-\pi_3+\pi_5-\pi_6 \end{cases}$

EXERCISE: Find the symmetry species, the symmetry adapted linear combunations, and the full symmetry classification of the π-orbits of planar, regular cyclopentadienyl (C_5H_5) .

7.15 Refer to Fig. 7.4. For simplicity, use the group O in Appendix 10 and adapt to the O_h classification later. It is sufficient to consider one operation of each class. Draw up the following Table :

Fig.7.4

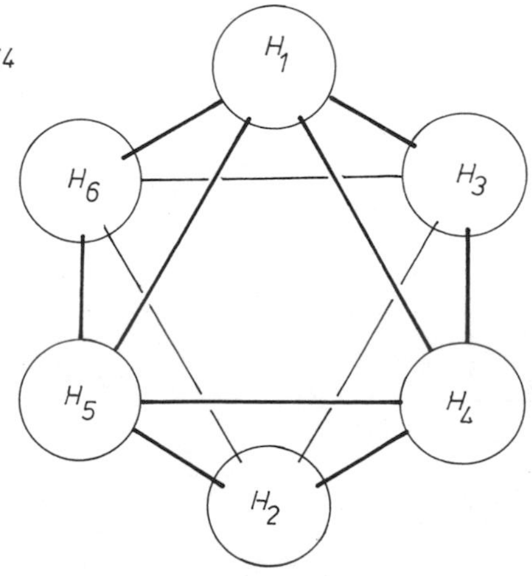

	H_1	H_2	H_3	H_4	H_5	H_6	χ
E	H_1	H_2	H_3	H_4	H_5	H_6	6
C_3	H_3	H_5	H_6	H_2	H_4	H_1	0
C_2	H_1	H_2	H_5	H_6	H_3	H_4	2
C_2'	H_2	H_1	H_6	H_5	H_4	H_3	0
C_4	H_1	H_2	H_6	H_3	H_4	H_5	2

The decomposition of 6 , 0 , 2 , 0 , 2 is obtained from :

$$a_l = (1/24)\{6\chi^{(l)}(E) + 6\chi^{(l)}(C_2) + 12\chi^{(l)}(C_4)\}$$
$$= \tfrac{1}{4}\{\chi^{(l)}(E) + \chi^{(l)}(C_2) + 2\chi^{(l)}(C_4)\}.$$

$a(A_1) = \tfrac{1}{4}\{1+1+2\} = 1$; $\quad a(A_2) = \tfrac{1}{4}\{1+1-2\} = 0$;

$a(E) = \tfrac{1}{4}\{2+2\} = 1$;

$a(T_1) = \tfrac{1}{4}\{3-1+2\} = 1$; $\quad a(T_2) = \tfrac{1}{4}\{3-1-2\} = 0$.

Therefore, the basis orbitals span $A_1 + E + T_1$. Under the group O_h we need the character of i. Since

$$i(H_1, \ldots, H_6) = (H_2, H_1, H_5, H_6, H_3, H_4), \quad \chi = 0.$$

The A_1 set is totally symmetric, and so it spans A_{1g}. Since $\chi^{(A_{1g})}(i) = 1$, the sum of the remaining two characters must be -1 [so that $\chi(i) = 0$ overall]. Clearly, we must have $E_g + T_{1u}$. Therefore, the orbitals must span $\underline{A_{1g} + E_g + T_{1u}}$.

Alternatively, use the full O_h character table. Note that the following numbers of orbitals remain unchanged :

	E	$8C_3$	$6C_2'$	$6C_4$	$3C_2$	i	$6C_4$	$8S_6$	$3\sigma_h$	$6\sigma_d$
$\chi =$	6	0	0	2	2	0	0	0	4	2

This decomposes into $A_{1g} + E_g + T_{1u}$.

EXERCISE: A hydrogen atom sits at each point of a regular icosahedron (I_h). What symmetry species do the 1s-orbitals span ?

7.16 (a) $a_1^2 b_1 b_2$: $A_1 \times A_1 \times B_1 \times B_2 = B_1 \times B_2 = A_2$; $\underline{{}^1A_2 \text{ and } {}^3A_2}$ may arise.

(b)(i) $a^2 e$: $A_2 \times E = E$; $\underline{{}^1E \text{ and } {}^3E}$ may arise.

(ii) e^2 : $E \times E = A_1 + [A_2] + E$; $\underline{{}^1A_1, {}^3A_2, {}^1E_2}$ may arise.

(c)(i) $a_2 e$: $A_2 \times E = E$; $\underline{{}^1E \text{ and } {}^3E}$ may arise.

(ii) et_1 : $E \times T_1 = T_1 + T_2$; $\underline{{}^1T_1, {}^3T_1, {}^1T_2, \text{ and } {}^3T_2}$ may arise.

(iii) $t_1 t_2$: $T_1 \times T_2 = A_2 + E + T_1 + T_2$;
$\underline{{}^1A_2, {}^3A_2, {}^1E, {}^3E, {}^1T_1, {}^3T_1, {}^1T_2, \text{ and } {}^3T_2}$ may arise.

(iv) t_1^2 : $T_1 \times T_1 = A_1 + E + [T_1] + T_2$; $\underline{{}^1A_1, {}^1E, {}^3T_1, \text{ and } {}^1T_2}$ may arise.

(v) t_2^2 : $T_2 \times T_2 = A_1 + E + [T_1] + T_2$ $\underline{{}^1A_1, {}^1E, {}^3T_1, \text{ and } {}^1T_2}$ may arise.

(d)(i) e^2 : $E \times E = A_1 + [A_2] + E$; $\underline{{}^1A_1, {}^3A_2, \text{ and } {}^1E}$ may arise.

(ii) et_1: $E \times T_1 = T_1 + T_2$; $\underline{{}^1T_1, {}^3T_1, {}^1T_2, \text{ and } {}^3T_2}$ may arise.

(iii) t_2^2 ; $T_2 \times T_2 = A_1 + E + [T_1] + T_2$; $\underline{{}^1A_1, {}^1E, {}^3T_1, \text{ and } {}^1T_2}$ may arise.

EXERCISE: Classify the term that may arise from d^2 in R_3, $\sigma\pi$ in $C_{\infty v}$, π^2 in $D_{\infty h}$, $e_g t_{1u}$ in O_h, and e_{1g}^2 in D_{6h}.

7.17 $O_h = O \times C_i$

$C_i = (E, i)$; $O = (E, 8C_3, 3C_2, 6C_4, 6C_2')$

$C_3 \times i = S_6$; $C_2 \times i = \sigma_h$; $C_4 \times i = S_4$; $C_2' \times i = \sigma_d$.

O	E		$8C_3$		$3C_2$		$6C_4$		$6C_2'$		$h = 24$
O_h	E	i	$8C_3$	$8S_6$	$3C_2$	$3\sigma_h$	$6C_4$	$6S_4$	$6C_2'$	$6\sigma_d$	$h = 48$
A_{1g}	1	1	1	1	1	1	1	1	1	1	
A_{2g}	1	1	1	1	1	1	-1	-1	-1	-1	
E_g	2	2	-1	-1	2	2	0	0	0	0	
T_{1g}	3	3	0	0	-1	-1	1	1	-1	-1	
T_{2g}	3	3	0	0	-1	-1	-1	-1	1	1	
A_{1u}	1	-1	1	-1	1	-1	1	-1	1	-1	
A_{2u}	1	-1	1	-1	1	-1	-1	1	-1	1	
E_u	2	-2	-1	1	2	-2	0	0	0	0	
T_{1u}	3	-3	0	0	-1	1	1	-1	-1	1	
T_{2u}	3	-3	0	0	-1	1	-1	1	1	-1	

$D_{6h} = D_6 \times C_i$

$C_i = (E, i)$; $D_6 = (E, 2C_6, 2C_3, C_2, 3C_2', 3C_2'')$

$C_6 \times i = S_3$, $C_3 \times i = S_6$, $C_2 \times i = \sigma_h$, $C_2' \times i = \sigma_d$, $C_2'' \times i = \sigma_v$

D_6	E		$2C_6$		$2C_3$		C_2		$3C_2'$		$3C_2''$		$h=12$
D_{6h}	E	i	$2C_6$	$2S_3$	$2C_3$	$2S_6$	C_2	σ_h	$3C_2'$	$3\sigma_d$	$3C_2''$	$3\sigma_v$	$h=24$
A_{1g}	1	1	1	1	1	1	1	1	1	1	1	1	
A_{2g}	1	1	1	1	1	1	1	1	-1	-1	-1	-1	
B_{1g}	1	1	-1	-1	1	1	-1	-1	1	1	-1	-1	
B_{2g}	1	1	-1	-1	1	1	-1	-1	-1	-1	1	1	
E_{1g}	2	2	1	1	-1	-1	-2	-2	0	0	0	0	
E_{2g}	2	2	-1	-1	-1	-1	2	2	0	0	0	0	
A_{1u}	1	-1	1	-1	1	-1	1	-1	1	-1	1	-1	
A_{2u}	1	-1	1	-1	1	-1	1	-1	-1	1	-1	1	
B_{1u}	1	-1	-1	1	1	-1	-1	1	1	-1	-1	1	
B_{2u}	1	-1	-1	1	1	-1	-1	1	-1	1	1	-1	
E_{1u}	2	-2	1	-1	-1	1	-2	2	0	0	0	0	
E_{2u}	2	-2	-1	1	-1	1	2	-2	0	0	0	0	

EXERCISE: Construct and identify the group $C_s \times C_5$.

7.18 H has the full symmetry of the system [definition of symmetry operation], and so it is a basis for A_1. Therefore, $\psi' H \psi$ spans $\Gamma' \times \Gamma$ if ψ' spans Γ' and ψ spans Γ. But $\Gamma' \times \Gamma$ contains A_1 only if $\Gamma' = \Gamma$. Therefore, the integral vanishes when ψ' and ψ belong to different symmetry species.

EXERCISE: Under what circumstances may a molecule possess a permanent electric dipole moment ?

7.19 (a) In C_{2v} translations span $A_1 + B_1 + B_2$; hence a $^2A_1$ term may make a transition to $A_1 \times {}^2A_1 = {}^2A_1$, $B_1 \times {}^2A_1 = {}^2B_1$, and $B_2 \times {}^2A_1 = {}^2B_2$ and a $^2B_1$ term may make a transitions to $A_1 \times {}^2B_1 = {}^2B_1$, $B_1 \times {}^2B_1 = {}^2A_1$, and $B_2 \times {}^2B_1 = {}^2A_2$. In $D_{\infty h}$ (as for O_2), translations span $\Sigma_u^+ + \Pi_u$. Therefore, since $\Sigma_u^+ \times \Sigma_g^- = \Sigma_u^-$ and $\Pi_u \times \Pi_g^- = \Pi_u^-$, transitions to $^3\Sigma_u^-$ and $^3\Pi_u^-$ are allowed.

(b) In C_{2v} rotations span $A_2 + B_1 + B_2$. Then, since $A_1 \times (A_2 + B_1 + B_2) = A_2 + B_1 + B_2$, transitions to $^2A_2$, $^2B_1$, and $^2B_2$ are allowed for NO_2. Since $B_1 \times (A_2 + B_1 + B_2) = B_2 + A_1 + A_2$, transitions to $^2B_2$, $^2A_1$, and $^2A_2$ are allowed for ClO_2. In $D_{\infty h}$, rotations transform as $\Sigma_g^- + \Pi_g$, and since $\Sigma_g^- \times (\Sigma_g^- + \Pi_g) = \Sigma_g^+ + \Pi_g$, transitions to $^3\Sigma_g^+$ and $^3\Pi_g$ are allowed in O_2.

EXERCISE: What electric and magnetic dipole transitions may take place from the E_{1g}, E_{2u}, and B_{2g} terms of benzene ?

7.20 The point group of a regular tetrahedron is T_d: three-dimensional irreps are allowed; therefore the maximum degeneracy is 3. (Accidental degeneracies could increase this number.)

EXERCISE: What is the maximum degeneracy of molecular orbitals of (a) benzene, (b) anthracene, (c) an icosahedral molecule ?

7.21 Under a translation, the point x changes to $x - \delta x$; hence

$$T_{\delta x} x = x - \delta x = \{1 - \delta x(\mathrm{d}/\mathrm{d}x)\}\, x \,.$$

Then, since $\mathrm{d}/\mathrm{d}x = (\mathrm{i}/\hbar)p$,

$$T_{\delta x} = 1 - (\mathrm{i}/\hbar)\,\delta x p \,.$$

Comparison with eqn (7.14.3) shows that p is the generator of the translation.

EXERCISE: Show that the transition operator can be regarded as the generator of displacements in time.

7.22 $\chi(C_{120°}) = \sin(\frac{3}{2}\times 120°)/\sin 60° = 0$; $\chi(E) = 3$.

$\chi(\sigma_v) = 1$ [since $p_y \rightarrow -p_y$, $p_x \rightarrow p_x$, $p_z \rightarrow p_z$] .

The characters for $(E, 2C_3, 3\sigma_v)$ are therefore $(3, 0, 1)$. Therefore, the orbitals span $\underline{A_1 + E}$.

EXERCISE: What symmetry species would be spanned if the p-orbitals were replaced by (a) f-orbitals, (b) g-orbitals ?

8 Techniques of approximation

8.1 $E_{\pm} = \frac{1}{2}(E_1 + E_2) \pm \frac{1}{2}\{(E_1 - E_2)^2 + 4\varepsilon^2\}^{\frac{1}{2}}$ [eqn (8.1.6)]

$$\left.\begin{aligned}\psi_+ &= \psi_1 \cos\beta + \psi_2 \sin\beta \\ \psi_- &= -\psi_1 \sin\beta + \psi_2 \cos\beta\end{aligned}\right\} \tan 2\beta = 2\varepsilon/(E_1 - E_2)$$ [eqn (8.1.8)]

$E_1 = 25\,739.86\ \text{cm}^{-1}$, $\quad E_2 = 50\,266.88\ \text{cm}^{-1}$.

$E_{\pm}/\text{cm}^{-1} = 38\,003.37 \pm \surd\{1.503\,93 \times 10^8 + \varepsilon^2\}$;

$\beta = \frac{1}{2}\ \text{arc tan}\ (-\varepsilon/12\,263.51)$.

Care is needed with the signs because $\surd\{\ldots\}$ may be positive or negative. When $\varepsilon = 0$, $\surd\{\ldots\} = E_1 - E_2$ which is *negative* if, as in this problem, $E_1 < E_2$. Therefore, take the negative root.

(a) $\varepsilon = 100\ \text{cm}^{-1}$;

$E_{\pm}/\text{cm}^{-1} = 38\,003.37 \pm (-12\,263.92) = 25\,739.45$ and $50\,267.29$.

$\beta = -4.077 \times 10^{-3}$ rad ;

$\psi_+ = 0.999\,992\,\psi_1 - 0.004\,077\,\psi_2$ (99.998 percent ψ_1)

$\psi_- = 0.004\,077\,\psi_1 + 0.999\,992\,\psi_2$ (99.998 percent ψ_2).

(b) $\varepsilon = 1000\ \text{cm}^{-1}$;

$E_{\pm}/\text{cm}^{-1} = 38\,003.37 \pm (-12\,304.21) = 25\,699.16$ and $50\,307.58$.

$\beta = -4.068 \times 10^{-2}$ rad;

$\psi_+ = 0.999\,173\,\psi_1 - 0.040\,670\,\psi_2$ (99.835 percent ψ_1)

$\psi_- = 0.040\,670\,\psi_1 + 0.999\,173\,\psi_2$ (99.835 percent ψ_2).

(c) $\varepsilon = 5000\ \text{cm}^{-1}$;

$E_{\pm}/\text{cm}^{-1} = 38\,003.37 \pm (-13\,243.63) = 25\,759.74$ and $51\,246.99$.

$\beta = -0.407\,72$ rad ;

$\psi_+ = 0.918\,030\,\psi_1 - 0.396\,511\,\psi_2$ (84.278 percent ψ_1)

$\psi_- = 0.396\,511\,\psi_1 + 0.918\,030\,\psi_2$ (84.278 percent ψ_2).

EXERCISE: An electron spin lies in a 10 T magnetic field. What are its energies and states when a perpendicular 2T field is also present ?

8.2 $E_{\pm} = \frac{1}{2}(E_1 + E_2) \pm \frac{1}{2}\{(E_1 - E_2)^2 + 4\varepsilon^2\}^{\frac{1}{2}}$ [eqn (8.1.6)]

$= E_{1s} \pm \varepsilon \quad [E_1 = E_2 = E_{1s}]$, $E_{1s} = -2\hbar^2/m_e a_0^2$ [eqn (1.5.3), $Z = 2$].

$H = H_1 + H_2 + e^2/4\pi\varepsilon_0 r_{12} = H^{(0)} + H^{(1)}$, with $H^{(0)} = H_1 + H_2$.

$\psi^{(0)} = \psi_{1s}(1)\,\psi_{1s}(2)$; $\quad H^{(0)}\psi^{(0)} = 2E_{1s}\psi^{(0)}$.

$$E^{(1)} = \langle \psi^{(0)} | H^{(1)} | \psi^{(0)} \rangle$$
$$= (e^2/4\pi\varepsilon_0) \int \psi_{1s}(1)\,\psi_{1s}(2)\,(1/r_{12})\,\psi_{1s}(1)\,\psi_{1s}(2)\,d\tau_1\,d\tau_2$$
$$= (5/4)(e^2/4\pi\varepsilon_0 a_0).$$

(a) $E \approx 2E_{1s} + E^{(1)} = -(4\hbar^2/m_e a_0^2) + (5/4)(e^2/4\pi\varepsilon_0 a_0)$
$\approx -(11/4)(\hbar^2/m_e a_0^2) = -(11/2)\,hcR_\infty \quad [\hbar^2/m_e a_0^2 = 2hcR_\infty]$.

Therefore, $E \approx -1.99 \times 10^{-17}$ J $\hat{=} -\underline{74.8 \text{ eV}}$.

(The experimental value is -78.98 eV.)

(b) $I = E_{1s} - E = (-52 + 11/4)(\hbar^2/m_e a_0^2) = \frac{3}{4}(\hbar^2/m_e a_0^2)$
$= \frac{3}{2}hcR_\infty = 3.27 \times 10^{-18}$ J $\hat{=}$ $\underline{20.4 \text{ eV}}$.

(The experimental value is 24.580 eV.)

EXERCISE : Find an expression for the dependence of the ionization energy on the atomic number of hydrogen-like one-electron ions.

8.3 Since $\tan 2\beta = 2\varepsilon/(E_1 - E_2)$ [eqn (8.1.8)],

$$E_\pm = \bar{E} \pm \tfrac{1}{2}\{(E_1 - E_2)^2 + (E_1 - E_2)^2 \tan^2 2\beta\}^{\frac{1}{2}}$$
$$= \bar{E} \pm \tfrac{1}{2}(E_1 - E_2)\{1 + \tan^2 2\beta\}^{\frac{1}{2}}$$
$$= \bar{E} \pm \tfrac{1}{2}(E_1 - E_2)/\cos 2\beta = \bar{E} \pm \tfrac{1}{2}(E_1 - E_2)\sec 2\beta .$$

For a graphical representation, use

$$E_\pm - \bar{E} = \pm\tfrac{1}{2}\{(E_1 - E_2)^2 + 4\varepsilon^2\}^{\frac{1}{2}}.$$

Then with a right-angled triangle of sides ε and $\frac{1}{2}(E_1 - E_2)$, the hypotenuse is given by $E_\pm - \bar{E}$.

EXERCISE: Sketch the dependence of $E - \bar{E}$ on β.

8.4 $H = -(\hbar^2/2m)(d^2/dx^2) + mgx$;

$H^{(0)} = -(\hbar^2/2m)(d^2/dx^2)$; $\quad H^{(1)} = mgx$.

$E^{(1)} = \langle 0 | H^{(1)} | 0 \rangle = mg\langle x \rangle = \underline{\tfrac{1}{2}mgl}$.

The first-order correction disregards the adjustment of the location of the particle in the gravitational field, and so $E^{(1)}$ is the potential energy of a particle at its average height $(\frac{1}{2}L)$. For $m = m_e$,

$$E^{(1)}/L = \tfrac{1}{2}mg = 4.47 \times 10^{-30}\ \mathrm{J\,m^{-1}} .$$

8.5 $E^{(2)} = -\sum_n \{H^{(1)}_{0n} H^{(1)}_{n0}/\Delta_{n0}\}$; $\quad \psi^{(1)} = -\sum_n \{H^{(1)}_{n0}/\Delta_{n0}\}\,\psi_n$.

$$H_{n0}^{(1)} = mg\,(2/L)\int_0^L x\sin(n\pi x/L)\sin(\pi x/L)\,dx$$

$$= (mg/L)(L/\pi)^2\left\{\frac{\cos(n-1)\pi-1}{(n-1)^2} - \frac{\cos(n+1)\pi-1}{(n+1)^2}\right\}$$

$$= mgL\,(4/\pi^2)\left\{1+(-1)^n\right\} n/(n^2-1)$$

$$= \begin{cases} -(8/\pi^2)mgLn/(n^2-1)\,, & n \text{ even} \\ 0, & n \text{ odd}\,. \end{cases}$$

$$\Delta_{n0} = E_n - E_0 = (n^2-1)(h^2/8mL^2)\,.$$

$$H_{n0}^{(1)}/\Delta_{n0} = -(8/\pi^2)\,a\,\{n/(n^2-1)^2\}\,,\quad a = mgL/(h^2/8mL^2)\,,\quad n \text{ even.}$$

$$H_n^{(1)2}/\Delta_n = (64/\pi^4)\,mgLa\{n^2/(n^2-1)^3\},\qquad n \text{ even}\,.$$

Therefore,

$$E^{(2)} = -(64/\pi^4)\,mgLa\sum_n^{\text{even}}\left\{n^2/(n^2-1)^3\right\}$$

$$= -(64/\pi^4)\,mgLa\{0.1481 + 0.0047 + 0.0008 + \ldots\}$$

$$= -(64/\pi^4)\,mgLa \times 0.1542 = \underline{-0.1013\,amgL}$$

$$\psi^{(1)} = (8/\pi^2)\,a\sum_n^{\text{even}}\left\{n\psi_n/(n^2-1)^2\right\}$$

$$= a\{0.1801\,\psi_2 + 0.0144\,\psi_4 + 0.0040\,\psi_6 + \ldots\}.$$

Note that all the ψ_n are positive close to $x = 0$, and so this sum represents an enhancement of amplitude at the bottom ($x \approx 0$) of the well, as is to be expected physically.

EXERCISE: A particle is confined to a box, but attached to one wall by a weak spring (so that its potential energy has a component of magnitude $\frac{1}{2}k(x-\frac{1}{2}L)^2$ inside the wall). Find the first and second-order corrections to the energy, and the first-order correction to the wavefunction.

8.6 Refer to Fig. 8.1. Take the modulation to be

$$H^{(1)} = V\sin^2(\pi x/R).$$

$$E_n^{(1)} = \langle n|H^{(1)}|n\rangle$$

$$= (2V/L)\int_0^L \sin^2(\pi x/R)\sin^2(n\pi x/L)dx = (2V/L)(L/4) = \tfrac{1}{2}V\,,$$

independent of n for $n \neq 1$. Hence, $E_{n+1} - E_n$ is unshifted from its unperturbed value (for $n \neq 1$).

Fig. 8.1

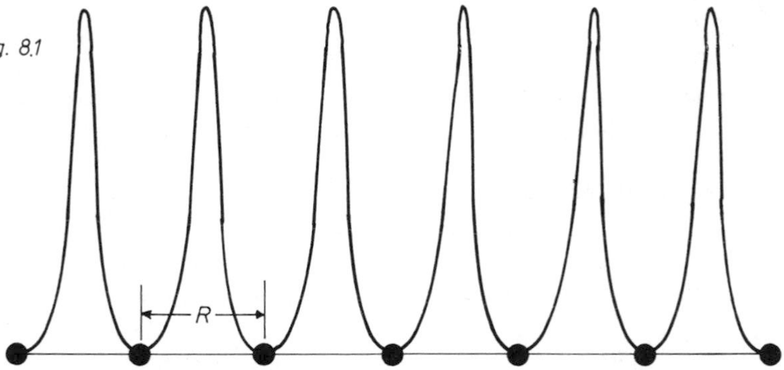

EXERCISE: Take the modulation to be $-V\cos^2(\pi x/R)$. Sketch this, and assess its first-order effects.

8.7 In the full rotational group (R_3), the ground state is fully symmetrical (S) while z transforms as a member of a basis spanning P. Therefore, $\int \psi_n z \psi_{1s} \, d\tau$ vanishes unless ψ_n is also a member of a basis spanning P. In the subgroup $C_{\infty v}$, z transforms as Σ (x, y span Π): ψ_{1s} spans Σ; therefore ψ_n must span Σ in order for the matrix elements to be non-zero. The parity of ψ_{1s} is g; that of z is u; therefore ψ_n must be u. The orbitals of Σ symmetry in $C_{\infty v}$, u in C_i, and P in R_3 are the $n\mathrm{p}_z$-orbitals.

(a) When $\psi_0 = \psi_{2\mathrm{p}_x}$, it transforms as P in R_3 and Π_u in $D_{\infty h}$; z spans P in R_3 and Σ in $D_{\infty h}$, and $P \times P = D + P + S$, $\Sigma_u \times \Pi_u = \Pi_g$. The first allows ψ_n to be d, p, and s. The second confines this list to d, and in particular to d_{xz} or d_{yz}. Consider the mirror plane σ_v under which $\mathrm{p}_x \to -\mathrm{p}_x$, $z \to z$, and $\mathrm{d}_{xz} \to -\mathrm{d}_{xz}$, $\mathrm{d}_{yz} \to \mathrm{d}_{yz}$. Hence only d_{xz}-orbitals can contribute

(b) When $\psi_0 = \psi_{2\mathrm{p}_z}$ it transforms as P (in R_3), Σ_u (in $D_{\infty h}$). Then $z\psi_{2\mathrm{p}_z}$ spans $P \times P = D + P + S$ (in R_3) and $\Sigma_u \times \Sigma_u = \Sigma_g$ (in $D_{\infty h}$). Therefore, only $n\mathrm{d}_{z^2}$-orbitals may contribute.

(c) When $\psi_0 = \psi_{3\mathrm{d}_{xy}}$, transforming as D (in R_3) and Δ_g (in $D_{\infty h}$), $z\psi_{3\mathrm{d}_{xy}}$ spans $D \times D = G + F + D + P + S$ (in R_3), $\Sigma_u \times \Delta_g = \Delta_u$ (in $D_{\infty h}$). Therefore, since f_{xyz} spans F (in R_3) and Δ_u (in $D_{\infty h}$), the $n\mathrm{f}_{xyz}$-orbitals may contribute.

EXERCISE: What orbitals contaminate s-orbitals when the perturbation has tthe form az^3? What orbitals contaminate $2p_y$-orbitals, $3d_{yz}$-orbitals, $3d_{x^2-y^2}$-orbitals, and f_{xyz}-orbitals when the perturbation has the form az?

8.8 (a) $x\psi_0$ spans $B_1 \times A_1 = B_1$ in $C_{\infty v}$; hence $\underline{B_1}$ states are admixed.
(b) $l_x\psi_0$ spans $B_2 \times A_1 = B_2$ in C_{2v}; hence $\underline{B_2}$ states are admixed.

EXERCISE: The symmetry of the ground state of ClO_2 is ${}^2B_1$. What symmetry species of excited states are admixed?

8.9 $\Delta\varepsilon^2 = \langle 0|H^{(1)2}|0\rangle - \langle 0|H^{(1)}|0\rangle^2$ [eqn (8.1.27)]

$\langle 0|H^{(1)}|0\rangle = \frac{1}{2}mgL$ [Problem 8.4].

$\langle 0|H^{(1)2}|0\rangle = m^2g^2(2/L)\int_0^L x^2\sin^2(\pi x/L)\,dx$

$= \frac{1}{3}m^2g^2L^2(1-3/2\pi^2)$ [Problem 3.11, $n=1$]

$\Delta\varepsilon^2 = \left\{\frac{1}{3}(1-3/2\pi^2)-\frac{1}{4}\right\}m^2g^2L^2 = (1/12)(1-6/\pi^2)(mgL)^2$

$= 0.032\,67\,(mgL)^2$.

$E^{(2)} \approx -\Delta\varepsilon^2/\Delta E\ [H^{(2)}=0] = -0.032\,67\,(mgL)^2/\Delta E$.

Write $\Delta E = \lambda(h^2/8mL^2)$ [the particle-in-a-box energies have the form $n^2(h^2/8mL^2)$, and so $h^2/8mL^2$ is a suitable 'scale factor' for the energy]. Then

$E^{(2)} = -0.032\,67\,(mgL)^2/\lambda(h^2/8mL^2) = -0.032\,67\,(a/\lambda)\,mgL$
[$a = mgL/(h^2/8mL^2)$, Problem 8.5]. But

$E^{(2)} = -0.1013\,amgL$ [Problem 8.5].

Therefore, the best choice of λ is

$$\lambda = 0.032\,67/0.1013 = 0.3225.$$

Consequently, take $\underline{\Delta E = 0.3225\,(h^2/8mL^2)}$. Notice that this value is significantly smaller than the energy separation of even the lowest (and closest) pair of levels contributing to the sum, $E_2 - E_1 = 3(h^2/8mL^2)$.

EXERCISE: Find the expression under the closure approximation for the *Exercise* following Problem 8.5, and determine the best value of ΔE.

8.10 $\langle 0|H^{(1)}|0\rangle = -\varepsilon(2/L)\int_0^L \sin^3(\pi x/L)\,dx = -(2\varepsilon/L)(4L/3\pi)$

$= -8\varepsilon/3\pi$.

$$\langle 0|H^{(1)2}|0\rangle = \varepsilon^2(2/L)\int_0^L \sin^4(\pi x/L)\,dx = \varepsilon^2(2/L)(3L/8) = 3\varepsilon^2/4.$$

$$\Delta\varepsilon^2 = 3\varepsilon^2/4 - 64\varepsilon^2/9\pi^2 = 0.029\,49\,\varepsilon^2$$

$$E^{(2)} \approx -\Delta\varepsilon^2/\Delta E = -0.029\,49\,\varepsilon^2/\Delta E\,.$$

Once again, write $\Delta E = \lambda'(h^2/8mL^2)$ for a suitable scaling expression, and set $b = \varepsilon/(h^2/8mL^2)$. Then

$$E^{(2)} \approx -0.029\,49\,\varepsilon b/\lambda'\,.$$

From the worked *Example* on p.177 in the text,

$$E^{(2)} = -0.003\,63\,\varepsilon b\,.$$

For these to be equal, we must choose $\lambda' = 8.12$. In this case, then,

$$\underline{\Delta E = 8.12\,(h^2/8mL^2)}.$$

EXERCISE: Consider a perturbation of the form $-\varepsilon\sin(2\pi x/L)$, and find the best value of ΔE.

8.11 $H^{(1)} = \varepsilon\sin^2\phi$.

Form the secular determinant using

$$H^{(1)}_{m_l' m_l} = (\varepsilon/2\pi)\int_0^{2\pi} e^{-i(m_l'-m_l)\phi}\left(e^{2i\phi}+e^{-2i\phi}-2\right)\Big/(-4)$$

$$= -(\varepsilon/8\pi)\int_0^{2\pi}\left\{e^{i(2-m_l'+m_l)\phi} + e^{i(-2-m_l'+m_l)\phi} - 2e^{-i(m_l'-m_l)\phi}\right\}d\phi$$

$$= -(\varepsilon/4)\left\{\delta_{m_l',m_l+2} + \delta_{m_l',m_l-2} - 2\delta_{m_l',m_l}\right\}.$$

Consequently,

$$H^{(1)}_{1,1} = \tfrac{1}{2}\varepsilon\,,\quad H^{(1)}_{-1,-1} = \tfrac{1}{2}\varepsilon\,,\quad H^{(1)}_{1,-1} = -\tfrac{1}{4}\varepsilon\,,\quad H^{(1)}_{-1,1} = -\tfrac{1}{4}\varepsilon\,;$$

$$S_{1,1} = S_{-1,-1} = 1;\quad S_{1,-1} = S_{-1,1} = 0\,.$$

$$\det\left|H^{(1)} - SE\right| = \begin{vmatrix}\tfrac{1}{2}\varepsilon - E & -\tfrac{1}{4}\varepsilon\\ -\tfrac{1}{4}\varepsilon & \tfrac{1}{2}\varepsilon - E\end{vmatrix} = (\tfrac{1}{2}\varepsilon - E)^2 - (\tfrac{1}{4}\varepsilon)^2 = 0\,.$$

Consequently, $E = \frac{1}{2}\varepsilon \pm \frac{1}{4}\varepsilon = \frac{3}{4}\varepsilon$ and $\frac{1}{4}\varepsilon$. Find the coefficients from the secular equations and $|c_1|^2 + |c_2|^2 = 1$ (or by intuition):

$$\left.\begin{aligned}(\tfrac{1}{2}\varepsilon - E)\,c_1 - \tfrac{1}{4}\varepsilon c_2 &= 0\\ -\tfrac{1}{4}\varepsilon c_1 + (\tfrac{1}{2}\varepsilon - E)\,c_2 &= 0\end{aligned}\right\}\quad\left\{\begin{aligned}\phi_{\frac{3}{4}} &= (\psi_1 - \psi_{-1})/\sqrt{2}\\ \phi_{\frac{1}{4}} &= (\psi_1 + \psi_{-1})/\sqrt{2}\,.\end{aligned}\right.$$

For the first-order energies we have $E = \frac{3}{4}\varepsilon$ and $\frac{1}{4}\varepsilon$. If desired, check this as follows:

$$H^{(1)}_{\frac{3}{4},\frac{3}{4}} = \tfrac{1}{2}\left(H^{(1)}_{1,1} + H^{(1)}_{-1,-1} - H^{(1)}_{1,-1} - H^{(1)}_{-1,1}\right) = \tfrac{1}{2}(\varepsilon + \tfrac{1}{2}\varepsilon) = \tfrac{3}{4}\varepsilon\,,$$

$$H^{(1)}_{\frac{1}{4},\frac{1}{4}} = \tfrac{1}{2}\left(H^{(1)}_{1,1} + H^{(1)}_{-1,-1} + H^{(1)}_{1,-1} + H^{(1)}_{-1,1}\right) = \tfrac{1}{2}(\varepsilon - \tfrac{1}{2}\varepsilon) = \tfrac{1}{4}\varepsilon\,,$$

$$H^{(1)}_{\frac{3}{4},\frac{1}{4}} = \tfrac{1}{2}\left(H^{(1)}_{1,1} - H^{(1)}_{-1,-1} + H^{(1)}_{1,-1} - H^{(1)}_{-1,1}\right) = 0\,.$$

This confirms that $H^{(1)}$ is diagonal in the $\phi_{\frac{3}{4}}$, $\phi_{\frac{1}{4}}$ basis, and that its eigenvalues are $\frac{3}{4}\varepsilon$ and $\frac{1}{4}\varepsilon$.

For the second-order energies we require the following matrix elements:

$$H^{(1)}_{m_l,\frac{3}{4}} = (1/\sqrt{2})\left\{H^{(1)}_{m_l,1} - H^{(1)}_{m_l,-1}\right\}$$

$$= \begin{cases} (1/\sqrt{2})\,H^{(1)}_{3,1} = -(\varepsilon/4\sqrt{2}) & \text{for } m_l = 3 \\ -(1/\sqrt{2})\,H^{(1)}_{-3,-1} = +(\varepsilon/4\sqrt{2}) & \text{for } m_l = -3, \text{ all others zero.} \end{cases}$$

$$H^{(1)}_{m_l,\frac{1}{4}} = (1/\sqrt{2})\left\{H^{(1)}_{m_l,1} + H^{(1)}_{m_l,-1}\right\}$$

$$= \begin{cases} (1/\sqrt{2})\,H^{(1)}_{3,1} = -(\varepsilon/4\sqrt{2}) & \text{for } m_l = 3 \\ (1/\sqrt{2})\,H^{(1)}_{-3,-1} = -(\varepsilon/4\sqrt{2}) & \text{for } m_l = -3, \text{ all others zero.} \end{cases}$$

$$E^{(0)}_{m_l} = m_l^2\hbar^2/2mr^2 = m_l^2 A\,, \text{ with } A = \hbar^2/2mr^2.$$

Both ϕ linear combinations correspond to $|m_l| = 1$, and so for them $E^{(0)} = A$. For the $\phi_{\frac{3}{4}}$ combination:

$$E^{(2)} = \sum_{m_l}{}' \left\{H^{(1)}_{\frac{3}{4},m_l} H^{(1)}_{m_l,\frac{3}{4}} \Big/ (1 - m_l^2)A\right\}$$

$$= \left|H^{(1)}_{\frac{3}{4},3}\right|^2 \Big/ (-8A) + \left|H^{(1)}_{\frac{3}{4},-3}\right|^2 \Big/ (-8A) = -\varepsilon^2/128A\,.$$

For the $\phi_{\frac{1}{4}}$ combination:

$$E^{(2)} = \sum_{m_l}{}' \left\{H^{(1)}_{\frac{1}{4},m_l} H^{(1)}_{m_l,\frac{1}{4}} \Big/ (1 - m_l^2)A\right\}$$

$$= \left|H^{(1)}_{\frac{1}{4},3}\right|^2 \Big/ (-8A) + \left|H^{(1)}_{\frac{3}{4},-3}\right|^2 \Big/ (-8A) = -\varepsilon^2/128A\,.$$

[The $m_l = 0$ does not in fact make a contribution to the sum.] The energies to second-order are therefore

$$\underline{E_{\frac{3}{4}} = A + \tfrac{3}{4}\varepsilon - \varepsilon^2/128A}\,, \qquad \underline{E_{\frac{1}{4}} = A + \tfrac{1}{4}\varepsilon - \varepsilon^2/128A}\,.$$

EXERCISE: Find the first- and second-order energy corrections for a particle subject to $H^{(1)} = \varepsilon\sin^4\phi$.

8.12 $H = -(\hbar^2/2\mu)\nabla^2 - (e^2/4\pi\varepsilon_0 r)$; $\psi = N e^{kr^2}$.

Since ψ is spherically symmetrical, use ∇^2 in the form

$$\nabla^2 = (1/r)\,(d^2/dr^2)\,r \qquad \text{[eqn (4.2.3)]}$$

$$\nabla^2\psi = N\,(1/r)(d^2/dr^2)\,r\,e^{kr^2} = N\,(6k + 4k^2 r^2)\,e^{kr^2}.$$

Anticipate that $k < 0$ [for an acceptable wavefunction], and so write $k = -\kappa$, with $\kappa > 0$. Then

$$\nabla^2\psi = -2N(3\kappa - 2\kappa^2 r^2)\,e^{-\kappa r^2}.$$

The Rayleigh ratio is then

$$\mathcal{E} = \frac{\int_0^\infty \left\{-(\hbar^2/\mu)(-3\kappa + 2\kappa^2 r^2)\,r^2 e^{-2\kappa r^2} - (e^2/4\pi\varepsilon_0)\,r\,e^{-2\kappa r^2}\right\} dr}{\int_0^\infty r^2 e^{-2\kappa r^2}\,dr}$$

[remember that $d\tau \propto r^2 dr$]. Therefore, since

$$\int_0^\infty r\,e^{-2\kappa r^2}\,dr = 1/4\kappa$$

$$\int_0^\infty r^2 e^{-2\kappa r^2}\,dr = \tfrac{1}{8}\,(\pi/2\kappa^3)^{\frac{1}{2}}$$

$$\int_0^\infty r^4 e^{-2\kappa r^2}\,dr = (3/32)(\pi/2\kappa^5)^{\frac{1}{2}},$$

$$\mathcal{E} = (3/2)(\hbar^2/\mu)\,\kappa - (e^2/\varepsilon_0\sqrt{(2\pi^3)})\sqrt{\kappa}.$$

Find κ such that $d\mathcal{E}/d\kappa = 0$:

$$\kappa^{\frac{1}{2}} = e^2\mu/3\varepsilon_0\hbar^2\sqrt{(2\pi^3)}, \qquad \kappa = e^4\mu^2/18\pi^3\varepsilon_0^2\hbar^4.$$

The best value of $\mathcal{E}$ is therefore

$$\mathcal{E} = \tfrac{3}{2}(\hbar^2/\mu)(e^4\mu^2/18\pi^3\varepsilon_0^2\hbar^4) - (e^2/\varepsilon_0\sqrt{(2\pi^3)})\,(e^2\mu/3\varepsilon_0\hbar^2\sqrt{(2\pi^3)})$$

$$= -e^4\mu/12\pi^3\varepsilon_0^2\hbar^2 = -(1/3\pi)(e^4\mu/\varepsilon_0^2 h^2)$$

$$= -(8/3\pi)\,hcR_H,$$

$-hcR_H$ being the exact ground state energy. This optimized wavefunction therefore accounts for $8/3\pi = 0.849$, or 84.9 percent, of the exact energy.

EXERCISE: Find the dependence of the best energy on the value of the atomic number of a hydrogen-like atom.

8.13 (a) This problem illustrates one precaution. The wavefunction must satisfy the boundary conditions, which in this case are sufficiently stringent to require $\sin kx = 0$ at $x = L$. Hence $k = n\pi/L$, n integral and greater than zero, must hold. Consequently the variation parameter is not a variable in the conventional sense, and we simply have to select the lowest value permitted by the boundary conditions:

$$\underline{k = \pi/L, \quad E = h^2/8mL^2}\ .$$

(b)
$$\int \psi H\psi\, d\tau = -(\hbar^2/2m)\int_0^L \psi(d^2/dx^2)\left\{(x-x^2/L) + k(x-x^2/L)^2\right\} dx$$

$$= -(\hbar^2/m)\int_0^L \left\{(x-x^2/L) + k(x-x^2/L)^2\right\}\left\{k - 1/L + 6kx^2/L^2 - 6kx/L\right\} dx$$

$$= (\hbar^2 L/m)\left\{1 + 7kL/5 - 19k^2L^2/35\right\}$$

$$\int \psi^2 d\tau = (L^2/30)\left\{1 + 3kL/7 - k^2L^2/21\right\}$$

$$\mathcal{E} = (5\hbar^2/mL^2)\left\{\frac{1 + 7kL/5 - 19k^2L^2/35}{1 + 3kL/7 - k^2L^2/21}\right\}.$$

The condition $d\mathcal{E}/dk = 0$ is satisfied if

$$(7L/5 - 38kL^2/35)(1 + 3kL/7 - k^2L^2/21)$$

$$- (1 + 7kL/5 - 19k^2L^2/35)(3L/7 - 2kL^2/21) = 0,$$

or

$$4 - (104/15)\,\underline{kL} - (57/35)\,\underline{k^2L^2} = 0.$$

The solutions are $kL = 0.5147$ and -4.7720. These give

$$\mathcal{E} = 1.3053\,(5\hbar^2/mL^2) = h^2/6.0489\,mL^2 \quad \text{for } kL = 0.5147$$

$$\mathcal{E} = 8.4727\,(5\hbar^2/mL^2) = h^2/0.9319\,mL^2 \quad \text{for } kL = -4.7720\ .$$

The former is the lower, and so we adopt $kL = 0.5147$. The best trial function is therefore

$$\underline{\psi = (x - x^2/L) + (0.5147/L)(x - x^2/L)^2}, \quad \underline{E = h^2/6.0489\,mL^2}\ .$$

(c) This problem illustrates another precaution.

$$\int \psi H\psi\, d\tau = -(\hbar^2/2m)\left\{\int_0^{\frac{1}{2}L} \psi\,(d^2/dx^2)\, e^{k(x-\frac{1}{2}L)}\, dx + \int_{\frac{1}{2}L}^{L} \psi\,(d^2/dx^2)\, e^{-k(x-\frac{1}{2}L)}\, dx\right\}$$

$$= -(k^2\hbar^2/2m)\left\{\int_0^{\frac{1}{2}L} \psi\, e^{k(x-\frac{1}{2}L)}\, dx + \int_{\frac{1}{2}L}^{L} \psi\, e^{-k(x-\frac{1}{2}L)}\, dx\right\}$$

$$=-(k^2\hbar^2/2m)\left\{\int_0^{\frac{1}{2}L}\left[e^{2k(x-\frac{1}{2}L)}-e^{k(x-L)}\right]dx + \int_{\frac{1}{2}L}^{L}\left[e^{-2k(x-\frac{1}{2}L)}-e^{-kx}\right]dx\right\}$$

$$=-(k^2\hbar^2/2m)\left\{e^{-kL}\int_0^{\frac{1}{2}L}(e^{2kx}-e^{kx})dx + \int_{\frac{1}{2}L}^{L}(e^{-2kx+kL}-e^{-kx})dx\right\}$$

$$=-(k^2\hbar^2/2m)\left\{\tfrac{1}{2}(1-e^{-kL})-(e^{-\frac{1}{2}kL}-e^{-kL}) + \tfrac{1}{2}(1-e^{-kL})-(e^{-\frac{1}{2}kL}-e^{-kL})\right\}$$

$$=k(\hbar^2/2m)(1-e^{-\frac{1}{2}kL})^2 .$$

$$\int\psi^2 d\tau = \int_0^{\frac{1}{2}L}\left\{e^{k(x-\frac{1}{2}L)}-e^{-\frac{1}{2}kL}\right\}^2 dx + \int_{\frac{1}{2}L}^{L}\left\{e^{-k(x-\frac{1}{2}L)}-e^{-\frac{1}{2}kL}\right\}^2 dx$$

$$=Le^{-kL}+(1/k)(1-e^{-\frac{1}{2}kL})(1-3e^{-\frac{1}{2}kL}) .$$

$$\mathcal{E}=\frac{-(\hbar^2/2m)k(1-e^{-\frac{1}{2}kL})^2}{Le^{-kL}+(1/k)(1-e^{-\frac{1}{2}kL})(1-3e^{-\frac{1}{2}kL})}.$$

Then, with $\mathcal{E}=-(\hbar^2/mL^2)\varepsilon$ and $z=\frac{1}{2}kL$,

$$\varepsilon=\frac{2z^2(1-e^{-z})^2e^{2z}}{2z+(1-e^{-z})(1-3e^{-z})e^{2z}} .$$

The condition $d\varepsilon/dz=0$ is satisfied for $z=0$ and for

$$3+z-(2z^2+7z+10)e^{z}+(2z^2+5z+2)e^{2z} -(5z+6)e^{3z}+(1+2z)e^{4z}=0 .$$

This must be solved numerically. One solution (I do not know if there are others) is $z=1.1486$, giving $\varepsilon=4.6565$. Therefore, at this stage, we conclude that $E=-h^2/8.4782\,mL^2$. This result exposes the difficulty: the E just calculated is not only *lower* than the true ground state energy $(\hbar^2/8mL^2)$ but it is also *negative*, whereas the particle possesses only kinetic energy, and $\langle p^2\rangle>0$ from the hermiticity of p. The resolution of the paradox is that the wavefunction contains a cusp (at $x=\frac{1}{2}L$), and so its first derivative is discontinuous, and its second derivative not defined — or, crudely, infinite — at $x=\frac{1}{2}L$. Therefore the trial function is inadmissible — or, crudely, it is necessary to add ∞ to the calculated energy.

EXERCISE: Take a gaussian trial function $\exp - k^2(x - \frac{1}{2}L)^2$ and find the optimum value of k and the best energy.

8.14 The symmetry adapted linear combinations for a C_s molecule (E, σ) are

$$A': \; a'_1 = s_{\mathbf{B}}\,, \; a'_2 = (s_{\mathbf{A}} + s_{\mathbf{C}})/\sqrt{2}\,,$$

$$A'': \; a'' = (s_{\mathbf{A}} - s_{\mathbf{C}})\sqrt{2}\,.$$

$$E(A'') = \int a''\, H\, a''\, \mathrm{d}\tau = \tfrac{1}{2}\int (s_{\mathbf{A}} + s_{\mathbf{C}})\, H\, (s_{\mathbf{A}} + s_{\mathbf{C}})\, \mathrm{d}\tau = \alpha\,.$$

$$a' = c_1 a'_1 + c_2 a'_2\,,$$

$$\int a'_1 H a'_1\, \mathrm{d}\tau = \alpha, \qquad \int a'_2 H a'_2\, \mathrm{d}\tau = \alpha$$

$$\int a'_1 H a'_2\, \mathrm{d}\tau = (1/\sqrt{2}) \int s_{\mathbf{B}} H (s_{\mathbf{A}} + s_{\mathbf{C}})\mathrm{d}\tau = \beta\sqrt{2} = \int a'_2 H a'_1\, \mathrm{d}\tau\,.$$

$$\det|H - ES| = \begin{vmatrix} \alpha - E & \beta\sqrt{2} \\ \beta\sqrt{2} & \alpha - E \end{vmatrix} = (\alpha - E)^2 - 2\beta^2 = 0\,.$$

Therefore, $E(A') = \alpha \pm \beta\sqrt{2}$.

The scalar equations are:

$$\left.\begin{aligned} (\alpha - E)c_1 + (\sqrt{2})\,\beta\, c_2 &= 0 \\ (\sqrt{2})\,\beta\, c_1 + (\alpha - E)\, c_1 &= 0 \end{aligned}\right\} \; c_1^2 + c_2^2 = 1\,.$$

When $E = \alpha + \beta\sqrt{2}$, $\quad -c_1 + c_2 = 0$; hence $c_1 = c_2 = 1/\sqrt{2}$.

When $E = \alpha - \beta\sqrt{2}$, $\quad c_1 + c_2 = 0$; hence $c_1 = -c_2 = 1/\sqrt{2}$.

Therefore, the energies and orbitals are

Highest: $\quad a'^{*} = (1/\sqrt{2})(a'_1 - a'_2) = \frac{1}{2}(s_{\mathbf{A}} - \sqrt{2}s_{\mathbf{B}} + s_{\mathbf{C}}),\quad E = \alpha - \beta\sqrt{2}$

Nonbonding: $a'' = (1/\sqrt{2})(s_{\mathbf{A}} - s_{\mathbf{C}}),\quad E = \alpha$.

Lowest: $\quad a' = (1/\sqrt{2})(a'_1 + a'_2) = \frac{1}{2}(s_{\mathbf{A}} + \sqrt{2}s_{\mathbf{B}} + s_{\mathbf{C}}),\quad E = \alpha + \beta\sqrt{2}$.

EXERCISE: Find the energies and orbitals of H_3 in the form of an equilateral triangle.

8.15 First, normalize the linear combinations to unity:

$$\int (a'_2)^2\, \mathrm{d}\tau = \tfrac{1}{2}\int (s_{\mathbf{A}} + s_{\mathbf{C}})^2\, \mathrm{d}\tau = \tfrac{1}{2}\int (s_{\mathbf{A}}^2 + s_{\mathbf{B}}^2 + 2s_{\mathbf{A}}s_{\mathbf{C}})\, \mathrm{d}\tau$$

$$= 1 + S_{\mathbf{AC}}\,.$$

$$\int (a'')^2\, \mathrm{d}\tau = 1 - S_{\mathbf{AC}}\,.$$

Therefore,

$$a_2' = (s_A + s_C)/\{2(1+S_{AC})\}^{\frac{1}{2}},$$
$$a'' = (s_A - s_C)/\{2(1-S_{AC})\}^{\frac{1}{2}}.$$

Now construct the matrix elements of H:

$$\int a_1' H a_1' \, d\tau = \alpha,$$
$$\int a_2' H a_2' \, d\tau = \int (s_A + s_C) H (s_A + s_C) \, d\tau / 2(1+S_{AC})$$
$$= (\alpha + \gamma)/(1 + S_{AC}),$$
$$\int a_2 H a_2 \, d\tau = (\alpha + \gamma)/(1 - S_{AC}),$$
$$\int a_1' H a_2' \, d\tau = \int s_B H (s_A + s_C) d\tau / \{2(1+S_{AC})\}^{\frac{1}{2}} = \{2/(1+S_{AC})\}^{\frac{1}{2}} \beta,$$
$$\int a_1' \, a_2' \, d\tau = \{2/(1+S_{AC})\}^{\frac{1}{2}} S_{AB} \qquad [S_{AB} = S_{BC}].$$

Hence, the 2×2 secular determinant is

$$\det|H - ES| = \begin{vmatrix} \alpha - E & \dfrac{(\beta - ES_{AB})\sqrt{2}}{\sqrt{(1+S_{AC})}} \\ \dfrac{(\beta - ES_{AB})\sqrt{2}}{\sqrt{(1+S_{AC})}} & \left(\dfrac{\alpha+\gamma}{1+S_{AC}}\right) - E \end{vmatrix}.$$

Set $\gamma = (S_{AC}/S_{AB})\beta$; then with $S_{AB} = 0.723$ and $S_{AC} = 0.345$,

$$\det|H - ES| = \begin{vmatrix} \alpha - E & 1.219(\beta - 0.723E) \\ 1.219(\beta - 0.723E) & (\alpha + 0.477\beta)/1.345 - E \end{vmatrix}$$
$$= 0.223E^2 + (1.794\beta - 1.744\alpha)E$$
$$+ (0.355\alpha\beta + 0.744\alpha^2 - 1.486\beta^2.$$

Therefore, we must solve

$$E^2 + (8.045\beta - 7.821\alpha)E + (1.592\alpha\beta + 3.336\alpha^2 - 6.664\beta^2) = 0.$$

Write $E/\alpha = \varepsilon$ and $\beta/\alpha = \lambda$; then

$$\varepsilon^2 + (8.045\lambda - 7.821)\varepsilon + (1.592\lambda + 3.336 - 6.664\lambda^2) = 0.$$
$$\varepsilon = 3.911 - 4.023\lambda \pm \sqrt{\{22.845\lambda^2 - 33.052\lambda + 11.956\}},$$

which can be plotted as a function of λ, Fig. 8.2. (The result from Problem 8.14, $\varepsilon = 1 \pm \lambda\sqrt{2}$, is also shown.)

EXERCISE: Include overlap in the *Exercise* attached to Problem 8.14.

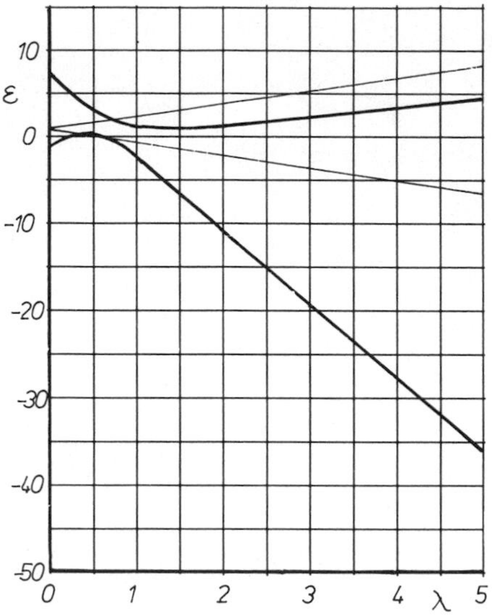

Fig. 8.2

8.16 $\Psi = a_1\psi_1 e^{-iE_1t/\hbar} + a_2\psi_2 e^{-iE_2t/\hbar}$ [eqn (8.5.6)]

$= \{a_{2s}\psi_{2s} + a_{2p}\psi_{2p}\} e^{-iEt/\hbar}$ $[E_1 = E_{2s},\ E_2 = E_{2p},\ E_{2s} = E_{2p} = E]$.

$\langle\Psi|\mu_z|\Psi\rangle = -e\langle\Psi|z|\Psi\rangle$

$= -e\{a^*_{2s}a_{2p}\langle 2s|z|2p\rangle + a_{2s}a^*_{2p}\langle 2p|z|2s\rangle\}$

$= -3ea_0(a^*_{2s}a_{2p} + a_{2s}a^*_{2p}) = -6ea_0\,\mathrm{re}\,(a_{2s}a^*_{2p})$.

$a_{2s} = \{\cos\Omega t - i(\omega_0/2\Omega)\sin\Omega t\} e^{\frac{1}{2}i\omega_0 t} = \cos Vt$ [eqn (8.5.9), $\omega_0 = 0$],

$a_{2p} = -i\sin Vt$,

$a^*_{2s}a_{2p} + a_{2s}a^*_{2p} = -i\sin Vt\cos Vt + i\sin Vt\cos Vt = 0$.

Therefore, $\langle\Psi|\mu_z|\Psi\rangle = \underline{0}$ at all times.

EXERCISE: Suppose there is a small energy difference between the 2s- and 2p-orbitals (as in real life there is, on account of relativistic effects). Find the electric dipole moment during and after its exposure.

8.17 $P_2 = \{4V^2/(\omega_0^2 + 4V^2)\}\sin^2\frac{1}{2}(\omega_0^2 + 4V^2)^{\frac{1}{2}}t$ [eqn (8.5.10)].

$\hbar V = \langle 1,0|H|0,0\rangle = \langle\alpha\beta + \beta\alpha|g_1 s_{z1} + g_2 s_{z2}|\alpha\beta - \beta\alpha\rangle\ (\mu_B/2\hbar)B$

$= \frac{1}{2}\hbar\langle\alpha\beta + \beta\alpha|g_1\alpha\beta + g_1\beta\alpha - g_2\alpha\beta - g_2\beta\alpha\rangle\ (\mu_B/2\hbar)B$

$= \frac{1}{2}(g_1 - g_2)\mu_B B$.

$\omega_0 = J$ [given]; therefore

$$P_{\mathrm{T}} = \frac{(g_1 - g_2)^2\,(\mu_{\mathrm{B}} B/\hbar)^2}{J^2 + (g_1 - g_2)^2(\mu_{\mathrm{B}} B/\hbar)^2}\ \sin^2\left\{\tfrac{1}{2}[J^2 + (g_1 - g_2)^2(\mu_{\mathrm{B}} B/\hbar)^2]^{\frac{1}{2}}\,t\right\}.$$

For $g_1 - g_2 = 10^{-3}$ and $J \approx 0$,

$$P_{\mathrm{T}} \approx \sin^2\left\{\tfrac{1}{2}(g_1 - g_2)\,(\mu_{\mathrm{B}} Bt/\hbar)\right\} \approx \sin^2(\mu_{\mathrm{B}} Bt/2000\,\hbar).$$

For $B = 1\,\mathrm{T}$, $\mu_{\mathrm{B}} = 9.274 \times 10^{-24}\,\mathrm{J\,T^{-1}}$, $\hbar = 1.055 \times 10^{-34}\,\mathrm{J\,s}$,

$$P_{\mathrm{T}} \approx \sin^2\{4.395 \times 10^7\,(t/\mathrm{s})\}.$$

Therefore, $P_{\mathrm{T}} \approx 1$ when $4.395 \times 10^7\,(t/\mathrm{s}) \approx \pi/2$, corresponding to $t \approx 3.6 \times 10^{-8}\,\mathrm{s}$, or <u>36 ns</u>.

EXERCISE: Consider the same biradical but with nuclear hyperfine interaction between an electron and a nucleus on centre 1 (so that there is an additional $a\boldsymbol{I}\cdot\boldsymbol{s}_1$ term in the hamiltonian). What is the *total* (i.e. including $T_{\pm 1}$ populations) probability of being in the triplet at time t? At what time does $P_{\mathrm{T}} \approx 1$ when the nucleus is a proton and $a \mathrel{\hat{=}} 50$ MHz?

8.18 $a_{\mathrm{f}}(t) = (1/\mathrm{i}\hbar)\int_0^t H^{(1)}_{\mathrm{fi}}(t)\,\mathrm{e}^{\mathrm{i}\omega_{\mathrm{fi}} t}\,\mathrm{d}t$ [eqn (8.6.6)]

$$a_{\mathrm{2p}}(t) = (1/\mathrm{i}\hbar)\,\langle 2\mathrm{p}_z|ez|1\mathrm{s}\rangle \int_0^t E(t)\,\mathrm{e}^{\mathrm{i}\omega_{\mathrm{2p,1s}} t}\,\mathrm{d}t \quad [H(t) = -\mu_z E(t) = ezE(t)].$$

Set $E(t) = \gamma t$; note that $\omega_{\mathrm{2p,1s}} = \frac{3}{4}hcR_{\mathrm{H}}/\hbar = (3\pi/2)cR_{\mathrm{H}}$. For simplicity of notation, write $\omega = \omega_{\mathrm{2p,1s}}$.

$$\begin{aligned} a_{\mathrm{2p}}(t) &= \gamma(e/\mathrm{i}\hbar)\,\langle 2\mathrm{p}_z|z|1\mathrm{s}\rangle \int_0^t t\,\mathrm{e}^{\mathrm{i}\omega t}\,\mathrm{d}t \\ &= (e\gamma/\mathrm{i}\hbar)\,\langle 2\mathrm{p}_z|z|1\mathrm{s}\rangle \left\{(t/\mathrm{i}\omega)\,\mathrm{e}^{\mathrm{i}\omega t} + (1/\omega^2)(\mathrm{e}^{\mathrm{i}\omega t} - 1)\right\} \end{aligned}$$

$$|a_{\mathrm{2p}}(t)|^2 = (e\gamma/\hbar)\,\langle 2\mathrm{p}_z|z|1\mathrm{s}\rangle^2\,(2/\omega^4)\left\{1 - \cos\omega t - \omega t\sin\omega t + \tfrac{1}{2}\omega^2 t^2\right\}.$$

EXERCISE: Find $|a_{\mathrm{2p}}(t)|^2$ in the case where the perturbation is turned on quadratically ($E \propto t^2$).

8.19 $a_{\mathrm{2p}}(T) = (e\gamma/\mathrm{i}\hbar)\langle 2\mathrm{p}_z|z|1\mathrm{s}\rangle\left\{\int_0^{\frac{1}{2}T} t\,\mathrm{e}^{\mathrm{i}\omega t}\,\mathrm{d}t + \int_{\frac{1}{2}T}^{T} (T - t)\,\mathrm{e}^{\mathrm{i}\omega t}\,\mathrm{d}t\right\}$

$$\int_{\frac{1}{2}T}^{T} (T - t)\,\mathrm{e}^{\mathrm{i}\omega t}\,\mathrm{d}t = \left\{\int_0^{\frac{1}{2}T} t\,\mathrm{e}^{-\mathrm{i}\omega t}\,\mathrm{d}t\right\}\mathrm{e}^{\mathrm{i}\omega T} \quad [\tau = T - t,\ \text{then}\ \tau \to t]$$

$$a_{2p}(T) = (e\gamma/i\hbar)\langle 2p_z|z|1s\rangle\Big\{\big[(T/2i\omega)\,e^{\frac{1}{2}i\omega T} + (1/\omega^2)\,(e^{\frac{1}{2}i\omega T} - 1)\big]$$
$$+\big[(-T/2i\omega)\,e^{-\frac{1}{2}i\omega T} + (1/\omega^2)(e^{-\frac{1}{2}i\omega T} - 1)\big]\,e^{i\omega T}\Big\}$$
$$= -(e\gamma/i\hbar\omega^2)\langle 2p_z|z|1s\rangle\,(1 - 2\,e^{\frac{1}{2}i\omega T} + e^{i\omega T})$$
$$= -(e\gamma/i\hbar\omega^2)\langle 2p_z|z|1s\rangle\,(1 - e^{\frac{1}{2}i\omega T})^2 .$$

$$|a_{2p}(T)|^2 = (e\gamma/\hbar\omega^2)^2\,\langle 2p_z|z|1s\rangle^2\,(4\sin^2\tfrac{1}{4}\omega T)^2$$
$$= (4e\gamma/\hbar\omega^2)^2\,\langle 2p_z|z|1s\rangle^2\,\sin^4\tfrac{1}{4}\omega T .$$

When $|1s\rangle$ is replaced by $|2s\rangle$, $\omega_{2p,1s}$ is replaced by $\omega_{2p,2s} = 0$. Form the limit of $(\sin\frac{1}{4}\omega T/\omega)^4$ as $\omega \to 0$:

$$\lim_{\omega\to 0}(\sin\tfrac{1}{4}\omega T/\omega)^4 = (\tfrac{1}{4}T)^4 .$$

Therefore,

$$|a_{2p}(T)|^2 = (e^2\gamma^2/16\hbar^2)\,\langle 2p_z|z|1s\rangle^2\,T^4 = 3e\gamma a_0/4\hbar)^2\,T^4 .$$

EXERCISE: The quadratic perturbation of the preceding *Exercise* was run until $\frac{1}{2}T$ and then extinguished quadratically. Find $|a_{2p}(T)|^2$ for initial 1s and 2s states.

8.20 $$a_f = (1/i\hbar)\left\{\int_0^T H^{(1)}_{fi}(t)\,e^{i\omega_{fi}t}\,dt + \int_T^t H^{(1)}_{fi}(t)\,e^{i\omega_{fi}t}\,dt\right\}.$$

Write

$$H^{(1)}_{fi}(t) = \begin{cases}\hbar V(1 - e^{-kt}) & \text{for } 0 \leqslant t \leqslant T\\ \hbar V\,e^{-k(t-T)} & \text{for } t \geqslant T ,\end{cases}$$

and put $\omega_{fi} = \omega$ for notational simplicity. Then

$$a_f = -iV\left\{\int_0^T (1 - e^{-kt})\,e^{i\omega t}\,dt + \int_T^t e^{-k(t-T)}\,e^{i\omega t}\,dt\right\}$$

$$= (V/\omega)\left\{1 - e^{i\omega T} - \left(\frac{i\omega}{k - i\omega}\right)\left(e^{i\omega T} - e^{-k(t-T) + i\omega t} + e^{-(k-i\omega)T} - 1\right)\right\}$$

$$\approx (V/\omega)\left\{1 - e^{i\omega T} - (i\omega/(-i\omega))(e^{i\omega T} - 1)\right\} \quad [k \ll \omega,\ kT \gg 1,\ k(t-T) \gg 1]$$

$$\approx (V/\omega)\left\{1 - e^{i\omega T} + e^{i\omega T} - 1\right\} = 0 .$$

Hence, $P_f(T) \approx 0$.

EXERCISE: Consider the case where an oscillating perturbation is switched on exponentially, and then switched off exponentially. Calculate $P_f(T)$ for slow switching.

8.21 $\mu_z = -e\int \psi_{2p_z} z \psi_{1s}\, d\tau = -e(Z/a)^4(1/3\sqrt{2})\int_0^\infty r^4 e^{-3Zr/2a}\, dr$.

[Use $\psi_{2p_z} = (Z/a)^{5/2}(1/4\sqrt{(2\pi)})\, r\cos\theta\, e^{-Zr/2a}$

$\psi_{1s} = (Z/a)^{3/2}(1/\sqrt{\pi})\, e^{-Zr/a}$.]

$\mu_z = -(2^8/3^5\sqrt{2})(ea/Z) = \mu_y^2 = \mu_x^2$ [for $2p_z$, $2p_y$, $2p_x \rightarrow 1s$].

$\mu^2 = \mu_x^2 + \mu_y^2 + \mu_z^2$ [for $2p \rightarrow 1s$ overall]

$= 3(2^8/3^5\sqrt{2})^2(ea/Z)^2$

$= (2^{15}/3^9)(ea/Z)^2 = (2^{15}/3^9)e^2a^2$ for $Z = 1$.

$B = \mu^2/6\varepsilon_0\hbar^2 = (2^{14}/3^{10})(e^2a^2/\varepsilon_0\hbar^2)$

$= (2^{14}/3^{10}) \times (7.300 \times 10^{20}\ \mathrm{J^{-1}\,m^3\,s^{-1}}) = \underline{2.025 \times 10^{20}\ \mathrm{J^{-1}\,m^3\,s^{-1}}}$.

$\nu(2p \rightarrow 1s) = \frac{3}{4}cR_{\mathrm{H}} = 2.467 \times 10^{15}$ Hz .

$A = 8\pi h(\nu/c)^3 B = \underline{1.880 \times 10^9\ \mathrm{s^{-1}}}$.

At 1000 K and $\nu = 2.467 \times 10^{15}$ Hz,

$\rho = (8\pi h\nu^3/c^3)(e^{h\nu/kT} - 1)^{-1} = 3.530 \times 10^{-63}\ \mathrm{J\,Hz^{-1}\,m^{-3}}$.

Therefore:

$W^{\text{stimulated}} = \rho B = \underline{7.148 \times 10^{-43}}$ $(1/W = 1.4 \times 10^{42}$ s$)$

$W^{\text{spontaneous}} = A = \underline{1.880 \times 10^9\ \mathrm{s^{-1}}}$ $(1/W = 5.3 \times 10^{-10}$ s$)$

EXERCISE: Calculate the rates of stimulated and spontaneous emission for $3p \rightarrow 1s$ in hydrogen on the surface of the sun (take $T \approx 6500$ K).

8.22 $\mu^2 = (2^{15}/3^9)(ea/Z)^2$ [Problem 8.21].

$B = \mu^2/6\varepsilon_0\hbar^2 = (2^{14}/3^{10})(e^2a^2/Z^2\varepsilon_0\hbar^2)$.

$\nu(2p \rightarrow 1s) = \frac{3}{4}Z^2Rc$ [$E \propto Z^2$].

$A = 8\pi h(\nu/c)^3 B = 8\pi h(3Z^2R/4)^3(2^{14}/3^{10})(e^2a^2/Z^2\varepsilon_0\hbar^2)$

$= (2^{11}/3^7)(\pi hR^3e^2a^2/\varepsilon_0\hbar^2)Z^4$.

Then, as $R \approx R_\infty = m_e e^4/8h^3\varepsilon_0^2 c$ and $a \approx a_0 = 4\pi\varepsilon_0\hbar^2/m_e e^2$,

$A = (2^4/3^7)(\pi m_e e^{10}/h^6\varepsilon_0^5 c^3)Z^4$

$= (2^9/3^7)\pi\alpha^5(m_e c^2/h)Z^4$ [$\alpha = e^2/4\pi\varepsilon_0\hbar c$]

$= \underline{(2^9/3^7)(\pi\alpha^5 c/\lambda_C)Z^4}$ [$\lambda_C = h/m_e c$, the Compton wavelength].

$$\rho(\nu_{2p,1s}) = 8\pi h\,(\tfrac{3}{2}Z^2R)^3\Big/\Big\{e^{\frac{3}{4}hcRZ^2/kT} - 1\Big\}$$

$$\approx 8\pi h(\tfrac{3}{4}Z^2R)^3\, e^{-\frac{3}{4}hcRZ^2/kT} \qquad [e^x \approx 1+x \text{ for } x \ll 1]$$

$$\approx 27\pi hR^3Z^6\, e^{-\frac{3}{4}hcRZ^2/kT}\,.$$

$$\rho B \approx (2/3)^7(\pi m_e e^{10}/h^6\varepsilon_0^5c^3)\,Z^4\, e^{-\frac{3}{4}hcRZ^2/kT}$$

$$\approx \underline{(2^{12}/3^7)(\pi\alpha^5 c/\lambda_C)\,Z^4\, e^{-\frac{3}{4}hcRZ^2/kT}}\,.$$

Note that $W^{\text{spontaneous}}/W^{\text{stimulated}} \approx \frac{1}{8}\,e^{\frac{3}{4}hcRZ^2/kT}$

and that $(2^9/3^7)\pi\alpha^5c/\lambda_C = 1.880\,49 \times 10^9\ \text{s}^{-1}$.

EXERCISE: Evaluate the Z-dependence for the $3p \rightarrow 1s$ transition.

8.23 $$\mu_x = -e\int_0^L \psi_n^* x\,\psi_{n+1}\,dx = -(2e/L)\int_0^L x\sin(n\pi x/L)\sin([n+1]\pi x/L)\,dx$$

$$= (2e/L)(L/\pi)^2\,\{4n(n+1)/(2n+1)^2\}$$

$$\left[\int x\sin ax\sin bx\,dx = \frac{\sin(a-b)x}{2(a-b)} - \frac{\sin(a+b)x}{2(a+b)} + \frac{\cos(a-b)x}{2(a-b)^2} - \frac{\cos(a+b)x}{2(a+b)^2}\,.\right]$$

$$= (8/\pi^2)\,eL\,\{n(n+1)/(2n+1)^2\}\,.$$

$$B = \mu_x^2/6\varepsilon_0\hbar^2$$

$$= (32/3\pi^4)(e^2L^2/\hbar^2\varepsilon_0)\{n(n+1)/(2n+1)^2\}\,.$$

$$\nu_{n+1,n} = (h/8mL^2)\,\{(n+1)^2 - n^2\} = (2n+1)h/8mL^2\,.$$

$$A = \left\{\frac{8\pi h^4(2n+1)^3}{(8mL^2c)^3}\right\}\left\{\frac{32}{3\pi^4}\right\}\left\{\frac{e^2L^2}{\hbar^2\varepsilon_0}\right\}\left\{\frac{n(n+1)}{(2n+1)^2}\right\}^2$$

$$= (2/3\pi)(h^2e^2/m^3\varepsilon_0c^3)(1/L^4)\{n^2(n+1)^2/(2n+1)\}$$

$$= (4/3\pi)\alpha(c\lambda_C^3/L^4)\{n^2(n+1)^2/(2n+1)\} \qquad [\lambda_C = h/m_ec].$$

Hence, $B \propto L^2$ while $A \propto 1/L^4$.

EXERCISE: (a) How do ρB and $W^{\text{spontaneous}}/W^{\text{stimulated}}$ depend on L?

(b) Evaluate A, B, and ρB for a one-dimensional harmonic oscillator.

9 Atomic spectra and atomic structure

9.1 $\tilde{\nu} = R_{He^+}\{(1/n_1^2) - (1/n_2^2)\}$ [eqn (9.1.5)].

$R_{He^+} = Z^2(\mu_{He^+}/m_e)R_\infty$ [eqn (9.1.3) and subsequent remarks].

$\mu_{He^+} = mm_e/(m+m_e)$, $m = m(He^{2+})$; $R_\infty = 109\,737.31\ cm^{-1}$.

$$m(He^{2+}) = m(^4_2He) - 2m_e$$
$$= 4.0026\times(1.660\,56\times10^{-27}\ kg) - 2(9.109\,53\times10^{-31}\ kg)$$
$$= 6.6447\times10^{-27}\ kg.$$
$$R_{He^+} = 4R_\infty/(1.000\,14) = 4.3889\times10^5\ cm^{-1}.$$

Therefore, the transitions are predicted to lie at

$$\tilde{\nu}/cm^{-1} = 4.3889\times10^5\{\tfrac{1}{4}-(1/n^2)\},\quad n = 3, 4, \ldots$$
$$= 6.0957\times10^4,\ 8.2292\times10^4,\ 9.2167\times10^4,\ \ldots\ 1.0972\times10^5.$$

EXERCISE: Find the positions of the corresponding lines in Li^{2+}.

9.2 $\mu/m_e = 1/(1+m_e/m_\mu) = 1/(1+1/207) = 0.995$.

$$\tilde{\nu} = R_\mu\{(1/n_1^2)-(1/n_2^2)\} = 0.995R_\infty\{(1/n_1^2)-(1/n_2^2)\}$$
$$= \underline{(1.092\times10^5\ cm^{-1})\times\{(1/n_1^2)-(1/n_2^2)\}},$$

with $n_1 = 1, 2, \ldots$ and $n_2 = n_1+1, n_1+2, \ldots$. The ionization limit (for $n_1 = 1$) lies at $1.092\times10^5\ cm^{-1}$ ($\lambda = 91.58$ nm).

EXERCISE: Predict the form of the spectrum of positronium (e^+, e^-).

9.3 Draw on $\Delta l = \pm 1$ [eqn (9.2.1)]. Then the allowed transitions are

<u>1s → 2p, 2p → 3d, and 3s → 5p</u>.

EXERCISE: Which of the following are electric-dipole allowed: 5s → 2p, 5s → 3d, 5s → 5f, 5s → 5p?

9.4 $E_n = -R/n^2$. For convenience, take the energy minimum as $E_1 \equiv 0$:

$$E_n = R\{1-1/n^2\} = 0,\ \tfrac{3}{4}R,\ \tfrac{8}{9}R,\ \tfrac{15}{16}R,\ \ldots.$$

The data fit this sequence with

$$R/cm^{-1} = 2743\,963,\quad 2744\,051,\quad 2744\,067$$

and mean value 2744 027. Write $R = Z^2R_\infty$ [ignore mass correction]; then $Z = 5.00$. The ion is therefore $\underline{B^{4+}}$.

The ionization energy is given by the series limit ($n = \infty$), which lies at $R = 2744\,027\ cm^{-1}$. Hence $I = hcR = 5.450\,95\times10^{-17}$ J, corresponding to $3.283\times10^4\ kJ\ mol^{-1}$ and 340.2 eV.

EXERCISE: Identify the one-electron ion giving rise to a spectrum showing that its np-orbitals were at 0, 6666 000 $\mathrm{cm^{-1}}$, 7901 000 $\mathrm{cm^{-1}}$, ... Predict its ionization energy (in $\mathrm{kJ\,mol^{-1}}$ and eV).

9.5 $\langle n'l'm_l'|\mu_m|nlm_l\rangle \propto \int R_{n'l'}Y^*_{l'm_l'}\, rY_{1m}R_{nl}Y_{lm_l}\, r^2\,\mathrm{d}r\,\sin\theta\,\mathrm{d}\theta\,\mathrm{d}\phi.$

[The proportionality factor is irrelevant; but since $z = 2(\pi/3)^{\frac{1}{2}}rY_{10}$ and $x \pm \mathrm{i}y = \mp 2(2\pi/3)^{\frac{1}{2}}rY_{1,\pm 1}$, with $\mu_0 = \mu_z$ and $\mu_{\pm 1} = \mp(\mu_x \pm \mathrm{i}\mu_y)/\sqrt{2}$, it is equal to $-2(\pi/3)^{\frac{1}{2}}e$.] Then

$$\langle n'l'm_l'|\mu_m|nlm_l\rangle \propto \int_0^\infty R_{n'l'}R_{nl}r^3\,\mathrm{d}r \int Y^*_{l'm_l'}Y_{1m}Y_{lm_l}\sin\theta\,\mathrm{d}\theta\,\mathrm{d}\phi.$$

Since

$$Y^*_{l'm_l'}Y_{1m}Y_{lm_l} \propto \exp\{\mathrm{i}(m_l + m - m_l')\phi\},$$

the integral vanishes unless $m_l' = m_l + m$; therefore $\Delta m_l = m = 0, \pm 1$. Since the spherical harmonics are bases for $\Gamma^{(l')}$, $\Gamma^{(1)}$, and $\Gamma^{(l)}$ in R_3, their product is a basis for the totally symmetric irrep only if $(l', 1, l)$ satisfies the triangle condition. Hence $l' = l, l \pm 1$. The parity of Y_{1m} under inversion is -1; Therefore $Y_{l'm_l'}$ and Y_{lm_l} must be of mutually opposite parity. Consequently $l' = l$ is excluded, and we conclude that $\Delta l = \pm 1$. There is no symmetry constraint on the r-integration, and the radial integral is non-zero for all values of $n' - n$. [For radial integrals, see Problem 13.4, and references there. There are symmetry properties of radial integrals — recall the high, hidden symmetry of the Coulomb potential: see M.J. Englefield, *Group theory and the Coulomb problem*, Wiley-Interscience (1972).]

EXERCISE: Establish the selection rules for the electric quadrupole transitions of atoms. [The transition operators are proportional to the quadratic forms xx, xy, etc., which themselves are proportional to the Y_{2m}.]

9.6 $hc\zeta_{nl} = \hbar^2 \int_0^\infty \xi(r)R^2_{nl}(r)\,r^2\,\mathrm{d}r$ [eqn (9.3.5)]

$$= (Ze^2\hbar^2/8\pi\varepsilon_0 m_e^2c^2)\int_0^\infty (1/r)R^2_{nl}(r)\,\mathrm{d}r \quad \text{[eqn (9.3.6)]}$$

$$= (Ze^2\hbar^2/8\pi\varepsilon_0 m_e^2c^2)\{Z^3/a_0^3n^3l(l+\tfrac{1}{2})(l+1)\}.$$

[Use the values of $\langle 1/r^3\rangle$ quoted on p. 145 of L. Pauling and E.B. Wilson, *Introduction to quantum mechanics*, McGraw-Hill (1935).]

EXERCISE: Find a relation between $\langle 1/r^3\rangle$ and $1/\langle r\rangle^3$ for an electron in a hydrogenic orbital.

9.7 $\psi_{nlm_l} = Nr^{n^*-1}\, e^{-Z^*\rho/n^*} Y_{lm_l} = R_{nl}\, Y_{lm_l}$.

$R_{nl} = Nr^{n^*-1}\, e^{-Z^*\rho/n^*}$; $\rho = r/a_0$. For 2p-orbitals, $n^* = n = 2$.

Then

$R_{2p} = Nre^{-Z^*/2a_0}$; $N = (Z^*/a_0)^{5/2}/2\sqrt{6}$.

$$hc\zeta_{2p} = (Ze^2\hbar^2/8\pi\varepsilon_0 m_e^2 c^2)\int_0^\infty (1/r)\, R_{2p}^2\, dr$$

$$= (Z^*/a_0)^5\,(Ze^2\hbar^2/192\pi\varepsilon_0 m_e^2 c^2)\int_0^\infty r e^{-Z^* r/a_0}\, dr$$

$$= (Z^*/a_0)^5 (Ze^2\hbar^2/\ 192\pi\varepsilon_0 m_e^2 c^2)(a_0/Z^*)^2$$

$$= Z^{*3} Z(e^2\hbar^2/192\pi\varepsilon_0 a_0^3 m_e^2 c^2) = (ZZ^{*3}/24)\,\alpha^2 hcR_\infty .$$

$\underline{\zeta_{2p} = (ZZ^{*3}/24)\,\alpha^2 R_\infty}$.

If the shielding comes into the form of $d\phi/dr$ [in the calculation of $\xi(r)$] then Z should be interpreted as Z^*. Then

$$\zeta_{2p} = (Z^{*4}/24)\,\alpha^2 R_\infty = (0.243\ \text{cm}^{-1}) \times (Z-\sigma)^4 .$$

Take $Z-\sigma$ from Table 9.1, and draw up the following Table:

ζ/cm^{-1}		B	C	N	O	F
Calculated	(a)	11	27	56	104	178
	(b)	21	50	101	183	308
Experimental		11	29	76	151	270

(a) using $Z^{*3}Z$; (b) using Z^{*4}.

EXERCISE: Evaluate ζ_{3p} for the elements Na to Cl.

9.8 $E_{so}(j) - E_{so}(j-1) = \tfrac{1}{2} hc\zeta_{nl}\,\{j(j+1) - (j-1)j\}$

$= jhc\zeta_{nl}$.

EXERCISE: Show that the difference of the squares of neighbouring level energies is proportional to j^3.

9.9 $E_{so}(J) - E_{so}(J-1) = Jhc\zeta_{nl}$; $\{E_{so}(J) - E_{so}(J-1)\}Jhc = \zeta_{3d}$.

$\{E_{so}(4) - E_{so}(3)\}/4hc = 104.0\ \text{cm}^{-1}$

$\{E_{so}(3) - E_{so}(2)\}/3hc = 96.0\ \text{cm}^{-1}$

$\{E_{so}(2) - E_{so}(1)\}/2hc = 92.1\ \text{cm}^{-1}$

$\{E_{so}(1) - E_{so}(0)\}/hc = 89.9\ \text{cm}^{-1}$

$\zeta_{3d,\text{mean}} = \underline{95.5\ \text{cm}^{-1}}$.

EXERCISE: The ground configuration of Mn is $\ldots 3d^5 4s^2$ and the ^6D term has five levels ($J = 4\frac{1}{2}, 3\frac{1}{2}, \ldots \frac{1}{2}$) at wavenumbers 17 052.29, 17 282.00, 17 451.52, 15 568.48 and 17 637.15 cm^{-1}. Investigate how well the Landé interval rule (and the rule deduced in the *Exercise*) is obeyed, and deduce a value of ζ.

9.10 $$E_{so}(l+s) - E_{so}(l-s) = \tfrac{1}{2}hc\zeta_{nl}\left\{(l+s)(l+s+1) - (l-s)(l-s+1)\right\}$$
$$= s(2l+1)\,hc\zeta_{nl}\,.$$

[This is the total width of a multiplet.]

$$\sum_j (2j+1)\,E_{so}(j) = \tfrac{1}{2}hc\zeta_{nl}\sum_j (2j+1)\left\{j(j+1) - l(l+1) - s(s+1)\right\}$$

$$\sum_j \equiv \sum_{j=l-s}^{l+s} = \sum_{j=0}^{l+s} - \sum_{j=0}^{l-s-1}\,.$$

$$\sum_j 1 = (l+s+1) - (l-s) = 2s+1\,,$$

$$\sum_j j = \tfrac{1}{2}(l+s)(l+s+1) - \tfrac{1}{2}(l-s-1)(l-s) = l(2s+1)\,,$$

$$\sum_j j^2 = \tfrac{1}{6}\left\{(l+s)(l+s+1)(2l+2s+1) - (l-s-1)(l-s)(2l-2s-1)\right\}$$
$$= \tfrac{1}{3}(2s+1)(s^2+s+3l^2),$$

$$\sum_j j^3 = \tfrac{1}{4}\left\{(l+s)^2(l+s+1)^2 - (l-s-1)^2(l-s)^2\right\} = (2s+1)\,l(l^2+s^2+s).$$

Then

$$\sum_j (2j+1)\,E_{so}(j) = \tfrac{1}{2}hc\zeta_{nl}\sum_j \left\{2j^3 + 3j^2 + j - (2j+1)[l(l+1) + s(s+1)]\right\}$$
$$= 0\,.$$

EXERCISE: Find an expression for the mean square energy.

9.11 Li : $1s^2 2s$, $^2S_{\frac{1}{2}}$.

Be : $1s^2 2s^2$, $^1S_0$.

B : $1s^2 2s^2 2p$, $^2P_{\frac{1}{2}}$, $^2P_{\frac{3}{2}}$ with $E(^2P_{\frac{1}{2}}) < E(^2P_{\frac{3}{2}})$.

C : $1s^2 2s^2 2p^2$, $^1D_2$; $^3P_2$, $^3P_1$, $^3P_0$; $^1S_0$

with $E(^3P_0) < E(^3P_1) < E(^3P_2) < E(^1D_2) < E(^1S_0)$

N : $1s^2 2s^2 2p^3$; $(l=1)\times(l=1) = (L=2,1,0)$

$(l=1)\times(L=2,1,0) = (L=3,2,1$ and $2,1,0$ and $1)$.

Hence, $2p^3$ gives rise to F, 2D, 3P, S terms. Three spin-$\frac{1}{2}$ species can give rise to $S = (\frac{1}{2}\times\frac{1}{2})\times\frac{1}{2} = (1+0)\times\frac{1}{2} = \frac{3}{2}+\frac{1}{2}+\frac{1}{2}$, or to two quartets and two doublets. Three *equivalent* p-electrons are constrained by the Pauli principle, and can give rise only to ^4S, ^2D, ^2P. Therefore, for N we predict $^4S < {}^2D < {}^2P$ (as observed).

O : $1s^22s^22p^4 \equiv 1s^22s^22p^62p^{-2}$ [particle-hole formalism]; this gives rise to $^1D_2$; $^3P_2$, $^3P_1$, $^3P_0$; $^1S_0$ with $^3P_2 < {}^3P_1 < {}^3P_0 < {}^1D_2 < {}^1S_0$.

F : $1s^22s^22p^5 \equiv 1s^22s^22p^62p^{-1}$; $^2P_{\frac{1}{2}}$, $^2P_{\frac{3}{2}}$ with $^2P_{\frac{3}{2}} < {}^2P_{\frac{1}{2}}$.

EXERCISE: Establish the terms for the atoms Na to Cl.

9.12 $$\langle nlm_l|H^{(1)}|nlm_l\rangle = -(1/2\mu c^2)\langle nlm_l|(E_{nlm}-V)^2|nlm_l\rangle$$
$$= -(1/2\mu c^2)\langle nlm_l|E^2_{nlm} + V^2 - 2VE_{nlm}|nlm_l\rangle$$
$$= -(1/2\mu c^2)\left\{E^2_{nlm} + \langle nlm_l|V^2|nlm_l\rangle - 2\langle nlm_l|V|nlm_l\rangle E_{nlm_l}\right\}.$$

From the virial theorem [Appendix 5],

$$\langle E\rangle = \langle T\rangle + \langle V\rangle = \tfrac{1}{2}(s+1)\langle V\rangle = \tfrac{1}{2}\langle V\rangle \quad [s=-1 \text{ for } V\propto 1/r].$$

Consequently

$$\langle nlm_l|H^{(1)}|nlm_l\rangle = -(1/2\mu c^2)\left\{\langle nlm_l|V^2|nlm_l\rangle - 3E^2_{nlm_l}\right\}.$$

$$V^2 = (e^2/4\pi\varepsilon_0)^2(1/r^2);\quad E^2_{nlm_l} = (hcR_H)^2/n^2.$$

$$\langle nlm_l|(1/r^2)|nlm_l\rangle = (1/a^2)\{1/(l+\tfrac{1}{2})n^3\}.$$

[See L. Pauling and E.B. Wilson, *Introduction to quantum mechanics*, McGraw-Hill (1935), p.145, for values of $\langle 1/r^p\rangle$.]

$$\langle nlm_l|(1/r^2)|nlm_l\rangle = -(1/2\mu c^2)\left\{\left(\frac{(e^2/4\pi\varepsilon_0 a)^2}{(l+\frac{1}{2})n^3}\right) - \left(\frac{3(hcR_H)^2}{n^2}\right)\right\}$$

$$= -\tfrac{1}{2}\alpha^4\mu c^2\left\{\left(\frac{1}{(l+\frac{1}{2})n^3}\right) - \left(\frac{3}{4n^2}\right)\right\}.$$

For the ground state $(n=1,\ l=0)$:

$$E^{(1)} = \langle 100|(1/r^2)|100\rangle = -\tfrac{5}{8}\alpha^4\mu c^2 = \underline{-1.451\times10^{-22}\ \text{J}}.$$

$(E^{(1)} \mathrel{\hat{=}} 7.305\ \text{cm}^{-1}.)$

EXERCISE: Find the first-order relativistic correction to the energy of a harmonic oscillator.

9.13 $H = -(\hbar^2/2m_e)(\nabla_1^2+\nabla_2^2+\nabla_3^2) - (3e^2/4\pi\varepsilon_0)\{(1/r_1)+(1/r_2)+(1/r_3)\}$
$+ (e^2/4\pi\varepsilon_0)\{(1/r_{12})+(1/r_{23})+(1/r_{13})\}$

$= H^{(0)} + H^{(1)}\ ;\ H^{(0)} = H_1 + H_2 + H_3\ ;$

$H_i = -(\hbar^2/2m)\nabla_i^2 - (3e^2/4\pi\varepsilon_0)(1/r_i); \qquad i = 1,2,3.$

$H^{(1)} = (e^2/4\pi\varepsilon_0)\{(1/r_{12})+(1/r_{23})+(1/r_{13})\}.$

$H^{(0)}\psi(1)\psi(2)\psi(3) = (H_1+H_2+H_3)\psi(1)\psi(2)\psi(3)$

$= \{H_1\psi(1)\}\psi(2)\psi(3) + \psi(1)\{H_2\psi(2)\}\psi(3) + \psi(1)\psi(2)\{H_3\psi(3)\}$

$= (E_1+E_2+E_3)\psi(1)\psi(2)\psi(3) = E^{(0)}\psi(1)\psi(2)\psi(3)\ .$

EXERCISE: Write down the general form of the hamiltonian for a Z-electron neutral atom, and show that $H^{(0)}\psi(1,\ldots,Z) = E^{(0)}\psi(1,\ldots,Z)$, with $\psi(1,\ldots,Z)$ a product of one-electron orbitals.

9.14 $H = -(\hbar^2/2m_e)\{\nabla_1^2+\nabla_2^2\} - (2e^2/4\pi\varepsilon_0)\{(1/r_1)+(1/r_2)\} + (e^2/4\pi\varepsilon_0)(1/r_{12}).$

Since the trial functions are isotropic, we have

$$\nabla^2\psi = (1/r)(d^2/dr^2)r\psi = (\zeta^3/\pi)^{\frac{1}{2}}(1/r)(d^2/dr^2)\,r\,e^{-\zeta r}$$
$$= (\zeta^3/\pi)^{\frac{1}{2}}(\zeta^2 - 2\zeta/r)\,e^{-\zeta r}.$$

The individual trial functions are normalized (to unity), and so

$$\mathcal{E} = \int \psi^* H\psi\,d\tau$$

$$= -(\hbar^2/2m_e)(\zeta^3/\pi)\left\{\int(\zeta^2-2\zeta/r_1)\,e^{-2\zeta r_1}d\tau_1 + \int(\zeta^2-2\zeta/r_2)\,e^{-2\zeta r_2}d\tau_2\right\}$$

$$- (2e^2/4\pi\varepsilon_0)(\zeta^3/\pi)\left\{\int(1/r_1)\,e^{-2\zeta r_1}d\tau_1 + \int(1/r_2)\,e^{-2\zeta r_2}d\tau_2\right\}$$

$$+ (e^2/4\pi\varepsilon_0)(\zeta^3/\pi)\int e^{-2\zeta r_1}(1/r_{12})\,e^{-2\zeta r_2}d\tau_1 d\tau_2$$

$$= -(\hbar^2/2m_e)8\zeta^3\int_0^\infty(\zeta^2-2\zeta/r)\,e^{-2\zeta r}r^2dr - (2e^2/4\pi\varepsilon_0)8\zeta^3\int_0^\infty r\,e^{-2\zeta r}dr$$

$$+ (e^2/4\pi\varepsilon_0)(\zeta^3/\pi)\int e^{-2\zeta r_1}(1/r_{12})e^{-2\zeta r_2}d\tau_1 d\tau_2$$

$$= \zeta^2\hbar^2/m_e - \zeta e^2/\pi\varepsilon_0 + \tfrac{5}{8}(e^2/4\pi\varepsilon_0)\zeta\,.$$

$d\mathcal{E}/d\zeta = 2\zeta\hbar^2/m_e - e^2/\pi\varepsilon_0 + \frac{5}{8}(e^2/4\pi\varepsilon_0) = 0$ for optimization. Then,

$\zeta = (27/64)(m_e e^2/\pi\varepsilon_0\hbar^2) = 27/16\,a_0 = \underline{1.69\,a_0}\,.$

(Note that the Slater rules give ζa_0 ($\equiv Z^*$) $= 1.70$.) The ground state energy is therefore

$$\mathcal{E} = (27/16)^2 \hbar^2/m_e a_0^2 - (e^2/\pi\varepsilon_0 a_0)(27/16) + \tfrac{5}{8}(e^2/4\pi\varepsilon_0 a_0)(27/16)$$
$$= (3^6/2^8 - 3^6/2^7)\hbar^2/m_e a_0^2 = -(3^6/2^8)\hbar^2/m_e a_0^2 = -(3^6/2^7)\,hcR_\infty .$$

The energy of He^+ in its ground state is

$$E(He^+) = -4hcR_\infty \; [R_{He} \approx R_\infty ; \; E \propto Z^2 = 4]$$
$$E(He^{2+}) = 0 .$$

The ionization energies are therefore

$$I_1 = E(He^+) - E(He) = (3^6/2^7 - 4)\,hcR_\infty = 1.695\,hcR_\infty$$
$$= 3.696 \times 10^{-18}\ \text{J} \mathrel{\hat{=}} 2.23 \times 10^3\ \text{kJ mol}^{-1} \mathrel{\hat{=}} \underline{23.1\ \text{eV}}.$$
$$I_2 = E(He^{2+}) - (He^+) = 4hcR_\infty$$
$$= 8.720 \times 10^{-18}\ \text{J} \mathrel{\hat{=}} 5.25 \times 10^3\ \text{kJ mol}^{-1} \mathrel{\hat{=}} \underline{54.4\ \text{eV}}.$$

The experimental values are $I_1 = 24.580$ eV, $I_2 = 54.400$ eV (for ^{4}He).

EXERCISE: Write $\psi = \psi_{1s} + \lambda\psi_{2s}$ for each electron, the ψ_{ns} being hydrogenic orbitals with general ζ_{1s}, ζ_{2s}. Take $\lambda, \zeta_{1s}, \zeta_{2s}$ as variation parameters and find expressions for the ionization energy of helium. (Adapt the *Example* on p.225 of the text to find J.)

9.15 The hamiltonian has $Ze^2/4\pi\varepsilon_0$ in place of $2e^2/4\pi\varepsilon_0$ in the preceding Problem. Hence,

$$\mathcal{E} = \zeta^2 \hbar^2/m_e - Ze^2\zeta/2\pi\varepsilon_0 + (5/8)(e^2/4\pi\varepsilon_0)\,\zeta .$$
$$d\mathcal{E}/d\zeta = 2\zeta\hbar^2/m_e - Ze^2/2\pi\varepsilon_0 + (5/8)(e^2/4\pi\varepsilon_0) = 0 .$$

This solves to

$$\zeta = (m_e e^2/4\pi\varepsilon_0 \hbar^2)(Z - 5/16) = (Z - 5/16)/a_0 .$$

$$\mathcal{E}_{\text{optimum}} = (Z-5/16)^2 \hbar^2/m_e a_0^2 - (Z-5/16)^2 e^2/2\pi\varepsilon_0 a_0$$
$$= -(Z-5/16)^2 \hbar^2/m_e a_0^2 = -2(Z-5/16)^2\,hcR_\infty .$$

Therefore,

$$E(X^{2+}) = 0 ,\; E(X^+) = -Z^2 hcR_\infty ,\; E(X) = -2(Z-5/16)^2 hcR_\infty .$$

$$I_1 = E(X^+) - E(X) = \{2(Z-5/16)^2 - Z^2\}\,hcR_\infty$$
$$I_2 = E(X^{2+}) - E(X^+) = Z^2 hcR_\infty .$$

$hcR_\infty \mathrel{\hat{=}} 13.61$ eV ; hence

	Li^+	Be^{2+}	B^{3+}	C^{4+}	
I_1/eV	98.2	202	342	518	calculated
I_2/eV	162	289	451	650	calculated
I_1/eV	73.5	153	258	389	experimental

EXERCISE: Are the ionization energies improved by allowing the ζ-values to be different for the two electrons ?

9.16 $\psi_{\parallel} = \{\psi_1(1)\psi_2(2) - \psi_2(1)\psi_1(2)\}/\sqrt{2}$

$$= (2^{\frac{1}{2}}/L)\{\sin(\pi x_1/L)\sin(2\pi x_2/L) - \sin(2\pi x_1/L)\sin(\pi x_2/L)\}$$

$$\psi_{\perp} = (2^{\frac{1}{2}}/L)\{\sin(\pi x_1/L)\sin(2\pi x_2/L) + \sin(2\pi x_1/L)\sin(\pi x_2/L)\}.$$

Plot $\psi_{\parallel}^2$ and $\psi_{\perp}^2$ on a surface with axes x_1, x_2. Notice that $\psi_{\parallel}^2 = 0$ and $\psi_{\perp}^2 = 0$ for $x_2 = \frac{1}{2}L$, and that $\psi_{\parallel}^2 = 0$ for all $x_1 = x_2$. A rough sketch is provided in Fig. 9.1(a) ($\psi_{\parallel}$) and (b) ($\psi_{\perp}$).

Fig. 9.1 (a) $\psi_{\parallel}^2$ (b) $\psi_{\perp}^2$

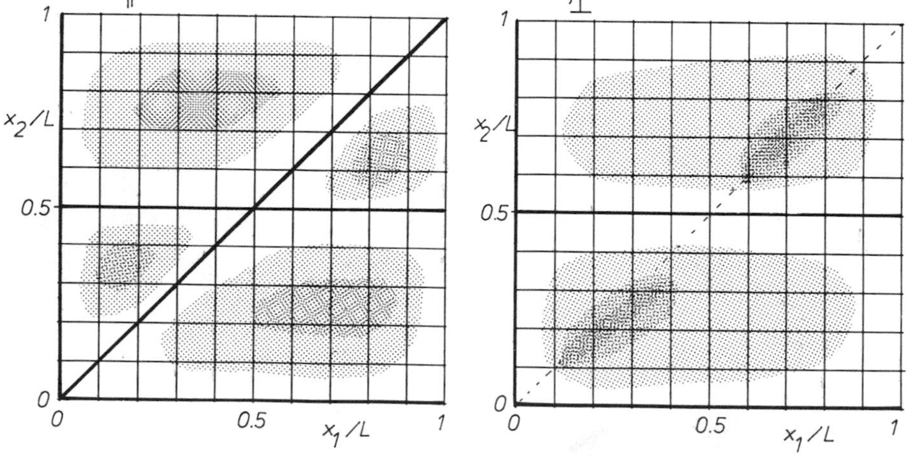

EXERCISE: Sketch (or generate computer graphics) for the singlet and triplet combinations of $\psi_2(x_1)\psi_3(x_2)$.

9.17 ${}^1E - {}^3E = (2E + J + K) - (2E + J - K) = 2K$ [eqn (9.6.10)]

$$\{{}^1E(1s\,2s) - {}^3E(1s\,2s)\}/hc = (166\,272 - 159\,850)\text{ cm}^{-1} = 6422\text{ cm}^{-1}.$$

Therefore, $K_{1s2s}/hc = 3211\text{ cm}^{-1} \mathrel{\hat{=}} \underline{0.3981\text{ eV}}$.

$$\{{}^1E(1s\,3s) - {}^3E(1s\,3s)\}/hc = (184\,859 - 183\,231)\text{ cm}^{-1} = 1628\text{ cm}^{-1}.$$

Therefore, $K_{1s3s}/hc = 814\text{ cm}^{-1} \mathrel{\hat{=}} \underline{0.1009\text{ eV}}$.

EXERCISE: The terms of Li^+ lie at 491 361 cm^{-1} (1s 2s ^1S), 476 046 cm^{-1} (1s 2s ^3S), 554 761 cm^{-1} (1s 3s ^3S), and 558 779 cm^{-1} (1s 2s ^1S). Find K_{1s2s} and K_{1s3s} and suggest reasons why they differ from those for He.

9.18 ^1S : $J = (0 \times 0) = 0$; hence $\underline{^1S_0}$.

^2P : $J = (\frac{1}{2} \times 1) = \frac{3}{2} + \frac{1}{2}$; hence $\underline{^2P_{\frac{3}{2}}, {}^2P_{\frac{1}{2}}}$.

^3D : $J = (1 \times 2) = 3 + 2 + 1$; hence $\underline{^3D_3, {}^3D_2, {}^3D_1}$.

^2D : $J = (\frac{1}{2} \times 2) = \frac{5}{2} + \frac{3}{2}$; hence $\underline{^2D_{\frac{5}{2}}, {}^2D_{\frac{3}{2}}}$.

^1D : $J = (0 \times 2) = 2$; hence $\underline{^1D_2}$.

^4D : $J = (\frac{3}{2} \times 2) = \frac{7}{2} + \frac{5}{2} + \frac{3}{2} + \frac{1}{2}$; hence $\underline{^4D_{\frac{7}{2}}, {}^4D_{\frac{5}{2}}, {}^4D_{\frac{3}{2}}, {}^4D_{\frac{1}{2}}}$.

1s2p : $L = 1$; $S = 0, 1$; hence $^1P, {}^3P$ with levels $^1P_1, {}^3P_2, {}^3P_1, {}^3P_0$.

Energies: $\underline{^3P_0 < {}^3P_1 < {}^3P_2 < {}^1P_1}$ [Hund rules].

2p3p : $L = 2, 1, 0$; $S = 0, 1$; hence $^3D, {}^1D, {}^3P, {}^1P, {}^3S, {}^1S$.

Energies : $\underline{^3D_1 < {}^3D_2 < {}^3D_3 < {}^3P_0 < {}^3P_1 < {}^3P_2 < {}^3S_1 < {}^1D_2 < {}^1P_1 < {}^1S_0}$.

3p3d : $L = 3, 2, 1$; $S = 0, 1$; hence $^3F, {}^1F, {}^3D, {}^1D, {}^3P, {}^1P$.

Energies: $\underline{^3F_2 < {}^3F_3 < {}^3F_4 < {}^3D_1 < {}^3D_2 < {}^3D_3 < {}^3P_0 < {}^3P_1 < {}^3P_2 < {}^1F_3 < {}^1D_2 < {}^1P_1}$.

(a) d^2 : $L = (2 \times 2) = 4 + 3 + 2 + 1 + 0$; $S = 1, 0$ but the Pauli principle forbids ^3G (and alternate triplets). Hence $\underline{^1G, {}^3F, {}^1D, {}^3P, {}^1S}$ arise.

(b) f^2 : $L = (3 \times 3) = 6 + 5 + \ldots + 0$; $S = 1, 0$. Pauli forbids ^3I etc.; hence $^1I, {}^3H, {}^1G, {}^3F, {}^1D, {}^3P, {}^1S$ arise.

EXERCISE: What levels may arise from the terms $^4F, {}^4D, {}^4P$? Arrange in order of increasing energy the levels and terms arising from 1s3p, $3p^2$, 2s4f, $5g^2$. What terms may arise from the general $(nl)^2$ configuration?

9.19

$$\psi = (1/4!)^{\frac{1}{2}} \begin{vmatrix} 1s(1) & 1\bar{s}(1) & 2s(1) & 2\bar{s}(1) \\ 1s(2) & 1\bar{s}(2) & 2s(2) & 2\bar{s}(2) \\ 1s(3) & 1\bar{s}(3) & 2s(3) & 2\bar{s}(3) \\ 1s(4) & 1\bar{s}(4) & 2s(4) & 2\bar{s}(4) \end{vmatrix},$$

where 1s and 2s denote α spin-orbitals and $1\bar{s}$ and $2\bar{s}$ denote β spin-orbitals.

$$\langle \| 1s(1) \ldots 2\bar{s}(4) \| \, |H| \, \| 1s(1) \ldots 2\bar{s}(4) \| \rangle$$

$$
\begin{aligned}
= (1/4!)^{\frac{1}{2}} \langle \, & [1s(1)\ 1\bar{s}(2)\ 2s(3)\ 2\bar{s}(4)] - [1s(1)\ 1\bar{s}(2)\ 2\bar{s}(3)\ 2s(4)] \\
& - [1s(1)\ 2s(2)\ 1\bar{s}(3)\ 2\bar{s}(4)] + [1s(1)\ 2s(2)\ 2\bar{s}(3)\ 1\bar{s}(4)] \\
& + [1s(1)\ 2\bar{s}(2)\ 1\bar{s}(3)\ 2s(4)] - [1s(1)\ 2\bar{s}(2)\ 2s(3)\ 1\bar{s}(4)] \\
& - [1\bar{s}(1)\ 1s(2)\ 2s(3)\ 2\bar{s}(4)] + [1\bar{s}(1)\ 1s(2)\ 2\bar{s}(3)\ 2s(4)] \\
& + [1\bar{s}(1)\ 2s(2)\ 1s(3)\ 2\bar{s}(4)] - [1\bar{s}(1)\ 2s(2)\ 2\bar{s}(3)\ 1s(4)] \\
& - [1\bar{s}(1)\ 2\bar{s}(2)\ 1s(3)\ 2s(4)] + [1\bar{s}(1)\ 2\bar{s}(2)\ 2s(3)\ 1s(4)] \\
& + [2s(1)\ 1s(2)\ 1\bar{s}(3)\ 2\bar{s}(4)] - [2s(1)\ 1s(2)\ 2\bar{s}(3)\ 1\bar{s}(4)] \\
& - [2s(1)\ 1\bar{s}(2)\ 1s(3)\ 2\bar{s}(4)] + [2s(1)\ 1\bar{s}(2)\ 2\bar{s}(3)\ 1\bar{s}(4)] \\
& + [2s(1)\ 2\bar{s}(2)\ 1s(3)\ 1\bar{s}(4)] - [2s(1)\ 2\bar{s}(2)\ 1\bar{s}(3)\ 1s(4)] \\
& - [2\bar{s}(1)\ 1s(2)\ 1\bar{s}(3)\ 2s(4)] + [2\bar{s}(1)\ 1s(2)\ 2s(3)\ 1\bar{s}(4)] \\
& + [2\bar{s}(1)\ 1\bar{s}(2)\ 1s(3)\ 2s(4)] - [2\bar{s}(1)\ 1\bar{s}(2)\ 2s(3)\ 1s(4)] \\
& - [2\bar{s}(1)\ 2s(2)\ 1s(3)\ 1\bar{s}(4)] + [2\bar{s}(1)\ 2s(2)\ 1\bar{s}(3)\ 1s(4)] \Big| \\
& \times [T_1 + T_2 + T_3 + T_4 + V_1 + V_2 + V_3 + V_4 \\
& \qquad + V_{12} + V_{13} + V_{14} + V_{23} + V_{24} + V_{34}] \\
& \times \Big| \det | 1s(1) \ldots 2\bar{s}(4) | \rangle
\end{aligned}
$$

$$
\begin{aligned}
= \underline{2E_{\mathbf{1s}} + 2E_{\mathbf{2s}}} & \underline{+ \langle 1s(1)\ 1s(2) | V_{12} | 1s(1)\ 1s(2) \rangle} \\
& \underline{+ 4 \langle 1s(1)\ 2s(2) | V_{12} | 1s(1)\ 2s(2) \rangle} \\
& \underline{+ \langle 2s(1)\ 2s(2) | V_{12} | 2s(1)\ 2s(2) \rangle} \\
& \underline{- 2 \langle 1s(1)\ 2s(2) | V_{12} | 2s(1)\ 1s(2) \rangle}
\end{aligned}
$$

[many terms are identical in value], with

$$\langle 1s(1)\ 1s(2) | V_{12} | 1s(1)\ 1s(2) \rangle = j_0 \int \psi^2_{\mathbf{1s}}(1)\ (1/r_{12})\ \psi^2_{\mathbf{1s}}(2)\ d\tau_1\ d\tau_2$$

$$[j_0 = -e^2/4\pi\varepsilon_0]$$

$$\langle 1s(1)\ 2s(2) | V_{12} | 1s(1)\ 2s(2) \rangle = j_0 \int \psi^2_{1s}(1)\ (1/r_{12})\ \psi^2_{\mathbf{2s}}(2)\ d\tau_1\ d\tau_2$$

$$\langle 2s(1)\ 2s(2) | V_{12} | 2s(1)\ 2s(2) \rangle = j_0 \int \psi^2_{\mathbf{2s}}(1)\ (1/r_{12})\ \psi^2_{\mathbf{2s}}(2)\ d\tau_1\ d\tau_2$$

$$\langle 1s(1)\ 2s(2) | V_{12} | 2s(1)\ 1s(2) \rangle = j_0 \int \psi_{\mathbf{1s}}(1)\ \psi_{\mathbf{2s}}(1)\ (1/r_{12})\ \psi_{\mathbf{1s}}(2)\ \psi_{\mathbf{2s}}(2)\ d\tau_1\ d\tau_2 .$$

In terms of the Hartree-Fock expressions [eqns (9.11.5) and (9.11.8)]:

$$E = 2E_{1s} + 2E_{2s} + \left\{(2J_{1s1s} - K_{1s1s}) + (2J_{2s2s} - K_{2s2s}) + (2J_{1s2s} - K_{1s2s}) + (2J_{2s1s} - K_{2s1s})\right\}$$

$$= 2E_{1s} + 2E_{2s} + \left\{J_{1s1s} + J_{2s2s} + 4J_{1s2s} - 2K_{1s2s}\right\}$$

$$= \underline{2E_{1s} + 2E_{2s} + \langle 1s(1)\,1s(1)|V_{12}|1s(1)\,1s(1)\rangle}$$
$$\underline{+ \langle 2s(1)\,2s(1)|V_{12}|2s(1)\,2s(1)\rangle}$$
$$\underline{+ 4\langle 1s(1)\,2s(2)|V_{12}|1s(1)\,2s(2)\rangle}$$
$$\underline{- 2\langle 1s(1)\,2s(2)|V_{12}|2s(1)\,1s(1)\rangle},$$

as before.

EXERCISE: Find an expression for the (Hartree-Fock) energy of Ne and for its first ionization energy.

9.20 $E^{(1)} = \mu_B B M_L$, [eqn (9.14.3)].

$\Delta E^{(1)} = \mu_B B$; $\Delta\tilde{\nu} = (\mu_B/hc)B$.

Therefore, $B = hc\tilde{\nu}/\mu_B = \underline{2.14\ \text{T}}$ when $\tilde{\nu} = 1\ \text{cm}^{-1}$.
(2.14 T ≙ 21.4 kG.)

$$g_J = 1 + \left\{\frac{J(J+1) + S(S+1) - L(L+1)}{2J(J+1)}\right\} \qquad \text{[eqn (9.14.8)]}$$

(a) $J_{max} = L + S$,

$$g_{L+S} = 1 + \left\{\frac{(L+S)(L+S+1) + S(S+1) - L(L+1)}{2(L+S)(L+S+1)}\right\} = \underline{1 + S/(L+S)}.$$

(b) $J_{min} = L - S$ (for $S \leqslant L$)

$$g_{L-S} = 1 + \left\{\frac{(L-S)(L-S+1) + S(S+1) - L(L+1)}{2(L-S)(L-S+1)}\right\} = \underline{1 - S/(L-S+1)}.$$

EXERCISE: Calculate the g-factor for a level in which J has its minimum value, but for which $L \leqslant S$. Evaluate $\sum_J J(J+1)\,g_J$ for a given S, L.

9.21 Since $^1F \rightarrow {}^1D$ is a transition between singlets, the normal Zeeman effect will be observed: the transition splits into three lines with separation $\Delta\tilde{\nu} = (\mu_B/hc)B = 1.87\ \text{cm}^{-1}$ for $B = 4\text{T}$ [use first part of Problem 9.20].

For the $^3P \rightarrow {}^3S$ transition we must distinguish the levels and calculate their respective g-factors:

$$g_J(L=S) = 1 + \left\{\frac{J(J+1)+S(S+1)-L(L+1)}{2J(J+1)}\right\}_{L=S} = 1+\tfrac{1}{2}=\tfrac{3}{2}\,.$$

Therefore, $g_J(^3\mathrm{P}) = \frac{3}{2}$ for $J = 0, 1, 2$.

$$g_J(^3\mathrm{S}_1) = 1 + \left\{\frac{J(J+1) + S(S+1)}{2J(J+1)}\right\}_{J=S} = 1 + 1 = 2\,.$$

At $B = 4\,\mathrm{T}$, $\tilde{\nu} = g_J\,\mu_\mathrm{B}B/hc = g_J \times (1.87\ \mathrm{cm}^{-1}) = \begin{cases} 2.80\ \mathrm{cm}^{-1} & \text{for } g_J = \frac{3}{2} \\ 3.74\ \mathrm{cm}^{-1} & \text{for } g_J = 2\,. \end{cases}$

Fig. 9.2

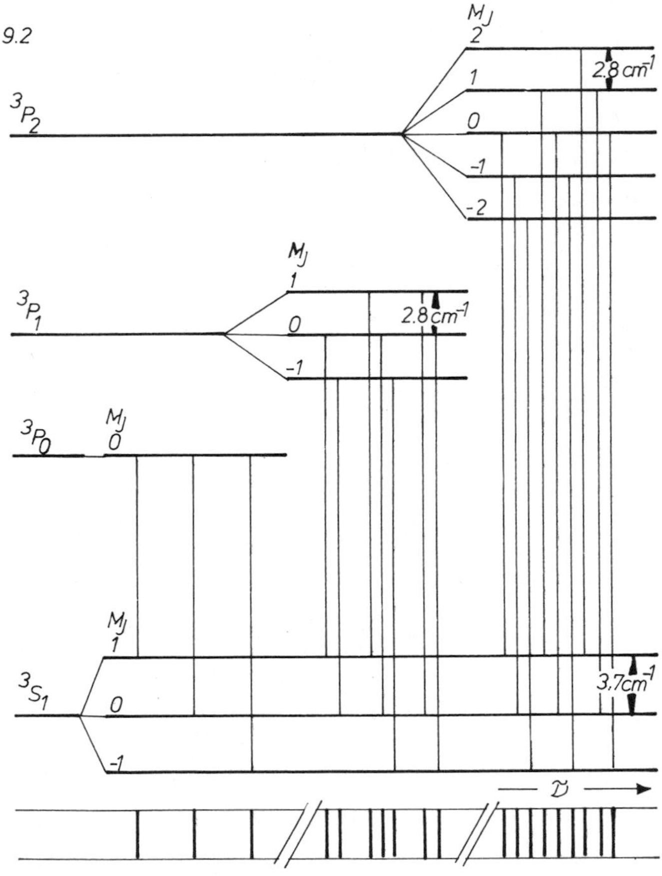

Construct the diagram in Fig. 9.2. The transitions are those for which $\Delta J = 0, \pm 1$ and $\Delta M_J = 0, \pm 1$.

EXERCISE: Construct the form of the Zeeman effect on a $^4\mathrm{F} \rightarrow {}^4\mathrm{D}$ transition.

10 Molecular structure

10.1 $H\psi = \left\{-(\hbar^2/2m)(\partial^2/\partial X^2) - (\hbar^2/2m)(\partial^2/\partial x^2) + V(X,x)\right\}\psi_{el}(x,X)\,\psi_{vib}(X)$

$$= -(\hbar^2/2M)\left\{\psi_{vib}(\partial^2\psi_{el}/\partial X^2) + 2(\partial\psi_{el}/\partial X)(\partial\psi_{vib}/\partial X) + \psi_{el}(\partial^2\psi_{vib}/\partial X^2)\right\}$$

$$-(\hbar^2/2m_e)(\partial^2\psi_{el}/\partial x^2)\,\psi_{vib} + V\psi_{el}\,\psi_{vib} = -(\hbar^2/2M)\,\psi_{el}(\partial^2\psi_{vib}/\partial X^2)$$

$$-(\hbar^2/2m_e)(\partial^2\psi_{el}/\partial x^2)\,\psi_{vib} + V\psi_{el}\,\psi_{vib} - (\hbar^2/2M)\left\{\psi_{vib}\psi''_{el} + 2\psi'_{el}\,\psi'_{vib}\right\}$$

$[\psi' = \partial\psi/\partial X]$.

For a fixed X solve

$$-(\hbar^2/2m_e)(\partial^2\psi_{el}/\partial x^2) + V(x,X)\psi_{el} = E_{el}(X)\,\psi_{el},$$

then

$$H\psi = -(\hbar^2/2M)\,\psi_{el}\psi''_{vib} + E_{el}(X)\,\psi_{el}\psi_{vib} - (\hbar^2/2M)\left\{\psi_{vib}\psi''_{el} + 2\psi'_{el}\psi'_{vib}\right\}.$$

Write $E_{el}(X) = E_{el}(X_0) + \tfrac{1}{2}(d^2E_{el}/dX^2)_0(X-X_0)^2 + \ldots$ $\quad[(dE/dX)_0 = 0]$

and note that

$$-(\hbar^2/2M)\psi''_{vib} + \tfrac{1}{2}(d^2E_{el}/dX^2)_0(X-X_0)^2\psi_{vib} = E_{vib}\psi_{vib}.$$

Therefore, if

$$-(\hbar^2/2M)(\psi_{vib}\psi''_{el} + 2\psi'_{el}\psi'_{vib}) = 0,$$

$H\psi = \{E_{vib} + E_{el}(X_0)\}\,\psi$, as required. Now,

$$(\hbar^2/2m_e)\,\psi''_{el} = (\hbar^2/2m_e)(\partial^2\psi_{el}/\partial X^2) \approx (\hbar^2/2m_e)(\partial^2\psi_{el}/\partial x^2)$$

because

$$\psi_{el}(x,X) \approx \psi_{el}(x-X), \quad \text{and} \quad (\hbar^2/2m_e)(\partial^2\psi_{el}/\partial x^2) \approx E_{el}\psi_{el}.$$

Therefore, $\quad (\hbar^2/2M)\,\psi''_{el} \approx (m_e/M)\,E_{el}\psi_{el} \ll E_{el}\psi_{el}$.

EXERCISE: Demonstrate the validity of the Born-Oppenheimer approximation for a general, three-dimensional molecule. [See *Quantum theory of molecules and solids*, vol. 1, J.C. Slater, McGraw-Hill (1963), p.252.]

10.2 $j'/j_0 = (1/R)\left\{1-(1+s)\,e^{-2s}\right\}$, $\quad s = R/a_0$, $\quad j_0 = e^2/4\pi\varepsilon_0$.

$k'/j_0 = (1/a_0)\left\{1+s\right\}e^{-s}$, $\quad S = \left\{1+s+\tfrac{1}{3}s^2\right\}$.

$E_+ - E_{1s} = j_0/R - (j'+k')/(1+S)$ $\quad$ [eqn (10.3.11a)],

$E_{\pm} - E_{1s} = j_0/R + (j'+k')/(1-S)$ $\quad$ [eqn (10.3.11b)].

$$(E_+ - E_{1s})/j_0 = (1/R) - \left\{\frac{(1/R)[1-(1+s)\,e^{-2s}] + (1/a_0)[1+s]\,e^{-s}}{1+[1+s+\frac{1}{3}s^2]\,e^{-s}}\right\}$$

$$= \left\{\frac{(1-\frac{2}{3}s^2) + (1+s)\,e^{-s}}{1+(1+s+\frac{1}{3}s^2)\,e^{-s}}\right\}\left(\frac{e^{-s}}{R}\right).$$

$$(E_+ - E_{1s})(a_0/j_0) = \left\{\frac{(1-\frac{2}{3}s^2) + (1+s)\,e^{-s}}{1+(1+s+\frac{1}{3}s^2)\,e^{-s}}\right\}\left(\frac{e^{-s}}{s}\right).$$

$$(E_- - E_{1s})(a_0/j_0) = \left\{\frac{2-[(1-\frac{2}{3}s^2) + (1+s)\,e^{-s}]\,e^{-s}}{1-(1+s+\frac{1}{3}s^2)\,e^{-s}}\right\}\left(\frac{1}{s}\right)$$

$$(\alpha - E_{1s})(a_0/j_0) = [-j' + (1/R)](a_0/j_0) \qquad [\text{eqn } (10.3.8)]$$

$$= (1+s)(e^{-2s}/s)\,.$$

$$(\beta - E_{1s})(a_0/j_0) = [-k' + (S/R)](a_0/j_0) \qquad [\text{eqn } (10.3.9)]$$

$$= (1-\tfrac{2}{3}s^2)(e^{-s}/s)\,.$$

The $E_\pm - E_{1s}$ values are plotted in Fig. 10.1 (there is a change of scale above and below zero energy); the α, β integrals are plotted in Fig. 10.2. The E_+ curve has a minimum at $R \approx 130$ pm ($s \approx 2.5$) corresponding to $E_+ - E_{1s} = -1.22 \times 10^{-3}\ j_0\ \text{pm}^{-1}$. Since $j_0 = 2.31 \times 10^{-28}$ J m, we have

$$E_+ - E_{1s} = -2.81 \times 10^{-19}\ \text{J} \mathrel{\hat{=}} 1.76\ \text{eV} \mathrel{\hat{=}} \underline{170\ \text{kJ mol}^{-1}}\,.$$

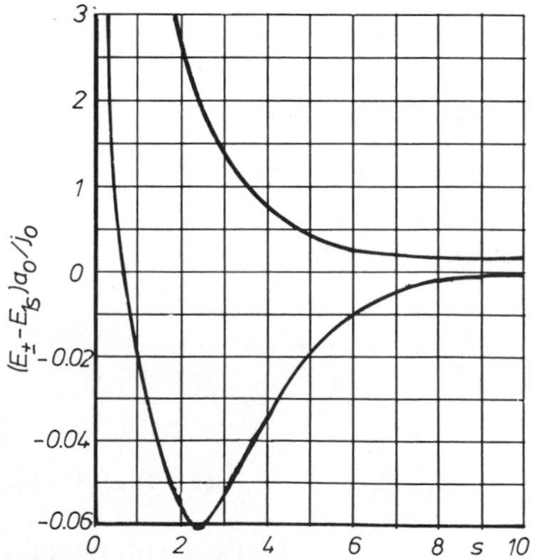

Fig. 10.1

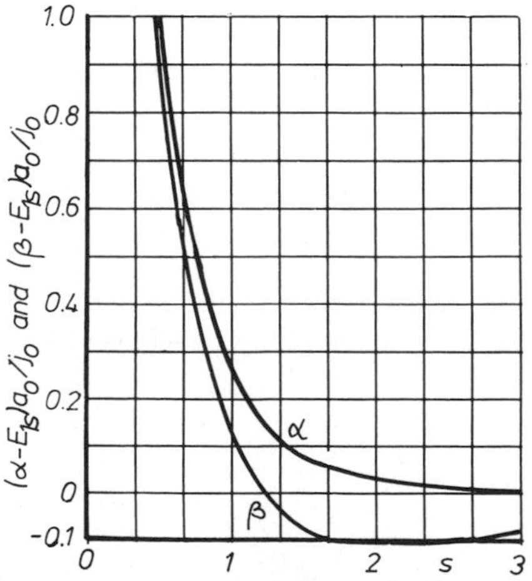

Fig. 10.2

Therefore the dissociation energy (neglecting the zero-point vibrational energy) is 170 kJ mol^{-1}.

EXERCISE: Plot the molecular potential energy curves for He_2^{3+} and estimate its bond length and dissociation energy if you find it to be stable. [$s \to ZR/a_0$.]

10.3 $\psi_\pm = (a \pm b)/[2(1 \pm S)]^{\frac{1}{2}}$,

$$|\psi_\pm|^2 = (a^2 + b^2 \pm 2ab)/2(1 \pm S)$$

$$\rho_\pm = \left\{a^2 + b^2 \pm 2ab)/2(1 \pm S)\right\} - \tfrac{1}{2}(a^2 + b^2)$$

$$= \pm\left[ab - \tfrac{1}{2}S(a^2 + b^2)\right]/(1 \pm S).$$

$S = (1 + s + \frac{1}{3}s^2)\,e^{-s}$, $s = R/a_0 = 2.46$ when $R = 130$ pm; hence $S = 0.469$.

$a \mathrel{\hat{=}} \psi_a = (1/\pi a^3)\,e^{-r_a/a}$; $b \mathrel{\hat{=}} \psi_b = (1/\pi a^3)^{\frac{1}{2}}\,e^{-r_b/a}$;

$(1/\pi a^3) = 2.148 \times 10^{-6}$ pm^{-3}.

$$\rho_+ = (1.462 \times 10^{-6}\ \text{pm}^{-3})\left\{e^{-(r_a + r_b)/a} - 0.235\left(e^{-2r_a/a} + e^{-2r_b/a}\right)\right\}.$$

$$\rho_- = -(4.045 \times 10^{-6}\ \text{pm}^{-3})\left\{e^{-(r_a + r_b)/a} - 0.235\left(e^{-2r_a/a} + e^{-2r_b/a}\right)\right\}.$$

Note that $\rho_- = -2.767\,\rho_+$, and so it is sufficient to plot one. We plot

$$\rho \equiv e^{-(r_a + r_b)/a} - 0.235\left(e^{-2r_a/a} + e^{-2r_b/a}\right).$$

For all points on a line joining the two nuclei and lying beyond b, $r_b = r_a - R$, we have

$$\rho = (e^s - 0.235 - 0.235\,e^{2s})\,e^{-2r_a/a} = -20.73\,e^{-2r_a/a}.$$

For points on the same line, but lying between the nuclei, $r_b = R - r_a$, and so

$$\rho = e^{-s} - 0.235\left(e^{-2r_a/a} + e^{-2s}\,e^{2r_a/a}\right)$$

$$= 0.0854 - 0.235\left(e^{-2r_a\ a} + 0.007\,30\,e^{2r_a/a}\right).$$

(Note that the two expressions for ρ are equal at $r_a = R$.) ρ is plotted in Fig. 10.3 (see p.110).

EXERCISE: Plot the difference density for points either side of $R = 130$ pm e.g., at $R = 80$ pm and 180 pm, and also for points in a plane bisecting and perpendicular to the internuclear distance (with $R = 130$ pm).

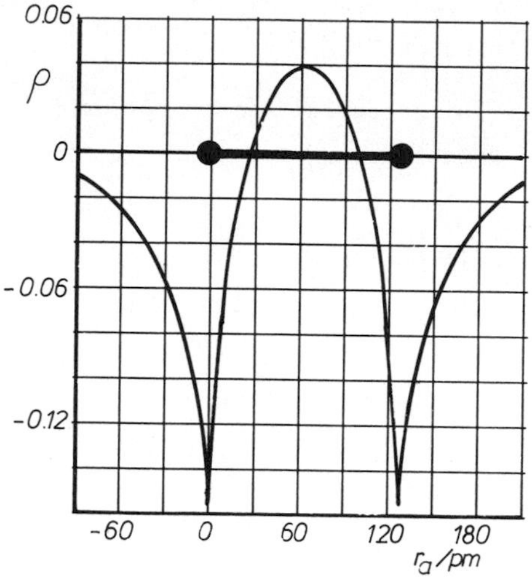

Fig. 10.3

10.4 $\psi(1,2) = \{1/2(1+S)\}\{a(1)+b(1)\}\{a(2)+b(2)\}$

[eqn (10.3.12), the space function].

$$H = T_1 + T_2 + (e^2/4\pi\varepsilon_0)\Big\{(1/R) + (1/r_{12}) - (1/r_{1a}) - (1/r_{1b}) - (1/r_{2a}) - (1/r_{2b})\Big\}$$

[eqn (10.2.1) with $T_i = -(\hbar^2/2m_e)\nabla_i^2$]. $\psi(1,2)$ is normalized to unity, and so

$$E = \Big\{1/2(1+S)\Big\}^2 \int d\tau_1\, d\tau_2 \Big\{a(1)\,a(2) + b(1)\,b(2) + a(1)\,b(2) + b(1)\,a(2)\Big\} \times H\Big\{a(1)\,a(2) + b(1)\,b(2) + a(1)\,b(2) + b(1)\,a(2)\Big\}$$

$$= \Big\{1/2(1+S)\Big\}^2 \int d\tau_1 d\tau_2 \{\ldots\}\Big\{[E_a + E_b + (j_0/R)]\,a(1)\,a(2)$$
$$+ j_0[(1/r_{12}) - (1/r_{1b}) - (1/r_{2b})]\,a(1)\,a(2) + [E_a + E_b + (j_0/R)]\,b(1)\,b(2)$$
$$+ j_0[(1/r_{12}) - (1/r_{1a}) - (1/r_{2a})]\,b(1)\,b(2) + [E_a + E_b + (j_0/R)]\,a(1)\,b(2)$$
$$+ j_0[(1/r_{12}) - (1/r_{1b}) - (1/r_{2a})]\,a(1)\,b(2) + [E_a + E_b + (j_0/R)]\,b(1)\,a(2)$$
$$+ j_0[(1/r_{12}) - (1/r_{1a}) - (1/r_{2b})]\,b(1)\,a(2)\Big\}$$

[$\{T_i - (j_0/r_{ia})\}\,a(i) = E_a a(i)$, etc; $j_0 = e^2/4\pi\varepsilon_0$].

$$= \{1/2(1+S)\}^2 \Big\{[E_a + E_b + (j_0/R)](1 + S^2 + 2S) \times 4 \qquad [\textstyle\int ab\, d\tau = S]$$
$$+ j_0 \int d\tau_1\, d\tau_2 \Big[2a(1)\,a(2)\,(1/r_{12})\,a(1)\,a(2) + 4a(1)\,a(2)\,(1/r_{12})\,b(1)\,b(2)$$
$$+ 2a(1)\,b(2)\,(1/r_{12})\,a(1)\,b(2) + 8a(1)\,a(2)\,(1/r_{12})\,a(1)\,b(2)\Big]$$

$$
\begin{aligned}
&- j_0 \int d\tau_1\, a(1)\,(1/r_{1b})\,a(1)(1+S) - j_0 \int d\tau_2\, a(2)(1/r_{2b})\,a(2)(1+S)\\
&- j_0 \int d\tau_1\, b(1)\,(1/r_{1b})\,a(1)(1+S) - j_0 \int d\tau_2\, b(2)\,(1/r_{2b})\,a(2)(1+S)\\
&- j_0 \int d\tau_1\, b(1)\,(1/r_{1a})\,b(1)(1+S) - j_0 \int d\tau_2\, b(2)\,(1/r_{2a})\,b(2)(1+S)\\
&- j_0 \int d\tau_1\, a(1)\,(1/r_{1a})\,b(1)(1+S) - j_0 \int d\tau_2\, b(2)\,(1/r_{2a})\,a(2)(1+S)\\
&- j_0 \int d\tau_1\, a(1)\,(1/r_{1b})\,a(1)(1+S) - j_0 \int d\tau_2\, a(2)\,(1/r_{2a})\,b(2)(1+S)\\
&- j_0 \int d\tau_1\, b(1)\,(1/r_{1b})\,a(1)(1+S) - j_0 \int d\tau_2\, b(2)\,(1/r_{2a})\,b(2)(1+S)\\
&- j_0 \int d\tau_1\, a(1)\,(1/r_{1a})\,b(1)(1+S) - j_0 \int d\tau_2\, b(2)\,(1/r_{2b})\,a(2)(1+S)\\
&- j_0 \int d\tau_1\, b(1)\,(1/r_{1a})\,b(1)(1+S) - j_0 \int d\tau_2\, a(2)\,(1/r_{2b})\,a(2)(1+S)\Big\}\\
&= E_a + E_b + (j_0/R)\\
&\quad + \left\{1/2(1+S)^2\right\} j_0 \int d\tau_1\, d\tau_2 \Big[a^2(1)(1/r_{12})\,a^2(2)\\
&\qquad\qquad + 2a(1)\,b(1)\,(1/r_{12})\,a(2)\,b(2)\\
&\qquad + a^2(1)\,(1/r_{12})\,b^2(2) + 4a^2(1)\,(1/r_{12})\,a(2)\,b(2)\Big]\\
&\quad - \left\{2/(1+S)\right\} j_0 \left\{\int d\tau_1\, a^2(1)\,(1/r_{1b}) + \int d\tau_1\, a(1)\,b(1)(1/r_{1b})\right\}.
\end{aligned}
$$

In Appendix 14 we define

$$
\begin{aligned}
j/j_0 &= \int a^2(1)(1/r_{12})\, b^2(2)\, d\tau_1\, d\tau_2 \qquad & j'/j_0 &= \int a^2(1)(1/r_{1b})\, d\tau_1\\
k/j_0 &= \int a(1)b(1)(1/r_{12})a(2)b(2)\, d\tau_1\, d\tau_2 \qquad & k'/j_0 &= \int a(1)b(1)(1/r_{1b})\, d\tau_1\\
l/j_0 &= \int a^2(1)(1/r_{12})\, a(2)\, b(2)\, d\tau_1\, d\tau_2 & &\\
m/j_0 &= \int a^2(1)(1/r_{12})\, a^2(2)\, d\tau_1\, d\tau_2 & &\\
V &= \int a^2(1)(1/r_{1a})\, d\tau_1 . & &
\end{aligned}
$$

Therefore

$$
\begin{aligned}
E = E_a + E_b + (j_0/R) &+ \{1/2(1+S)^2\}\{j + 2k + 4l + m\}\\
&- \{2/(1+S)\}\{j' + k'\} .
\end{aligned}
$$

We have taken a, b equivalent; therefore $E_a = E_b$, and so

$$
\underline{E = 2E_a + (j_0/R) + \{1/2\,(1+S)^2\}\{j + 2k + 4l + m\}}
$$
$$
\underline{- \{2/(1+S)\}\{j' + k'\}} .
$$

EXERCISE: Find the corresponding expression for a heteronuclear two-electron molecule.

10.5 Use
$$a_0 j/j_0 = 1/s - \tfrac{1}{2}\{(2/s) + 11/4 + \tfrac{3}{2}s + \tfrac{1}{3}s^2\}\,e^{-2s}$$
$$a_0 k/j_0 = (1/5)(A-B)$$
$$a_0 l/j_0 = \tfrac{1}{2}\left\{(2s+\tfrac{1}{4}+5/8s)\,e^{-s} - (\tfrac{1}{4}+5/8s)\,e^{-3s}\right\}$$
$$a_0 m/j_0 = 5/8$$
$$A = (6/s)\left\{S^2(\gamma + \ln s) - S'^2 E_1(4s) + 2SS'E_1(2s)\right\}$$
$$B = \left\{-25/8 + 23s/4 + 3s^2 + \tfrac{1}{3}s^3\right\}e^{-2s}$$
$$S = \left\{1 + s + \tfrac{1}{3}s^2\right\}e^{-s}$$
$$S'(s) = S(-s)$$
$$a_0 j'/j_0 = (1/S)\left\{1 - (1+s)\,e^{-2s}\right\}$$
$$a_0 k'/j_0 = \{1+s\}\,e^{-s}.$$

For the exponential integrals ($E_1(z)$) see *Handbook of mathematical functions*, M. Abramowitz and I.A. Stegun, Dover (1965).

EXERCISE: Now write the program.

10.6
$$\begin{aligned}\psi^2(1,2) &= \left\{1/4(1+S)^2\right\}\left\{a(1)\,a(2) + b(1)\,b(2) + a(1)\,b(2) + b(1)\,a(2)\right\}^2\\ &= \left\{1/4(1+S)^2\right\}\left\{a^2(1)a^2(2) + b^2(1)b^2(2) + a^2(1)b^2(2) + b^2(1)a^2(2)\right.\\ &\quad + 2a(1)b(1)a(2)b(2) + 2a^2(1)a(2)b(2) + 2a(1)b(1)a^2(2)\\ &\quad \left. + 2b(1)a(1)b^2(2) + 2b^2(1)a(2)b(2) + 2a(1)b(1)b(2)a(2)\right\}\end{aligned}$$
$$\begin{aligned}\rho_1 &\equiv \int \psi^2(1,2)\,d\tau_2\\ &= \left\{1/4(1+S)^2\right\}\left\{2a^2(1) + 2b^2(1) + 2a(1)\,b(1)S + 2a^2(1)\,S\right.\\ &\quad \left. + 2a(1)\,b(1) + 2a(1)\,b(1) + 2b^2(1)S + 2a(1)\,b(1)S\right\}\\ &= \left\{1/2(1+S)\right\}\left\{a^2(1) + b^2(1) + 2a(1)\,b(1)\right\}.\end{aligned}$$

It follows that
$$\begin{aligned}\delta\rho &\equiv \rho_1 - \tfrac{1}{2}\left\{a^2(1) + b^2(1)\right\}\\ &= \left\{1/2(1+S)\right\}\left\{a^2(1) + b^2(1) + 2a(1)\,b(1) - a^2(1)(1+S) - b^2(1)(1+S)\right\}\\ &= \left\{1/2(1+S)\right\}\left\{2a(1)\,b(1) - a^2(1)\,S - b^2(1)\,S\right\}.\end{aligned}$$

As in Problem 10.2, outside the nuclei but on the line of centres, $r_b = r_a - R$, while between the nuclei, $r_b = R - r_a$. Since
$$a = (1/\pi a_0^3)^{\frac{1}{2}}\,e^{-r_a/a_0},$$
and likewise for b, and since

$$S = (1+s+\tfrac{1}{2}s^2)\,e^{-s} = 0.75 \text{ when } R = 74 \text{ pm},$$

we have:

Outside the nuclei:

$$2\pi a_0^3\,\delta\rho = \left\{2e^{-2r_a/a_0+s} - 0.75\,e^{-2r_a/a_0}(1+e^{2s}\right\}\Big/1.75$$

$$= -2.82\,e^{-2r_a/a_0}$$

Between the nuclei:

$$2\pi a_0^3\,\delta\rho = \left\{2\,e^{-s} - 0.75\,e^{-2r_a/a_0} - 0.75\,e^{2r_a/a_0-2s}\right\}\Big/1.75$$

$$= 0.282 - 0.429\,e^{-2r_a/a_0} - 0.026\,e^{2r_a/a_0}.$$

The two components (which coincide at $r_a = R$) are plotted in Fig. 10.4.

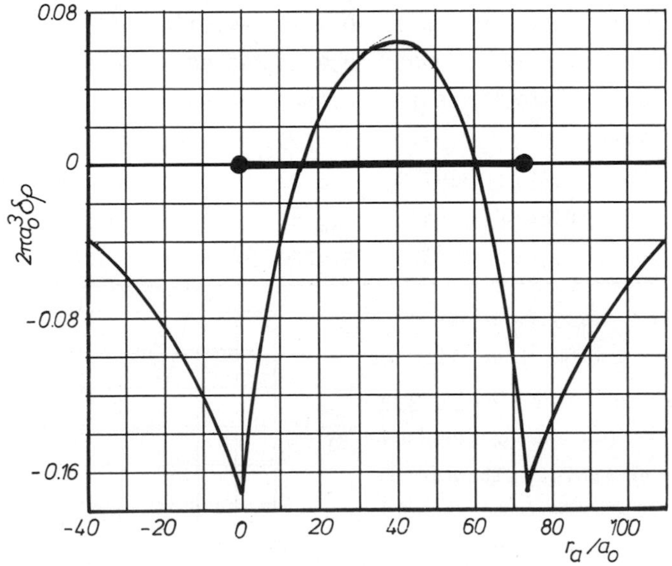

Fig. 10.4

EXERCISE: Plot the difference density for different bond lengths (e.g., $R = 50$ pm, 100 pm).

10.7 $(E_+ - 2E_{1s})a_0/j_0 = 1/s + a_0(J+K)/(1+S^2)j_0$ [eqn (10.4.4)].

$$J = j - 2j', \quad K = k - 2k'S.$$

$$a_0J/j_0 = 1/s - \tfrac{1}{2}\left\{(2/s) + 11/4 + \tfrac{3}{2}s + \tfrac{1}{3}s^2\right\}e^{-2s}$$

$$- (2/s)\left\{1-(1+s)\,e^{-2s}\right\}$$

$$= -(1/s) - [-(1/s) - (5/8) + (3/4)s + (1/6)s^2]\,e^{-2s}.$$

$$a_0K/j_0 = \tfrac{1}{5}(A-B) - 2(1+s)(1+s+\tfrac{1}{3}s^2)\,e^{-2s}.$$

$$A = (6/s)\left\{S^2(\gamma + \ln s) - S'^2 E_1(4s) + 2SS'E_1(2s)\right\}$$

$$B = \left\{-(25/8) + 23s/4 + 3s^2 + \tfrac{1}{3}s^3\right\} e^{-2s} .$$

$$\gamma = 0.577\,22 .$$

$E_1(x)$ is listed in *Handbook of mathematical functions*, M. Abramowitz and I.A. Stegun, Dover (1965).

These expressions enable us to draw up the following table.

	$E_1(2s)$	$E_1(4s)$	S	A	B	a_0J/j_0	a_0K/j_0	$\frac{(E_+ - 2E_{1s})}{(j_0/a_0)}$
0.4	0.3106	0.0863	0.9742	2.7911	−0.1454	−1.2426	−1.2412	1.2256
0.6	0.1584	0.0284	0.9440	3.1703	0.4449	−1.1300	−1.1127	0.4808
0.8	0.0863	0.0101	0.9046	3.1862	0.7199	−1.0141	−0.9700	0.1588
1.0	0.0490	0.0038	0.8584	2.9906	0.8064	−0.9041	−0.8263	0.0037
1.2	0.0284	0.0015	0.8072	2.6811	0.7866	−0.8045	−0.6908	−0.0720
1.4	0.0169	0.0009	0.7529	2.3277	0.7127	−0.7166	−0.5682	−0.1057
1.6	0.0101	0.0002	0.6972	1.9740	0.6163	−0.6404	−0.4605	−0.1158
1.8	0.0062	—	0.6414	1.6425	0.5161	−0.5749	−0.3684	−0.1128
2.0	0.0038	—	0.5865	1.3336	0.4220	−0.5191	−0.2939	−0.1049

The average energy is plotted in Fig. 10.5. The minimum lies at $s = 1.7$, corresponding to <u>$R = 90$ pm</u>. The dissociation energy (neglecting the zero-point vibrational energy) is

$$0.12\,j_0/a_0 = 0.52 \times 10^{-18}\ \text{J},$$

corresponding to 3.3 eV or <u>310 kJ mol^{-1}</u>.

EXERCISE: Determine whether He_2^{2+} is a stable species, and, if so, its bond length and dissociation energy.

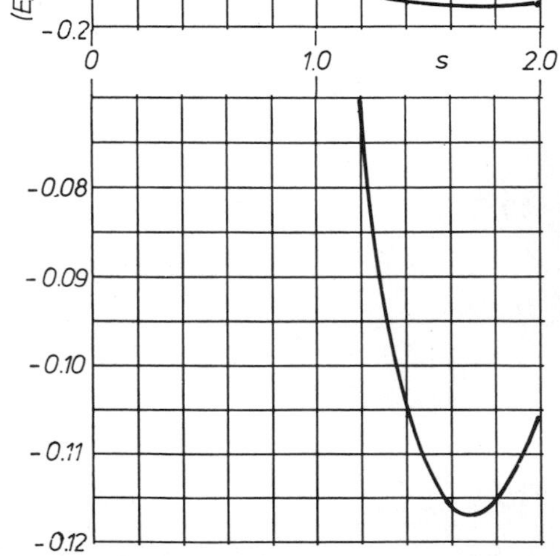

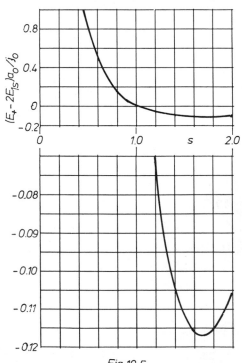

Fig. 10.5

10.8 $\psi = a(1)b(2)+b(1)a(2)+\lambda\{a(1)a(2)+b(1)b(2)\}$, unnormalized.

$$\begin{aligned}\langle\psi|\psi\rangle &= \langle a(1)b(2)|a(1)b(2)\rangle + \langle b(1)a(2)|b(1)a(2)\rangle \\ &\quad + \langle a(1)b(2)|b(1)a(2)\rangle + \langle b(1)a(2)|a(1)b(2)\rangle \\ &\quad + \lambda\{\langle a(1)b(2)|a(1)a(2)\rangle + \langle a(1)b(2)|b(1)b(2)\rangle \\ &\qquad + \langle b(1)a(2)|a(1)a(2)\rangle + \langle b(1)a(2)|b(1)b(2)\rangle\} \\ &\quad + \lambda^2\{\langle a(1)a(2)|a(1)a(2)\rangle + \langle b(1)b(2)|b(1)b(2)\rangle \\ &\qquad + \langle a(1)a(2)|b(1)b(2)\rangle + \langle b(1)b(2)|a(1)a(2)\rangle\} \\ &= 1+1+S^2+S^2 + \lambda(S+S+S+S) + \lambda^2(1+1+S^2+S^2) \\ &= 2(1+S^2) + 4\lambda S + 2\lambda^2(1+S^2) = 2(1+S^2)(1+\lambda^2) + 4\lambda S.\end{aligned}$$

Therefore, the normalized function is

$$\psi = \frac{a(1)b(2)+b(1)a(2) + \lambda\{a(1)a(2)+b(1)b(2)\}}{\sqrt{[2(1+\lambda^2)(1+S^2)+4\lambda S]}}.$$

$$\begin{aligned}\psi_{\mathrm{CI}}^{\mathrm{MO}} &= \sigma_{\mathbf{g}}(1)\sigma_{\mathbf{g}}(2) + \lambda'\sigma_{\mathbf{u}}^*(1)\sigma_{\mathbf{u}}^*(2) \\ &= [a(1)+b(1)][a(2)+b(2)]/2(1+S) \\ &\quad +\lambda'[a(1)-b(1)][a(2)-b(2)]/2(1-S) \\ &= \{a(1)a(2)+b(1)b(2)+a(1)b(2)+b(1)a(2)\}/2(1+S) \\ &\quad + \lambda'\{a(1)a(2)+b(1)b(2) - a(1)b(2)-b(1)a(2)\}/2(1-S)\end{aligned}$$

$$\psi_{\mathrm{CI}}^{\mathrm{MO}} = a(1)b(2)+b(1)a(2) + \lambda''a(1)a(2) + \lambda''b(1)b(2).$$

Consequently, for the equality of these two expressions,

$$\lambda'' = \frac{1}{2(1+S)} + \frac{\lambda'}{2(1-S)}, \qquad 1 = \frac{1}{2(1+S)} - \frac{\lambda'}{2(1-S)}.$$

These equations are sufficient to determine the value of λ' and λ'':

$$\lambda'' = \{1/(1+S)\} - 1 = -S/(1+S), \quad \lambda' = -(1-S)(1+2S)/(1+S).$$

For the VB wavefunction in its normalized form,

$$\begin{aligned}\mathcal{E} &= \tfrac{1}{2}\{\langle a(1)b(2)|H|a(1)b(2)\rangle + \langle a(1)b(2)|H|b(1)a(2)\rangle \\ &\qquad + \langle b(1)a(2)|H|b(1)a(2)\rangle + \langle b(1)a(2)|H|a(1)b(2)\rangle \\ &\quad + \lambda[\langle a(1)b(2)|H|a(1)a(2)\rangle + \langle a(1)b(2)|H|b(1)b(2)\rangle \\ &\qquad + \langle b(1)a(2)|H|a(1)a(2)\rangle + \langle b(1)a(2)|H|b(1)b(2)\rangle] \\ &\quad + \lambda^2[\langle a(1)a(2)|H|a(1)a(2)\rangle + \langle b(1)b(2)|H|b(1)b(2)\rangle \\ &\qquad + \langle a(1)a(2)|H|b(1)b(2)\rangle + \langle b(1)b(2)|H|a(1)a(2)\rangle]\} \\ &\quad \times \{(1+\lambda^2)(1+S^2) + 2\lambda S\}^{-1} \\ &= \frac{H_{\mathbf{ab,ab}} + H_{\mathbf{ab,ba}} + 2\lambda H_{\mathbf{ab,aa}} + \lambda^2(H_{\mathbf{aa,aa}} + H_{\mathbf{aa,bb}})}{(1+\lambda^2)(1+S^2) + 2\lambda S}.\end{aligned}$$

$$\begin{aligned} d\mathcal{E}/d\lambda = \Big\{ & 4(H_{aa,aa}+H_{aa,bb})(1+S^2)\lambda^3 \\ & + [6(H_{aa,aa}+H_{aa,bb})S - 2H_{ab,aa}(1+S^2)]\lambda^2 \\ & + [2(H_{aa,aa}+H_{aa,bb})(1+S^2) \\ & \qquad +8H_{ab,aa}S - 2(H_{ab,ab}+H_{ab,ba})(1+S^2)]\lambda \\ & + [2H_{ab,aa} - 2(H_{ab,ab}+H_{ab,ba})S]\Big\} \\ & \qquad \times \Big\{(1+\lambda^2)(1+S^2)+2\lambda S\Big\}^{-2} = 0 . \end{aligned}$$

That is, solve $a_3\lambda^3 + a_2\lambda^2 + a_1\lambda + a_0 = 0$ for λ, where

$$a_3 = 4(1+S^2)(H_{aa,aa}+H_{aa,bb})$$

$$a_2 = 6S(H_{aa,aa}+H_{aa,bb}) - 2(1+S^2)H_{ab,aa}$$

$$a_1 = 2(1+S^2)(H_{aa,aa}+H_{aa,bb}) + 8SH_{ab,aa} - 2(1+S^2)(H_{ab,ab}+H_{ab,ba})$$

$$a_0 = 2H_{ab,aa} - 2S(H_{ab,ab}+H_{ab,ba}).$$

EXERCISE: Find a similar expression for the molecular orbital function.

10.9 Refer to Fig. 10.12 of the text (p. 264).

C_2: $1s\sigma_g^2\, 1s\sigma_u^{*2}\, 2s\sigma_g^2\, 2s\sigma_u^{*2}\, 2p\pi_u^4$, $^1\Sigma_g^+$ [but see the discussion in the text, p. 265].

C_2^+: $1s\sigma_g^2\, 1s\sigma_u^{*2}\, 2s\sigma_g^2\, 2s\sigma_u^{*}\, 2p\pi_u^3$, $^2\Pi_u$

C_2^-: $\ldots 2p\pi_u^4\, 2p\sigma_g$, $^2\Sigma_g^+$

N_2^+: $\ldots 2p\pi_u^4\, 2p\sigma_g$, $^2\Sigma_g^+$

N_2^-: $\ldots 2p\pi_u^4\, 2p\sigma_g^2\, 2p\pi_g^*$, $^2\Pi_g$

O_2: $\ldots 2p\pi_u^4\, 2p\sigma_g^2\, 2p\pi_g^{*2}$, $^1\Delta_g$ and $^3\Sigma_g^+$, with $E(^3\Sigma) < E(^1\Delta)$.

O_2^+: $\ldots 2p\sigma_g^2\, 2p\pi_g^*$, $^2\Pi_g$

CO: $1s\sigma^2\, 1s\sigma^{*2}\, 2s\sigma^2\, 2s\sigma^{*2}\, 2p\pi^4\, 2p\sigma^2$, $^1\Sigma$ [isoelectronic with N_2]

NO: $\ldots 2p\pi^4\, 2p\sigma^2\, 2p\pi^*$, $^2\Pi$ [isoelectronic with O_2^+]

F_2^+: $\ldots 2p\pi_u^4\, 2p\sigma_g^2\, 2p\pi_g^{*3}$, $^2\Pi_g$

Ne_2^+: $\ldots 2p\pi_g^{*4}\, 2p\sigma_u^*$, $^2\Sigma_u^+$.

EXERCISE: Predict the ground configurations of Na_2, S_2, HCl, O_2^- and decide which terms lie lowest.

10.10 Use the basis $H1s_a$, $H1s_b$, O2s, $O2p_x$, $O2p_y$, $O2p_z$, and proceed as described in Chapter 7 (using C_{2v}). Draw up the following Table:

	$H1s_a$	$H1s_b$	O2s	$O2p_x$	$O2p_y$	$O2p_z$	χ(R)
E	$H1s_a$	$H1s_b$	O2s	$O2p_x$	$O2p_y$	$O2p_z$	6
C_2	$H1s_b$	$H1s_a$	O2s	$-O2p_x$	$-O2p_y$	$O2p_z$	0
σ_v	$H1s_a$	$H1s_b$	O2s	$-O2p_x$	$O2p_y$	$O2p_z$	4
σ'_v	$H1s_b$	$H1s_a$	O2s	$O2p_x$	$-O2p_y$	$O2p_z$	2

[The characters are equal to the net number of unchanged orbitals under each operation.] The decomposition of 6,0,4,2 is $3A_1+B_1+2B_2$. The symmetry-adapted bases are:

A_1: (O2s, $O2p_z$, $H1s_a+H1s_b$)

B_1: $O2p_x$

B_2: ($O2p_y$, $H1s_a-H1s_b$).

Therefore, the secular determinant factorizes into 3×3 (A_1), 1×1 (B_1), and 2×2 (B_2) determinants. Note that $p_z = (p+p')\surd2$, $p_y = (p-p')\surd2$ [eqn (10.8.1)], and that $\langle p|H|s_b\rangle = \langle p'|H|s_a\rangle = \beta$. Now ignore O2s and take A_1: ($O2p_z$, $[H1s_a+H1s_b]/\surd2$. The secular determinant (with $S=0$) is

$$A_1: \quad \begin{vmatrix} \alpha_0-E & \beta \\ \beta & \alpha_H-E \end{vmatrix} = 0.$$

The eigenvalues are therefore [p. 100]:

$$\underline{E_\pm = \tfrac{1}{2}(\alpha_0+\alpha_H) \pm \tfrac{1}{2}\Delta}, \quad \Delta^2 = (\alpha_0-\alpha_H)^2 + 4\beta^2,$$

and the eigenfunctions are

$$\left.\begin{aligned} \psi_+ &= (O2p_z)\cos\theta + (1/\surd2)(H1s_a+H1s_b)\sin\theta \\ \psi_- &= -(O2p_z)\sin\theta + (1/\surd2)(H1s_a+H1s_b)\cos\theta \end{aligned}\right\} \tan2\theta = -2\beta/(\alpha_H-\alpha_0).$$

In the case of B_1 (a non-bonding orbital)

$$\underline{E = \alpha_0}, \quad \psi = O2p_x.$$

In the case of B_2 the secular determinant for the basis ($O2p_y$, $[H1s_a-H1s_b]/\surd2$) is

$$B_2: \quad \begin{vmatrix} \alpha_0-E & -\beta \\ -\beta & \alpha_H-E \end{vmatrix} = 0.$$

The eigenvalues are therefore [p. 100]

$$\underline{E_\pm = \tfrac{1}{2}.(\alpha_0+\alpha_H) \pm \tfrac{1}{2}\Delta}, \quad \Delta^2 = (\alpha_0-\alpha_H)^2 + 4\beta^2,$$

and the eigenfunctions are

$$\left.\begin{aligned}\psi_+ &= (\mathrm{O2p}_y)\cos\theta + (1/\sqrt{2})(\mathrm{H1s_a}-\mathrm{H1s_b})\sin\theta\\ \psi_- &= -(\mathrm{O2p}_y)\sin\theta + (1/\sqrt{2})(\mathrm{H1s_a}-\mathrm{H1s_b})\cos\theta\end{aligned}\right\} \tan 2\theta = 2\beta/(\alpha_{\mathrm{H}}-\alpha_0).$$

Consequently, $E_\pm(A_1) = E_\pm(B_2)$ when the bond angle is 90°.

EXERCISE: Find the energies of the orbitals of NH_3 under the same set of assumptions (but using C_{3v}) and for a 90° pyramidal molecule.

10.11 $h_1 = as + bp$, $h_2 = as + bp'$ [eqn (10.8.2)]

$h' = a's + b'p_z$ [eqn (10.8.4)].

$p = p_z\cos\tfrac{1}{2}\Theta + p_z\sin\tfrac{1}{2}\Theta$, $p' = p_z\cos\tfrac{1}{2}\Theta - p_y\sin\tfrac{1}{2}\Theta$ [eqn (10.8.1)].

The symmetry-adapted linear combinations are

$$h(A_1) = (h_1+h_2)/\sqrt{2},\quad h(B_2) = (h_1-h_2)/\sqrt{2},\quad h'(A_1) = h'.$$

$$\mathrm{H1s}(A_1) = (\mathrm{H1s_a}+\mathrm{H1s_b})/\sqrt{2},\quad \mathrm{H1s}(B_2) = (\mathrm{H1s_a}-\mathrm{H1s_b})/\sqrt{2}.$$

The B_2 scalar determinant is based on the elements

$$\langle \mathrm{H1s}(B_2)|H|\mathrm{H1s}(B_2)\rangle = \alpha_{\mathrm{H}}$$

$$\begin{aligned}\langle h(B_2)|H|h(B_2)\rangle &= \tfrac{1}{2}\{\langle h_1|H|h_1\rangle + \langle h_2|H|h_2\rangle\\ &= a^2\alpha(\mathrm{O2s}) + b^2\alpha(\mathrm{O2p}) \approx \alpha_0\ [\text{for } \alpha(\mathrm{O2s}) = \alpha(\mathrm{O2p})]\end{aligned}$$

$$\begin{aligned}\langle \mathrm{H1s}(B_2)|H|h(B)\rangle &= \tfrac{1}{2}\Big\{\langle \mathrm{H1s_a}|H|h_1\rangle - \langle \mathrm{H1s_a}|H|h_2\rangle\\ &\qquad - \langle \mathrm{H1s_b}|H|h_1\rangle + \langle \mathrm{H1s_b}|H|h_2\rangle\Big\}\\ &= \langle \mathrm{H1s_a}|H|h_1\rangle - \langle \mathrm{H1s_a}|H|h_2\rangle \qquad [\text{symmetry}]\\ &= a\langle \mathrm{H1s_a}|H|s\rangle + b\langle \mathrm{H1s_a}|H|p\rangle\\ &\qquad - a\langle \mathrm{H1s_a}|H|s\rangle - b\langle \mathrm{H1s_a}|H|p'\rangle\\ &= b\Big\{\langle \mathrm{H1s_a}|H|p\rangle - \langle \mathrm{H1s_a}|H|p'\rangle\Big\}.\end{aligned}$$

Now, p points directly at $\mathrm{H1s_a}$, and so $\langle \mathrm{H1s}|H|p\rangle = \beta$. On the other hand, p′ does not point at $\mathrm{H1s_a}$ (but at $\mathrm{H1s_b}$) unless $\Theta = 0$. We write

$$\langle \mathrm{H1s_a}|H|p'\rangle = \langle \mathrm{H1s_a}|H|p_z\rangle\cos\tfrac{1}{2}\Theta - \langle \mathrm{H1s_a}|H|p_y\rangle\sin\tfrac{1}{2}\Theta,$$

and use

$$\langle \mathrm{H1s_a}|H|p_z\rangle = \beta\cos\tfrac{1}{2}\Theta,\quad \langle \mathrm{H1s_a}|H|p_y\rangle = \beta\sin\tfrac{1}{2}\Theta.$$

Then,

$$\langle \mathrm{H1s_a}|H|p'\rangle = \beta(\cos^2\tfrac{1}{2}\Theta - \sin^2\tfrac{1}{2}\Theta) = \beta\cos\Theta.$$

It follows that $\langle \mathrm{H1s}(B_2)|H|h(B_2)\rangle = b\beta(1-\cos\Theta)$.

The B_2 secular determinant is therefore:

$$\begin{vmatrix} a_0 - E & b\beta(1-\cos\Theta)\\ b\beta(1-\cos\Theta) & \alpha_{\mathrm{H}} - E\end{vmatrix} = 0,$$

and the eigenvalues and eigenfunctions are:

$$E_{\pm} = \tfrac{1}{2}(\alpha_0 + \alpha_H) \pm \tfrac{1}{2}\Delta, \quad \Delta^2 = (\alpha_0 - \alpha_H)^2 + 4b^2\beta^2(1-\cos\Theta)^2,$$

$$\left.\begin{aligned}\psi_+ &= h(B_2)\cos\theta + [\mathrm{H1s}(B_2)]\sin\theta \\ \psi_- &= -h(B_2)\sin\theta + [\mathrm{H1s}(B_2)]\cos\theta\end{aligned}\right\} \tan 2\theta = -2b\beta(1-\cos\Theta)/(\alpha_H - \alpha_0).$$

We also know that $b^2 = 1 - a^2$, $a^2 = \cos\Theta/(\cos\Theta - 1)$ [eqn (10.8.3)], and so $b^2 = 1/(1-\cos\Theta)$. Consequently

$$\underline{E_{\pm} = \tfrac{1}{2}(\alpha_0 + \alpha_H) \pm \tfrac{1}{2}\Delta}, \quad \underline{\Delta^2 = (\alpha_0 - \alpha_H)^2 + 4\beta^2(1-\cos\Theta)}$$

$$\underline{\tan 2\theta = -2\beta(1-\cos\Theta)^{\frac{1}{2}}/(\alpha_H - \alpha_0)}.$$

The A_1 secular determinant is based on the elements

$$\langle \mathrm{H1s}(A_1)|H|\mathrm{1s}(A_1)\rangle = \alpha_H$$

$$\langle h(A_1)|H|h(A_1)\rangle = a^2\alpha(\mathrm{O2s}) + b^2\alpha(\mathrm{O2p}) \approx \alpha_0$$

$$\langle h'(A_1)|H|h'(A_1)\rangle = a'^2\alpha(\mathrm{O2s}) + b'^2\alpha(\mathrm{O2p}) \approx \alpha_0$$

$$\begin{aligned}\langle \mathrm{H1s}(A_1)|H|h'(A_1)\rangle &= \{\langle \mathrm{H1s_a}|H|h'(A_1)\rangle + \langle \mathrm{H1s_b}|H|h'(A_1)\rangle\}/\sqrt{2} \\ &= \langle \mathrm{H1s_a}|H|h'(A_1)\rangle\sqrt{2} \qquad [\text{symmetry}] \\ &= \{\langle \mathrm{H1s_a}|H|s\rangle a' + \langle \mathrm{H1s_a}|H|p_z\rangle b'\}\sqrt{2} \\ &= \{\beta_s a' + \beta b'\cos\tfrac{1}{2}\Theta\}\sqrt{2} \\ &\approx (a' + b'\cos\tfrac{1}{2}\Theta)\beta\sqrt{2} \qquad [\beta_s \approx \beta].\end{aligned}$$

Write $A = (a' + b'\cos\tfrac{1}{2})\sqrt{2}$.

$$\begin{aligned}\langle \mathrm{H1s}(A_1)|H|h(A_1)\rangle &= \langle \mathrm{H1s_a}|H|h_1\rangle + \langle \mathrm{H1s_a}|H|h_2\rangle \\ &= a\langle \mathrm{H1s_a}|H|s\rangle + b\langle \mathrm{H1s_a}|H|p\rangle \\ &\quad + a\langle \mathrm{H1s_a}|H|s\rangle + b\langle \mathrm{H1s_a}|H|p''\rangle \\ &= 2a\beta_s + b\beta + b\beta\cos\Theta \qquad [\text{from above}] \\ &\approx \{2a + b(1+\cos\Theta)\}\beta \qquad [\beta_s \approx \beta].\end{aligned}$$

Write $B = 2a + b(1+\cos\Theta)$.

$\langle h(A_1)|H|h'(A_1)\rangle = 0$ [orthogonal in this approximation (but there are actually repulsions)].

The secular determinant is therefore

$$\begin{vmatrix}\alpha_0 - E & 0 & A\beta \\ 0 & \alpha_0 - E & B\beta \\ A\beta & B\beta & \alpha_H - E\end{vmatrix} \approx \begin{vmatrix}\alpha - E & 0 & A\beta \\ 0 & \alpha - E & B\beta \\ A\beta & B\beta & \alpha - E\end{vmatrix}$$

$$= (\alpha - E)[(\alpha - E)^2 - B^2\beta^2] - (\alpha - E)A^2\beta^2 = 0.$$

Consequently, $E = \alpha$ and $\alpha \pm \beta\sqrt{(A^2+B^2)}$.

$$A^2+B^2 = 2(a'+b'\cos\tfrac{1}{2}\Theta)^2 + [2a+b(1+\cos\Theta)]^2$$
$$= 2(a'^2+b'^2\cos^2\tfrac{1}{2}\Theta + 2a'b'\cos\tfrac{1}{2}\Theta)$$
$$+ 4a^2+b^2(1+\cos\Theta)^2+4ab(1+\cos\Theta)$$
$$= 2\left\{\cos\tfrac{1}{2}\Theta + \left(\frac{2\sqrt{(-\cos\Theta)}}{1-\cos\Theta}\right)\right\}\cos\tfrac{1}{2}\Theta + \left\{\frac{1+2\cos\Theta+5\cos^2\Theta}{1-\cos\Theta}\right\}.$$

EXERCISE: Find (numerically) the value of Θ corresponding to the energy minimum of the molecule. Relax the approximation $\alpha_0 \approx \alpha_H$, $\beta_s \approx \beta$ and write instead $\alpha(\text{O2s}) = \lambda\alpha(\text{O2p}) = \lambda\alpha$, $\alpha(\text{H1s}) = \mu\alpha$, $\beta_s = \nu\beta$. Find the corresponding energies.

10.12 A planar, equilateral triangular molecule belongs to D_{3h}. In this group (s, p_x, p_y) spans $A_1' + E'$. Three planar hybrids pointing towards the corners span A_1+E in C_{3v} [just like s_A, s_B, s_C at each corner], and hence span $A_1'+E'$ in D_{3h} [even under σ_h]. Hence the two basis sets span the same symmetry species, and so one may be expressed as linear combinations of the other.

$\left(d_{x^2-y^2}, d_{xy}\right)$ also spans E'; hence sd^2 may also be used. d_{z^2} spans A_1', and so dp^2 may also be used.

$\left(d_{z^2}, d_{x^2-y^2}, d_{xy}\right)$ spans $A_1'+E'$, and so d^3 may also be used.

EXERCISE: Show that sp and dp hybrids may arise for linear species, and that d^2p hybridization may arise for trigonal pyramidal molecules.

10.13 The corners of the tetrahedron (T_d) span A_1+T_2 [in exactly the same way that s-orbitals attached to them would]. Therefore, four equivalent hybrids span A_1+T_2. In T_d, s spans A_1 and both (p_x, p_y, p_z) and (d_{xy}, d_{xz}, d_{yz}) span T_2. Hence sp^3 and sd^3 hybridization may arise.

For tetragonal planar molecules, consider D_{4h}. In this group s, d_{z^2} span A_{1g}, $d_{x^2-y^2}$ spans B_{1g}, d_{xy} spans B_{2g}, and (d_{xz}, d_{yz}) spans E_g, and (p_x, p_y) spans E_u. Four equivalent hybrids span $A_{1g}+B_{2g}+E_u$. These irreps are also spanned by both (s, d_{xy}, p_x, p_y) and $(d_{z^2}, d_{xy}, p_x, p_y)$, and so either sdp^2 or d^2p^2 hybridization may arise.

For pyramidal molecules, use D_{3h}. The five equivalent hybrids span $2A_1'+A_2''+E'$. In this group s, d_{z^2} each span A_1' while (p_x, p_y) and $d_{x^2-y^2}, d_{xy})$ each span E', and p_z spans A_2''. Therefore, we may have

(s, d_{z^2}, p_z) with either (p_x, p_y), to give sdp^3, or $(d_{x^2-y^2}, d_{xy})$, to give spd^3.

In O_h the six equivalent octahedral bonds span $A_{1g}+E_g+T_{1u}$. These species are also spanned by s, $(d_{z^2}, d_{x^2-y^2})$, (p_x, p_y, p_z) respectively, and so sd^2p^3 hybridization may arise.

EXERCISE: Show that d^2sp^2, d^4s, d^2p^3, and d^4p may each participate in tetragonal pyramidal hybridization, and that d^3p^3 gives rise to trigonal antiprismatic hybridization.

10.14 (a) Under C_2, $\pi_1 \leftrightarrow \pi_4$, $\pi_2 \leftrightarrow \pi_3$; therefore, use the symmetry-adapted linear combinations based on

$$\pi(A_1): (\pi_1+\pi_4)/\sqrt{2}, \quad (\pi_2+\pi_3)/\sqrt{2};$$

$$\pi(B_2): (\pi_1-\pi_4)/\sqrt{2}, \quad (\pi_2-\pi_3)/\sqrt{2}.$$

For $\pi(A_1) = (c_1/\sqrt{2})(\pi_1+\pi_4) + (c_2/\sqrt{2})(\pi_2+\pi_2)$

$\pi(B_2) = (c_1'/\sqrt{2})(\pi_1-\pi_4) + (c_2'/\sqrt{2})(\pi_2-\pi_3)$.

The secular determinant factorizes into two 2×2 determinants. Use

$$\tfrac{1}{2}\langle\pi_1+\pi_4|H|\pi_1+\pi_4\rangle = \tfrac{1}{2}\{\langle\pi_1|H|\pi_1\rangle + \langle\pi_1|H|\pi_4\rangle + \langle\pi_4|H|\pi_1\rangle + \langle\pi_4|H|\pi_4\rangle\}$$
$$= \tfrac{1}{2}\{\alpha+0+0+\alpha\} = \alpha.$$

$$\tfrac{1}{2}\langle\pi_1-\pi_4|H|\pi_1-\pi_4\rangle = \alpha.$$

$$\tfrac{1}{2}\langle\pi_2+\pi_3|H|\pi_2+\pi_3\rangle = \tfrac{1}{2}\{\langle\pi_2|H|\pi_2\rangle + \langle\pi_3|H|\pi_3\rangle + \langle\pi_2|H|\pi_3\rangle + \langle\pi_3|H|\pi_2\rangle\}$$
$$= \tfrac{1}{2}\{\alpha+\alpha+\beta+\beta\} = \alpha+\beta.$$

$$\tfrac{1}{2}\langle\pi_2-\pi_3|H|\pi_2-\pi_3\rangle = \alpha-\beta.$$

$$\tfrac{1}{2}\langle\pi_1+\pi_4|H|\pi_2+\pi_3\rangle = \tfrac{1}{2}\{\langle\pi_1|H|\pi_2\rangle + \langle\pi_1|H|\pi_3\rangle + \langle\pi_4|H|\pi_2\rangle + \langle\pi_4|H|\pi_3\rangle\}$$
$$= \tfrac{1}{2}\{\beta+0+0+\beta\} = \beta.$$

$$\tfrac{1}{2}\langle\pi_1-\pi_4|H|\pi_2-\pi_3\rangle = \tfrac{1}{2}\{\langle\pi_1|H|\pi_2\rangle - \langle\pi_1|H|\pi_3\rangle - \langle\pi_4|H|\pi_2\rangle + \langle\pi_4|H|\pi_3\rangle\}$$
$$= \tfrac{1}{2}\{\beta-0-0+\beta\} = \beta.$$

$$\Delta(A_1) = \begin{vmatrix}\alpha-E & \beta \\ \beta & \alpha+\beta-E\end{vmatrix} = (\alpha-E)(\alpha+\beta-E)-\beta^2 = 0,$$

$$\Delta(B_2) = \begin{vmatrix}\alpha-E & \beta \\ \beta & \alpha-\beta-E\end{vmatrix} = (\alpha-E)(\alpha-\beta-E)-\beta^2 = 0.$$

The solutions are

$$E_\pm(A_1) = \alpha+\tfrac{1}{2}(1\pm\sqrt{5})\beta, \quad E_\pm(B_2) = \alpha-\tfrac{1}{2}(1\mp\sqrt{5})\beta.$$

The wavefunctions are formed from [p. 100]

$$\left.\begin{aligned} A_1:\ \psi_+ &= [(\pi_1+\pi_4)/\sqrt{2}]\cos\theta + [(\pi_2+\pi_3)/\sqrt{2}]\sin\theta \\ \psi_- &= -[(\pi_1+\pi_4)/\sqrt{2}]\sin\theta + [(\pi_2+\pi_3)/\sqrt{2}]\cos\theta \end{aligned}\right\} \tan 2\theta = -2\beta/\beta = -2\,.$$

Therefore, $\theta = -31.72^\circ$; $\cos\theta = 0.851$, $\sin\theta = -0.526$.

$$\left.\begin{aligned} B_2:\ \psi_+ &= [(\pi_1-\pi_4)/\sqrt{2}]\cos\theta + [(\pi_2-\pi_3)/\sqrt{2}]\sin\theta \\ \psi_- &= -[(\pi_1-\pi_4)/\sqrt{2}]\sin\theta + [(\pi_2-\pi_3)/\sqrt{2}]\cos\theta \end{aligned}\right\} \tan 2\theta = -2\beta/(-\beta) = 2\,.$$

Therefore, $\theta = 31.72^\circ$; $\cos\theta = 0.851$, $\sin\theta = 0.526$.

It follows that the explicit forms of the wavefunctions and their energies are:

$$\underline{\pi^*(A_1) = 0.602(\pi_1+\pi_4) - 0.372(\pi_2+\pi_3)\,, \quad E = \alpha - 1.62\,\beta}$$

$$\underline{\pi^*(B_2) = 0.602(\pi_1-\pi_4) - 0.372(\pi_2-\pi_3)\,, \quad E = \alpha - 0.62\,\beta}$$

$$\underline{\pi^*(B_2) = 0.372(\pi_1-\pi_4) + 0.602(\pi_2-\pi_3)\,, \quad E = \alpha + 0.62\,\beta}$$

$$\underline{\pi^*(A_1) = 0.372(\pi_1+\pi_4) + 0.602(\pi_2+\pi_3)\,, \quad E = \alpha + 1.62\,\beta}\,.$$

The delocalization energy is

$$\left\{2[\alpha+\tfrac{1}{2}(1+\sqrt{5})\beta] + 2[\alpha-\tfrac{1}{2}(1-\sqrt{5})\beta]\right\} - \left\{2(\alpha+\beta) + 2(\alpha+\beta)\right\}$$

$$= (2\sqrt{5}-4)\beta = \underline{0.472\,\beta}\,.$$

(b) For cyclobutene use C_{4v}; the symmetry-adapted combinations are

$$\pi(A_1) = \tfrac{1}{2}(\pi_1+\pi_2+\pi_3+\pi_4)$$

$$\pi(B_1) = \tfrac{1}{2}(\pi_1-\pi_2+\pi_3-\pi_4)$$

$$\pi(E) = \tfrac{1}{2}\left\{(\pi_1+\pi_2)-(\pi_3+\pi_4)\right\},\quad \tfrac{1}{2}\left\{(\pi_1+\pi_4)-(\pi_2+\pi_3)\right\}.$$

The relevant matrix elements are therefore

$$\begin{aligned} \langle\pi(A_1)|H|\pi(A_1)\rangle = \tfrac{1}{2}\Big\{&\langle\pi_1|H|\pi_1\rangle + \langle\pi_2|H|\pi_2\rangle + \langle\pi_3|H|\pi_3\rangle + \langle\pi_4|H|\pi_4\rangle \\ &+ 2\langle\pi_1|H|\pi_2\rangle + 2\langle\pi_1|H|\pi_3\rangle + 2\langle\pi_1|H|\pi_4\rangle \\ &+ 2\langle\pi_2|H|\pi_3\rangle + 2\langle\pi_2|H|\pi_4\rangle + 2\langle\pi_2|H|\pi_1\rangle \\ &+ 2\langle\pi_3|H|\pi_4\rangle + 2\langle\pi_3|H|\pi_1\rangle + 2\langle\pi_3|H|\pi_2\rangle \\ &+ 2\langle\pi_4|H|\pi_1\rangle + 2\langle\pi_4|H|\pi_2\rangle + 2\langle\pi_4|H|\pi_3\rangle\Big\} \\ = \tfrac{1}{4}\Big\{&\alpha+\alpha+\alpha+\alpha + 2\beta+0+2\beta+2\beta+0+2\beta \\ &+ 2\beta+0+2\beta+2\beta+0+2\beta\Big\} \end{aligned}$$

$= \alpha + 2\beta$; hence, $E(A_1) = \underline{\alpha + 2\beta}$.

$$\langle\pi(B_1)|H|\pi(B_1)\rangle = \tfrac{1}{4}\{\alpha+\alpha+\alpha+\alpha-2\beta+0-2\beta-\ldots\}$$
$$= \alpha-2\beta\,; \quad \text{hence,}\ E(B_1) = \underline{\alpha-2\beta}\,.$$
$$\langle\pi_1(E)|H|\pi_1(E)\rangle = \alpha+\tfrac{1}{4}\{2\langle\pi_1|H|\pi_2\rangle - 2\langle\pi_1|H|\pi_3\rangle+\ldots\}$$
$$= \alpha = \langle\pi_2(E)|H|\pi_2(E)\rangle\,.$$
$$\langle\pi_1(E)|H|\pi_2(E)\rangle = 0\,.$$

The E secular determinant is therefore diagonal, and $E(E)=\underline{\alpha}$. The four energy levels are therefore

$$E=\alpha+2\beta\,,\ \alpha,\ \alpha,\ \alpha-2\beta,$$

and the delocalization energy is

$$\{2(\alpha+2\beta)+2\alpha\} - \{2(\alpha+\beta)+2(\alpha+\beta)\} = \underline{0}\,.$$

EXERCISE: Find the energies and orbitals of the molecule HN = CH − CH = NH and its cyclic isomer, using $\alpha_N \neq \alpha_C$.

10.15 Use the group C_{6v} initially. Draw up the following Table :

	π_1	π_2	π_3	π_4	π_5	π_6	χ
E	π_1	π_2	π_3	π_4	π_5	π_6	6
C_6^-	π_2	π_3	π_4	π_5	π_6	π_1	0
C_6^+	π_6	π_1	π_2	π_3	π_4	π_5	0
C_3^-	π_3	π_4	π_5	π_6	π_1	π_2	0
C_3^+	π_5	π_6	π_1	π_2	π_3	π_4	0
C_2	π_4	π_5	π_6	π_1	π_2	π_3	0
σ_v	π_1	π_6	π_5	π_4	π_3	π_2	2
σ_v'	π_3	π_2	π_1	π_6	π_5	π_4	2
σ_v''	π_5	π_4	π_3	π_2	π_1	π_6	2
σ_d	π_2	π_1	π_6	π_5	π_4	π_3	0
σ_d'	π_4	π_3	π_2	π_1	π_6	π_5	0
σ_d''	π_6	π_5	π_4	π_3	π_2	π_1	0

[The final column is formed by counting the total number of orbitals left unchanged under the operation.] The π_i-orbitals therefore span a reducible representation with characters 6,0,0,0,2,0 for C_{6v}. This decomposes into $a(A_1)=1$, $a(B_1)=1$, $a(E_1)=1$, $a(E_2)=1$ [eqn (7.8.4)]. The basis therefore spans $A_1+B_1+E_1+E_2$. Now use the projection operator [eqn (7.9.6)] to generate the symmetry-adapted combinations. Then, ignoring normalization factors,

$$p_{A_1}\pi_1 = \pi_1+\pi_2+\pi_3+\pi_4+\pi_5+\pi_6\,; \qquad p_{E_1}\pi_2 = 2\pi_2+\pi_3+\pi_1-\pi_4-\pi_6-2\pi_5\,;$$

$$p_{B_1}\pi_1 = \pi_1-\pi_2-\pi_6+\pi_3+\pi_5-\pi_4\,; \qquad p_{E_2}\pi_2 = 2\pi_2-\pi_3-\pi_1-\pi_4-\pi_6+2\pi_5\,;$$

$$p_{E_1}\pi_1 = 2\pi_1+\pi_2+\pi_6-\pi_3-\pi_5-2\pi_4\,; \qquad p_{E_1}\pi_3 = 2\pi_3+\pi_4+\pi_2-\pi_5-\pi_1-2\pi_6\,;$$

$$p_{E_1}\pi_1 = 2\pi_1-\pi_2-\pi_6-\pi_3-\pi_5+2\pi_4\,; \qquad p_{E_2}\pi_3 = 2\pi_3-\pi_4-\pi_2-\pi_5-\pi_1+2\pi_6\,.$$

$p_{E_1}\pi_1$ is symmetric under σ_v. The linear combination of $p_{E_1}\pi_2$ and $p_{E_1}\pi_3$ that is antisymmetric under σ_v is

$$p_{E_1}\pi_2 + p_{E_2}\pi_2 = 3(\pi_2+\pi_3-\pi_6-\pi_5)\,.$$

The corresponding σ_v-antisymmetric linear combination of $p_{E_2}\pi_2$, $p_{E_2}\pi_3$ is

$$p_{E_2}\pi_2 - p_{E_2}\pi_3 = 3(\pi_2-\pi_3+\pi_5-\pi_6)\,.$$

These combinations (when normalized to unity) conform with those set out on p. 276 of the text.

For naphthalene use C_{2v}, and draw up the following Table :

	π_1	π_2	π_3	π_4	π_5	π_6	π_7	π_8	π_9	π_{10}	χ
E	π_1	π_2	π_3	π_4	π_5	π_6	π_7	π_8	π_9	π_{10}	10
C_2	π_6	π_7	π_8	π_9	π_{10}	π_1	π_2	π_3	π_4	π_5	0
σ_v	π_4	π_3	π_2	π_1	π_{10}	π_9	π_8	π_7	π_6	π_5	0
σ'_v	π_9	π_8	π_7	π_6	π_5	π_4	π_3	π_2	π_1	π_{10}	2

The characters 10 , 0 , 0 , 2 span $3A_1+2A_2+2B_1+3B_2$ [use eqn (7.8.4)]. The symmetry-adapted linear combinations are

$$\pi(A_1) = p_{A_1}\pi_1 = \pi_1+\pi_4+\pi_6+\pi_9\,, \qquad \pi(B_1) = p_{B_1}\pi_1 = \pi_1+\pi_4-\pi_6-\pi_9\,,$$

$$\pi(A_1) = p_{A_1}\pi_2 = \pi_2+\pi_3+\pi_7+\pi_8\,, \qquad \pi(B_1) = p_{B_1}\pi_2 = \pi_2+\pi_3-\pi_7-\pi_8\,,$$

$$\pi(A_1) = p_{A_1}\pi_5 = \pi_5+\pi_{10}\,, \qquad \pi(B_2) = p_{B_2}\pi_1 = \pi_1-\pi_4-\pi_6+\pi_9\,,$$

$$\pi(A_2) = p_{A_2}\pi_1 = \pi_1-\pi_4+\pi_6-\pi_9\,, \qquad \pi(B_2) = p_{B_2}\pi_2 = \pi_2-\pi_3-\pi_7+\pi_8\,,$$

$$\pi(A_2) = p_{A_2}\pi_2 = \pi_2-\pi_3+\pi_7-\pi_8\,, \qquad \pi(B_2) = p_{B_2}\pi_5 = \pi_5-\pi_{10}\,.$$

EXERCISE: Find the symmetry-adapted linear combinations for the π-orbitals of anthracene in C_{2v} and D_{2h}.

10.16 The basis $\pi_1, \pi_2, \pi_3, \pi_4 \equiv \pi_N, \pi_5, \pi_6$ transforms as follows in C_{2v} (write the C_2 axis cutting through π_1 and π_N):

	π_1	π_2	π_3	π_N	π_5	π_6	χ
E	π_1	π_2	π_3	π_N	π_5	π_6	6
C_2	$-\pi_1$	$-\pi_6$	$-\pi_5$	$-\pi_N$	$-\pi_3$	$-\pi_2$	-2
σ_v	$-\pi_1$	$-\pi_2$	$-\pi_3$	$-\pi_N$	$-\pi_5$	$-\pi_6$	-6
σ_v'	π_1	π_6	π_5	π_N	π_3	π_2	2

The characters $6,-2,-6,2$ span $2A_2+4B_2$. The unnormalized symmetry-adapted linear combinations are

$$\pi(A_2) = p_{A_2}\pi_2 = \pi_2-\pi_6\,; \quad (\pi_2-\pi_6)/\sqrt{2} \ \text{when normalized,}$$

$$\pi(A_2) = p_{A_2}\pi_3 = \pi_3-\pi_5\,, \quad (\pi_3-\pi_5)/\sqrt{2} \ \text{when normalized,}$$

$$\pi(B_2) = p_{B_2}\pi_1 = \pi_1,$$

$$\pi(B_2) = p_{B_2}\pi_2 = \pi_2+\pi_6\,, \quad (\pi_2+\pi_6)/\sqrt{2} \ \text{when normalized,}$$

$$\pi(B_2) = p_{B_2}\pi_3 = \pi_3+\pi_5\,, \quad (\pi_3+\pi_5)/\sqrt{2} \ \text{when normalized,}$$

$$\pi(B_2) = p_{B_2}\pi_N = \pi_N\,.$$

The A_2 determinant in the Hückel approximation involves the matrix elements

$$\tfrac{1}{2}\langle\pi_2-\pi_6|H|\pi_2-\pi_6\rangle = \tfrac{1}{2}\langle\pi_3-\pi_5|H|\pi_3-\pi_5\rangle = \alpha$$

$$\tfrac{1}{2}\langle\pi_2-\pi_6|H|\pi_3-\pi_5\rangle = \tfrac{1}{2}\{\langle\pi_2|H|\pi_3\rangle + \langle\pi_6|H|\pi_5\rangle - \langle\pi_6|H|\pi_3\rangle - \langle\pi_2|H|\pi_5\rangle\}$$

$$= \tfrac{1}{2}\{\beta+\beta-0-0\} = \beta\,.$$

The A_2 secular determinant is therefore

$$\begin{vmatrix} \alpha-E & \beta \\ \beta & \alpha-E \end{vmatrix} = (\alpha-E)^2-\beta^2 = 0; \quad \text{consequently } \underline{E=\alpha\pm\beta}\,.$$

The B_2 determinant involves

$$\langle\pi_1|H|\pi_1\rangle = \alpha$$

$$\tfrac{1}{2}\langle\pi_2+\pi_6|H|\pi_2+\pi_6\rangle = \alpha = \tfrac{1}{2}\langle\pi_3+\pi_5|H|\pi_3+\pi_5\rangle$$

$$\langle\pi_N|H|\pi_N\rangle = \alpha_N \approx \alpha + \tfrac{1}{2}\beta$$

$$\langle\pi_1|H|\pi_2+\pi_6\rangle/\sqrt{2} = (\beta+\beta)/\sqrt{2} = \beta\sqrt{2}\,.$$

$$\langle\pi_1|H|\text{all others}\rangle = 0\,.$$

$$\tfrac{1}{2}\langle\pi_2+\pi_6|H|\pi_3+\pi_5\rangle = \tfrac{1}{2}(\beta+\beta) = \beta.$$
$$\langle\pi_2+\pi_6|H|\pi_N\rangle/\sqrt{2} = 0$$
$$\langle\pi_3+\pi_5|H|\pi_N\rangle/\sqrt{2} = \beta_N\sqrt{2} \approx \beta\sqrt{2}.$$

The determinant itself is therefore

$$\begin{vmatrix} \alpha-E & \beta\sqrt{2} & 0 & 0 \\ \beta\sqrt{2} & \alpha-E & \beta & 0 \\ 0 & \beta & \alpha-E & \beta\sqrt{2} \\ 0 & 0 & \beta\sqrt{2} & \alpha+\tfrac{1}{2}\beta-E \end{vmatrix}$$
$$= (\alpha-E)^3(\alpha+\tfrac{1}{2}\beta-E) - 2(\alpha-E)^2\beta^2 - (\alpha-E)(\alpha+\tfrac{1}{2}\beta-E)\beta^2$$
$$- 2\beta^2(\alpha-E)(\alpha+\tfrac{1}{2}\beta-E) + 4\beta^4 = 0.$$

Write $(\alpha-E)/\beta=x$; then solve

$$x^3(x+\tfrac{1}{2}) - 2x^2 - x(x+\tfrac{1}{2}) - 2x(x+\tfrac{1}{2}) + 4 = 0,$$

or

$$x^4+\tfrac{1}{2}x^3 - 5x^2 - \tfrac{3}{2}x + 4 = 0.$$

The roots of this equation (determined numerically) are

$$x = 0.8410,\ 1.9337,\ -1.1672,\ -2.1074,$$

and so, in this approximation, the energies of the B_2 orbitals lie at

$$\underline{E = \alpha+1.9337\beta,\ \alpha+0.8410\beta,\ \alpha-1.1672\beta,\ \alpha-2.1074\beta}.$$

The π-electron energy is therefore

$$E_\pi = 2(\alpha+1.9337\beta) + 2(\alpha+\beta) + 2(\alpha+0.8410\beta) = \underline{6\alpha+7.5494\beta}.$$

EXERCISE: Find the Hückel molecular orbital energies of pyrazine using the same set of approximations.

10.17 If we do not deal with symmetry-adapted linear combinations the elements of the determinant are

$\langle\pi_i|H|\pi_i\rangle = \alpha$, $i = 1,2,3,4$;

$\langle\pi_1|H|\pi_2\rangle = \beta$, $\langle\pi_1|\pi_2\rangle = S$, and so on for neighbours;

$\langle\pi_1|H|\pi_3\rangle = 0$, $\langle\pi_1|\pi_3\rangle = 0$, and so on for non-neighbours.

This gives

$$\begin{vmatrix} \alpha-E & \beta-ES & 0 & \beta-ES \\ \beta-ES & \alpha-E & \beta-ES & 0 \\ 0 & \beta-ES & \alpha-E & \beta-ES \\ \beta-ES & 0 & \beta-ES & \alpha-E \end{vmatrix} = 0.$$

Divide through by $\beta-ES$ and set $w = (\alpha-E)/(\beta-ES)$; then the determinant becomes

$$\begin{vmatrix} w & 1 & 0 & 1 \\ 1 & w & 1 & 0 \\ 0 & 1 & w & 1 \\ 1 & 0 & 1 & w \end{vmatrix} = 0 ,$$

whereas with $S = 0$ we would have $x = (\alpha - E)/\beta$ in place of w. We must therefore solve $w^4 - 4w^2 = 0$. The roots are $w = 0, 0, \pm 2$ (where before they were $x = 0, 0, \pm 2$), and so the eigenvalues are

$$\underline{E = \alpha, \alpha, (\alpha \mp 2\beta)/(1 \mp 2S)}$$

in place of $\alpha, \alpha, \alpha \mp 2\beta$.

EXERCISE: Include overlap in the calculation of butadiene. What is its delocalization energy ?

10.18 The secular determinant with non-zero nearest neighbour overlap is

$$\begin{vmatrix} \alpha - E & \beta - ES & 0 & 0 & 0 & \beta' - ES' \\ \beta - ES & \alpha - E & \beta' - ES' & 0 & 0 & 0 \\ 0 & \beta' - ES' & \alpha - E & \beta - ES & 0 & 0 \\ 0 & 0 & \beta - ES & \alpha - E & \beta' - ES' & 0 \\ 0 & 0 & 0 & \beta' - ES' & \alpha - E & \beta - ES \\ \beta' - ES' & 0 & 0 & 0 & \beta - ES & \alpha - E \end{vmatrix} = 0$$

where we have allowed for the (1,2), (3,4), (5,6) bond lengths to be different from the other three. Then, with $\beta = \beta_0 S$ and $\beta' = \beta_0 S'$,

$$\begin{vmatrix} \alpha - E & (\beta_0 - E)S & 0 & 0 & 0 & (\beta_0 - E)S' \\ (\beta_0 - E)S & \alpha - E & (\beta_0 - E)S' & 0 & 0 & 0 \\ 0 & (\beta_0 - E)S' & \alpha - E & (\beta_0 - E)S & 0 & 0 \\ 0 & 0 & (\beta_0 - E)S & \alpha - E & (\beta_0 - E)S' & 0 \\ 0 & 0 & 0 & (\beta - E)S' & \alpha - E & (\beta_0 - E)S \\ (\beta_0 - E)S' & 0 & 0 & 0 & (\beta_0 - E)S & \alpha - E \end{vmatrix} = 0 .$$

Write $w = (\alpha - E)/(\beta_0 - E)$; then

$$\begin{vmatrix} w & S & 0 & 0 & 0 & S' \\ S & w & S' & 0 & 0 & 0 \\ 0 & S' & w & S & 0 & 0 \\ 0 & 0 & S & w & S' & 0 \\ 0 & 0 & 0 & S' & w & S \\ S' & 0 & 0 & 0 & S & w \end{vmatrix} = 0 .$$

(a) When $S' = S$ the determinant has the form of a *circulant* (where successive rows of the determinant differ by cyclic permutation of the elements). A circulant, written $\mathrm{cir}(a_1, a_2, \ldots, a_n)$ and defined as

$$\mathrm{cir}(a_1, a_2, \ldots, a_n) = \begin{vmatrix} a_1 & a_2 & a_3 \cdots & a_n \\ a_n & a_1 & a_2 \cdots & a_{n-1} \\ \cdot & \cdot & \cdot & \cdot \\ \cdot & \cdot & \cdot & \cdot \\ \cdot & \cdot & \cdot & \cdot \\ a_2 & a_3 & a_4 \cdots & a_1 \end{vmatrix}$$

can be expressed as

$$\mathrm{cir}(a_1, a_2, \ldots, a_n) = \prod_{k=1}^{n} \left(a_1 + w_k a_2 + w_k^2 a_3 + \ldots w_k^{n-1} a_n\right); \quad w_k = e^{2\pi i k/n}.$$

This lets us find the roots readily (by setting each factor equal to zero in turn, and solving for $a_1 = w$ in the present case).

In the case of benzene with $S' = S$, $a_1 = w$, $a_2 = S$, $a_3 = a_4 = a_5 = 0$, $a_6 = S$; then

$$\mathrm{cir}(w, S, 0, 0, 0, S) = \prod_{k=1}^{n} \left\{w + S\left(e^{2\pi i k/6} + e^{-2\pi i k/6}\right)\right\}$$

$$= \prod_{k=1}^{n} \left\{w + 2S\cos(\pi k/3)\right\} = 0, \quad k = 1, 2, \ldots 6.$$

It follows that $w = -2S\cos(\pi k/3)$, and as $E = (\alpha - \beta_0 w)/(1-w)$, we find

$$E = \frac{\alpha + 2\beta_0 S\cos(\pi k/3)}{1 + 2S\cos(\pi k/3)} = \frac{\alpha + 2\beta\cos(\pi k/3)}{1 + 2S\cos(\pi k/3)}.$$

Since $\cos(\pi k/3) = \frac{1}{2}, -\frac{1}{2}, -1, -\frac{1}{2}, \frac{1}{2}, 1$ for $k = 1, 2, \ldots 6$, we have

$$E = \frac{\alpha + 2\beta}{1 + 2S}, \frac{\alpha + \beta}{1 + S}, \frac{\alpha + \beta}{1 + S}, \frac{\alpha - \beta}{1 - S}, \frac{\alpha - \beta}{1 - S}, \frac{\alpha - 2\beta}{1 - 2S}.$$

For carbon, $Z^* = 3.25$ [Table 9.1]; for $R = 140$ pm, $s = Z^*R/na_0 = 4.30$ and so $S = 0.244$. Consequently,

$$E = (\alpha + 2\beta)/1.488, \ (\alpha + \beta)/1.244, \ (\alpha + \beta)/1.244,$$

$$(\alpha - \beta)/0.756, \ (\alpha - \beta)/0.756, \ (\alpha - 2\beta)/0.512.$$

The total energy of the ground configuration is therefore

$$E = 2(\alpha + 2\beta)/1.488 + 4(\alpha + \beta)/1.244 = \underline{4.560\alpha + 5.904\beta}.$$

The energy of a single π-bond with $R = 140$ pm (which is artificially long) is obtained from the root of the determinant (which is also a circulant)

$$\begin{vmatrix} \alpha - E & \beta - ES \\ \alpha - ES & \alpha - E \end{vmatrix} = (\alpha - E)^2 - (\beta - ES)^2 = 0,$$

or $\quad E = (\alpha+\beta)/(1+S)$ and $= (\alpha-\beta)/(1-S)$.

Hence

$$E_\pi = 2(\alpha+\beta)/(1+S) = 2(\alpha+\beta)/1.244 = 1.608\,\alpha + 1.608\,\beta\ .$$

The delocalization energy is therefore

$$E_{\mathbf{deloc}} = E - 3E_\pi = (4.560\,\alpha + 5.904\,\beta) - 3(1.608\,\alpha + 1.608\,\beta)$$
$$= \underline{-0.264\,\alpha + 1.080\,\beta}$$

(in place of $E_{\mathbf{deloc}} = \beta$ when overlap is neglected).

(b) When $S' \neq S$ the determinant is no longer a circulant, and an expansion leads to

$$w^6 - 2w^4(S^2 + S'^2) + 3w^2(S^4 + S^2 S'^2 + S'^4) - (S^3 + S'^3)^2 = 0.$$

When $R(\mathrm{C-C}) = 153$ pm, $s = 4.70$ and $S = 0.195$.

When $R(\mathrm{C=C}) = 133$ pm, $s = 4.08$ and $S = 0.275$.

Write $w^2 = u$, then solve

$$u^3 - 0.341u^2 + 3.01 \times 10^{-2}u - 7.96 \times 10^{-4} = 0\ .$$

The roots (obtained numerically) lie at $u = 0.0600$ ($w = \pm\,0.2449$) and at $u = 0.2209$ ($w = \pm\,0.4700$); the former is doubly degenerate. Therefore the orbital energies are

$$E = (\alpha - w\beta_0)/(1-w)$$
$$= \underline{0.68\,\alpha + 0.22\,\beta_0,\ 0.80\,\alpha + 0.20\,\beta_0,\ 1.32\,\alpha - 0.32\,\beta_0,\ 1.89\,\alpha - 0.89\,\beta_0}\ .$$

The total energy of the molecule is therefore

$$E = 2(0.68\,\alpha + 0.22\,\beta_0) + 4(0.80\,\alpha + 0.20\,\beta_0) = 4.56\alpha + 1.24\,\beta_0$$
$$= 4.56\alpha + 4.51\,\beta_0 \qquad [\beta_0 = \beta/S].$$

The delocalization energy is therefore

$$E_{\mathbf{deloc}} = E - 3E_\pi = (4.56\,\alpha + 4.51\,\beta) - 3(1.608\,\alpha + 1.608\,\beta)$$
$$= -\underline{0.26\,\alpha - 0.31\,\beta}\ ,$$

which is positive ('antiaromatic').

EXERCISE: Repeat the calculation for planar cyclo-octatetraene.

10.19 The secular determinant (neglecting overlap) is

$$\Delta = \begin{vmatrix} \alpha-E & \beta & 0 & 0 & \cdots \\ \beta & \alpha-E & \beta & 0 & \cdots \\ 0 & \beta & \alpha-E & \beta & \cdots \\ 0 & 0 & \beta & \alpha-E & \cdots \\ \vdots & \vdots & \vdots & \vdots & \end{vmatrix} = 0\ .$$

N rows and columns

Such *tridiagonal* determinants (which have non-zero entries only on the principal diagonal and its two immediate neighbours) are called *continuants* [see *The theory of determinants*, T. Muir; Dover (1960)].

Their general form is

$$\mathrm{con}(a, a', a'') = \begin{vmatrix} a & a' & 0 & 0 & \cdots \\ a'' & a & a' & 0 & \cdots \\ 0 & a'' & a & a' & \cdots \\ 0 & 0 & a'' & a & \cdots \\ \vdots & \vdots & \vdots & \vdots & \end{vmatrix}_{n\times n}$$

They can be expressed in the form

$$\mathrm{con}\,(a\,,\,a'\,,a'') = \prod_{k=1}^{n}\left\{a - 2(a'a'')^{\frac{1}{2}}\cos\left[k\pi/(n+1)\right]\right\} .$$

The solutions of $\mathrm{con} = 0$ can therefore be picked out at once. In the present case,

$$\begin{aligned}\Delta &= \mathrm{con}\;(\alpha - E\,,\beta\,,\beta) \\ &= \prod_{k=1}^{n}\left\{\alpha - E - 2\beta\cos\left[k\pi/(N+1)\right]\right\} = 0 .\end{aligned}$$

Therefore,

$$E_k = \alpha + 2\beta\cos\left[k\pi/(N+1)\right] , \qquad k = 1, 2, \ldots N .$$

This, with a slight change of notation, is the required result. The energies are plotted in Fig. 10.6(a).

Fig. 10.6 (a) (b)

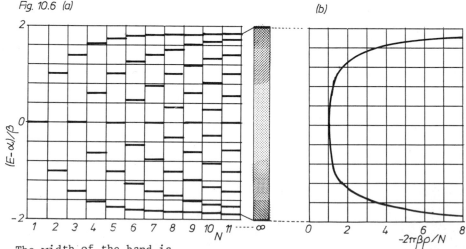

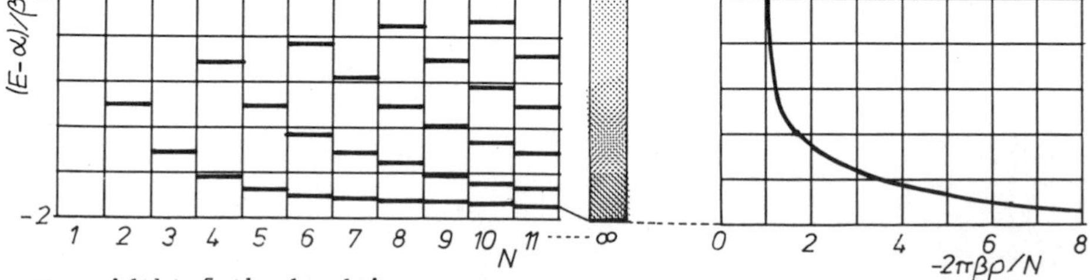

The width of the band is

$$\left|E_N - E_1\right| = \left|2\beta\left\{\cos\left[N\pi/(N+1)\right] - \cos\left[\pi/(N+1)\right]\right\}\right| ,$$

$$\lim_{N\to\infty}\left|E_N - E_1\right| = \left|2\beta\left\{\cos\pi - \cos 0\right\}\right| = 4\beta ,$$

which is finite. In passing, note that for finite N,

$$|E_N - E_1| = 4|\beta| \sin\left\{(N-1)\pi/2(N+1)\right\}.$$

The *density of states* is

$$\rho = \mathrm{d}n/\mathrm{d}E = (\mathrm{d}/\mathrm{d}E)\left\{[(N+1)/\pi] \operatorname{arc\,cos}[(E-\alpha)/2\beta]\right\}$$

$$= -[(N+1)/\pi](1/2\beta)/\left\{1-[(E-\alpha)/2\beta]^2\right\}^{\frac{1}{2}} \qquad [\mathrm{d}\operatorname{arc\,cos} z/\mathrm{d}z = -1/(1-z^2)^{\frac{1}{2}}]$$

$$= -[(N+1)/2\pi\beta]/\left\{1-\cos^2[n\pi/(N+1)]\right\}^{\frac{1}{2}}$$

$$= -[(N+1)/2\pi\beta] \operatorname{cosec}[n\pi/(N+1)].$$

For large N, $\rho \approx -(N/2\pi\beta)\operatorname{cosec}(n\pi/N)$,

and $-2\pi\beta\rho/N$ is plotted against $x = n/N$ in Fig. 10.6(b).

EXERCISE: Find expressions for the energy levels of a linear chain when

(a) overlap is not neglected between nearest neighbours,

(b) there is a bond length alternation along the chain.

10.20 In a tetrahedral environment (symmetry group T_{d}), the d-orbitals span $E(\mathrm{d}_{z^2}, \mathrm{d}_{x^2-y^2})$ and $T_2(\mathrm{d}_{xy}, \mathrm{d}_{xz}, \mathrm{d}_{yz})$. [Refer to the T_{d} character table; $\mathrm{d}_{xy} \propto xy$ etc.]

EXERCISE: Determine which symmetry species are spanned by the f-orbitals in a tetrahedral complex.

10.21 The Clebsch-Gordan series for f^2 $(l = 3)$ is

$$3 \times 3 = 6 + 5 + \ldots + 0,$$

and so $\mathrm{f}^2 \rightarrow$ I, H, G, F, D, P, S. Since the orbitals are equivalent, I must be ^{1}I [Pauli principle], and so the permitted terms are

$$^1\mathrm{I},\ ^3\mathrm{H},\ ^1\mathrm{G},\ ^3\mathrm{F},\ ^1\mathrm{D},\ ^3\mathrm{P},\ ^1\mathrm{S}$$

[either note that terms alternate in general, or else evaluate the symmetrized and antisymmetrized direct products].

For the second part, use eqn (7.14.6):

$$\chi(C_\alpha) = \sin[(L+\tfrac{1}{2})\alpha]/\sin\tfrac{1}{2}\alpha$$

with $\alpha = 0(E)$, $2\pi/3(C_3)$, $\pi(C_2)$, $\pi/2(C_4)$, $\pi(C_2')$ [see *Example* on p. 166]. Draw up the following Table:

Term	E	C_3	C_2	C_4	C_2'	Decomposition
I	13	1	1	−1	1	$A_1 + A_2 + E + T_1 + 2T_2$
H	11	−1	−1	1	−1	$E + 2T_1 + T_2$
G	9	0	1	1	1	$A_1 + E + T_1 + T_2$
F	7	1	−1	−1	−1	$A_2 + T_1 + T_2$
D	5	−1	1	−1	1	$E + T_2$
P	3	0	−1	1	−1	T_1
S	1	1	1	1	1	A_1

For the decompositions use

$$a_l = (1/h)\sum_c g(c)\,\chi^{(l)}(c)\,\chi(c) \qquad \text{[eqn (7.8.5)]}$$

$$= (1/24)\left\{\chi^{(l)}(E)\,\chi(E) + 8\chi^{(l)}(C_3)\,\chi(C_3) + 3\chi^{(l)}(C_2)\,\chi(C_2) + 6\chi^{(l)}(C_4)\,\chi(C_4) + 6\chi^{(l)}(C_2')\,\chi(C_2')\right\}$$

in conjunction with the O character table. The multiplicities carry over. Therefore:

$$^1\mathrm{I} \rightarrow \underline{^1A_1 + {}^1A_2 + {}^1E + {}^1T_1 + 2\,{}^1T_2},$$

$$^3\mathrm{H} \rightarrow \underline{^3E + 2\,{}^3T_1 + {}^3T_2},$$

$$^1\mathrm{G} \rightarrow \underline{^1A_1 + {}^1E + {}^1T_1 + {}^1T_2},$$

$$^3\mathrm{F} \rightarrow \underline{^3A_2 + {}^3T_1 + {}^3T_2},$$

$$^1\mathrm{D} \rightarrow \underline{^1E + {}^1T_2},$$

$$^3\mathrm{P} \rightarrow \underline{^3T_1},$$

$$^1\mathrm{S} \rightarrow \underline{^1A_1}.$$

EXERCISE: What terms does a g^2 configuration give rise to, (a) in a free atom, (b) an octahedral complex?

10.22 Use the direct product tables in Appendix 11. Triplets arise from antisymmetrized products, singlets from symmetrized products.

$e_g^2 \rightarrow A_{1g} + [A_{2g}] + E_g$; hence $^1A_{1g}$, $^3A_{2g}$, ^1E_g

$t_{2g}e_g \rightarrow T_{1g} + T_{2g}$; hence $^1T_{1g}$, $^3T_{1g}$, $^1T_{2g}$, $^3T_{2g}$

$t_{2g}^2 \rightarrow A_{1g} + E_g + [T_{1g}] + T_{2g}$; hence $^1A_{1g}$, ^1E_g, $^3T_{1g}$, $^1T_{2g}$

The d^2 configuration of the free ion gives $^1\mathrm{G} + {}^3\mathrm{F} + {}^1\mathrm{D} + {}^3\mathrm{P} + {}^1\mathrm{S}$. Then, as in Problem 10.21, $\mathrm{G} \rightarrow A_1 + E + T_1 + T_2$, $\mathrm{D} \rightarrow E + T_2$, $\mathrm{S} \rightarrow A_1$, the multiplicites are preserved, and all parities are g. Hence

$$^1\mathrm{G} \rightarrow \underline{^1A_{1g} + {}^1E_g + {}^1T_{1g} + {}^1T_{2g}}$$

$$^1\mathrm{D} \rightarrow \underline{^1E_g + {}^1T_{2g}}$$

$$^1\mathrm{S} \rightarrow \underline{^1A_{1g}}$$

In a tetrahedral complex use

$$\chi(C_\alpha) = \sin\left[(L+\tfrac{1}{2})\,\alpha\right]/\sin\tfrac{1}{2}\alpha \qquad \text{[eqn (7.14.6)]}$$

with $\alpha = 0\,(E)$, $2\pi/3(C_3)$, $\pi(C_2)$. The entries in the following Table can be calculated in this way (or simply taken from the relevant entries in the Table accompanying the solution to Problem 10.21; in the descent in symmetry $T_1 \rightarrow T$, $T_2 \rightarrow T$, $A_1 \rightarrow A$):

	E	C_3	C_2	Decomposition
G	9	0	1	$A+E+2T$
F	7	1	-1	$A+2T$
D	5	-1	1	$E+T$
P	3	0	-1	T
S	1	1	1	A

Therefore

$$^1\mathrm{G} \longrightarrow \underline{^1A + {}^1E + 2\,{}^1T}\ ,$$

$$^3\mathrm{F} \longrightarrow \underline{^3A + 2\,{}^3T}\ ,$$

$$^1\mathrm{D} \longrightarrow \underline{^1E + {}^1T}\ ,$$

$$^3\mathrm{P} \longrightarrow \underline{^3T}\ ,$$

$$^1\mathrm{S} \longrightarrow \underline{^1A}\ .$$

EXERCISE: Repeat the calculation for the various d^2 configurations in a complex of (a) D_{6d} symmetry, (b) D_{3d} symmetry.

10.23 (a) Refer to Fig. 10.7. Under the operations of the group (O), σ_1 becomes

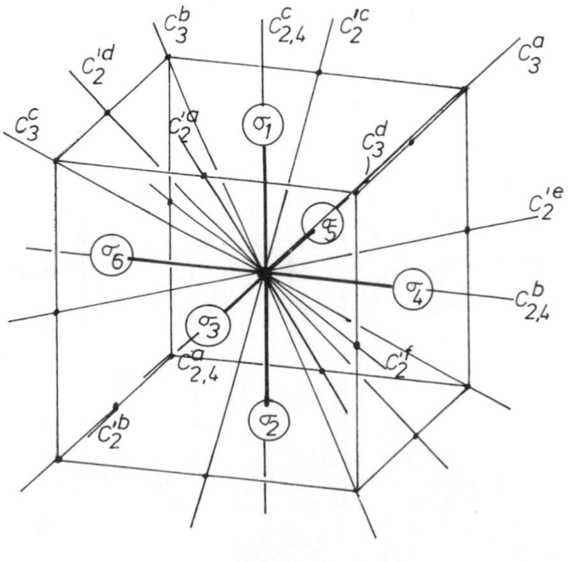

Fig. 10.7

R	E	C_3^{+a}	C_3^{+b}	C_3^{+c}	C_3^{+d}	C_3^{-a}	C_3^{-b}	C_3^{-c}	C_3^{-d}	C_2^{a}	C_2^{b}	C_2^{c}
$R\sigma_1$	σ_1	σ_1	σ_4	σ_5	σ_6	σ_3	σ_5	σ_6	σ_3	σ_2	σ_2	σ_1
R	$C_2'^{a}$	$C_2'^{b}$	$C_2'^{c}$	$C_2'^{d}$	$C_2'^{e}$	$C_2'^{f}$	C_4^{+a}	C_4^{+b}	C_4^{+c}	C_4^{-a}	C_4^{-b}	C_4^{-c}
$R\sigma_1$	σ_3	σ_4	σ_5	σ_6	σ_2	σ_2	σ_6	σ_3	σ_1	σ_4	σ_5	σ_1

The projection operators then generate

$$p_{A_1}\sigma_1 = (1/24)\left\{\sigma_1+\sigma_4+\sigma_5+\sigma_6+\sigma_3+\sigma_5+\ldots+\sigma_1\right\}$$
$$= (1/6)\left\{\sigma_1+\sigma_2+\sigma_3+\sigma_4+\sigma_5+\sigma_6\right\}.$$

$$p_{A_2}\sigma_1 = (1/24)\left\{\sigma_1+\sigma_4+\ldots+\sigma_2+\sigma_1-\sigma_3-\sigma_4-\;\ldots\;-\sigma_5-\sigma_1\right\} = 0$$

$$p_E\sigma_1 = (2/24)\left\{2\sigma_1-\sigma_4-\sigma_5-\sigma_6-\sigma_3-\sigma_5-\sigma_6-\sigma_3-\sigma_4+2\sigma_2+2\sigma_2+2\sigma_1\right\}$$
$$= (1/6)\left\{2\sigma_1+2\sigma_2-(\sigma_3+\sigma_4+\sigma_5+\sigma_6)\right\}.$$

$$p_E\sigma_3 = (1/6)\left\{2\sigma_3+2\sigma_5-(\sigma_1+\sigma_4+\sigma_2+\sigma_6)\right\}.$$

$$p_E\sigma_4 = (1/6)\left\{2\sigma_4+2\sigma_6-(\sigma_1+\sigma_2+\sigma_3+\sigma_5)\right\}.$$

The linear combination

$$\tfrac{1}{2}(p_E\sigma_3-p_E\sigma_4) = \tfrac{1}{4}\{\sigma_3+\sigma_5-\sigma_4-\sigma_6\}$$

has the same symmetry as $d_{x^2-y^2}$ (while $p_E\sigma_1$ has that of d_{z^2}).

$$p_{T_1}\sigma_1 = (3/24)\left\{3\sigma_1-\sigma_2-\sigma_2-\sigma_1-\sigma_3-\sigma_4-\ldots+\sigma_5+\sigma_6\right\} = \tfrac{1}{2}\{\sigma_1-\sigma_2\}.$$

Likewise $p_{T_1}\sigma_3 = \tfrac{1}{2}\{\sigma_3-\sigma_5\}$, $p_{T_1}\sigma_4 = \tfrac{1}{2}\{\sigma_5-\sigma_6\}$.

Similarly $p_{T_2}\sigma_i = 0$ as there are no T_2 combinations.

(b) Refer to Fig. 10.8 (it is a good idea to build a paper model). Under the operations of the group π_1 becomes :

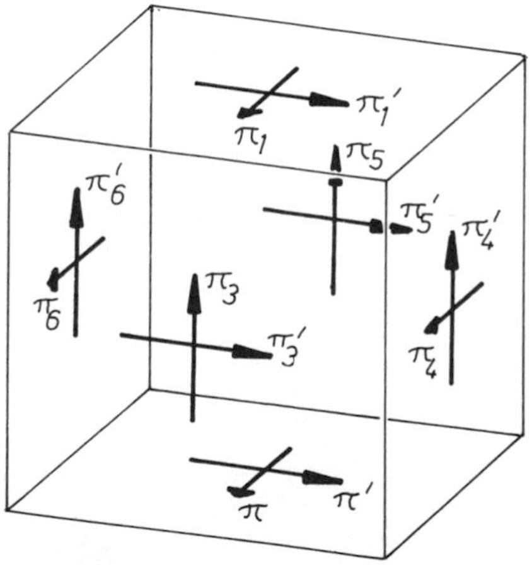

Fig. 10.8

R	E	$C_3^{\mathbf{a}}$	$C_3^{\mathbf{b}}$	$C_3^{\mathbf{c}}$	$C_3^{\mathbf{d}}$	$C_3^{-\mathbf{a}}$	$C_3^{-\mathbf{b}}$	$C_3^{-\mathbf{c}}$	$C_3^{-\mathbf{d}}$	$C_2^{\mathbf{a}}$	$C_2^{\mathbf{b}}$	$C_2^{\mathbf{c}}$
$R\pi_1$	π_1	$-\pi_4'$	π_5'	π_6'	π_3'	$-\pi_5'$	$-\pi_6'$	$-\pi_3'$	π_4'	π_2	$-\pi_2$	$-\pi_1$
R	$C_2'^{\mathbf{a}}$	$C_2'^{\mathbf{b}}$	$C_2'^{\mathbf{c}}$	$C_2'^{\mathbf{d}}$	$C_2'^{\mathbf{e}}$	$C_2'^{\mathbf{f}}$	$C_4^{\mathbf{a}}$	$C_4^{\mathbf{b}}$	$C_4^{\mathbf{c}}$	$C_4^{-\mathbf{a}}$	$C_4^{-\mathbf{b}}$	$C_4^{-\mathbf{c}}$
$R\pi_1$	π_3	$-\pi_4$	$-\pi_5$	$-\pi_6$	$-\pi_2'$	π_2'	π_6	$-\pi_3$	π_1'	π_4	π_5	$-\pi_1'$

The projection operators then generate

$$p_{A_1}\pi_1 = 0$$

$$p_{A_2}\pi_1 = 0$$

$$p_E\pi_1 = (2/24)\left\{2\pi_1+\pi_4'-\pi_5'-\pi_6'-\pi_3'+\pi_5'+\pi_6'+\pi_3'-\pi_4'+2\pi_2-2\pi_2-2\pi_1\right\} = 0$$

$$p_{T_1}\pi_1 = (3/24)\left\{3\pi_1-\pi_2+\pi_2+\pi_1-\pi_3+\pi_4+\pi_5+\pi_6+\pi_2'-\pi_2' + \pi_6-\pi_3+\pi_1'+\pi_4+\pi_5-\pi_1'\right\}$$

$$= (1/6)\left\{2\pi_1-\pi_3+\pi_4+\pi_5+\pi_6\right\}$$

$$p_{T_1}\pi_2 = (1/6)\left\{2\pi_2+\pi_3+\pi_6-\pi_5+\pi_4\right\}$$

[by symmetry; if using a paper model, rotate π_2 into π_1's position and note the relabelling of orbitals].

$$p_{T_1}\pi_3 = (1/6)\left\{2\pi_3-\pi_1+\pi_6'+\pi_2+\pi_4'\right\} \qquad p_{T_1}\pi_5 = (1/6)\left\{2\pi_5+\pi_1+\pi_4'-\pi_2+\pi_6'\right\}$$

$$p_{T_1}\pi_4 = (1/6)\left\{2\pi_4-\pi_3'+\pi_2+\pi_5'+\pi_1\right\} \qquad p_{T_1}\pi_6 = (1/6)\left\{2\pi_6+\pi_3'+\pi_1-\pi_5'+\pi_2\right\}$$

$$p_{T_1}\pi_1' = (1/6)\left\{2\pi_1'-\pi_4'+\pi_5'+\pi_6'+\pi_3'\right\} \qquad p_{T_1}\pi_4' = (1/6)\left\{2\pi_4'-\pi_1'+\pi_3+\pi_2'+\pi_5\right\}$$

$$p_{T_1}\pi_2' = (1/6)\left\{2\pi_2'+\pi_4'+\pi_3'-\pi_6'+\pi_5'\right\} \qquad p_{T_1}\pi_5' = (1/6)\left\{2\pi_5'+\pi_4+\pi_2'-\pi_6+\pi_1'\right\}$$

$$p_{T_1}\pi_3' = (1/6)\left\{2\pi_3'-\pi_4+\pi_1'+\pi_6+\pi_2'\right\} \qquad p_{T_1}\pi_6' = (1/6)\left\{2\pi_6'+\pi_1'+\pi_5-\pi_2'+\pi_3\right\}.$$

We select the following linear combinations (and neglect normalization factors):

$$\left.\begin{aligned} p_{T_1}\pi_1 + p_{T_1}\pi_2 &= \underline{\pi_1+\pi_2+\pi_4+\pi_6} \\ p_{T_1}\pi_1' + p_{T_1}\pi_2' &= \underline{\pi_1'+\pi_2'+\pi_3'+\pi_5'} \\ p_{T_1}\pi_4' + p_{T_1}\pi_6' &= \underline{\pi_3+\pi_4'+\pi_5+\pi_6'} \end{aligned}\right\} \text{ parity u; } T_{\mathbf{1u}}.$$

$$\left.\begin{aligned} p_{T_1}\pi_3 - p_{T_1}\pi_5 &= \underline{\pi_1-\pi_2-\pi_3+\pi_5} \\ p_{T_1}\pi_5' - p_{T_1}\pi_6' &= \underline{\pi_1'-\pi_2'-\pi_4'+\pi_6'} \\ p_{T_1}\pi_4 - p_{T_1}\pi_6 &= \underline{\pi_3'-\pi_4-\pi_5'+\pi_6} \end{aligned}\right\} \text{ parity g; } T_{\mathbf{1g}}.$$

Fig. 10.9

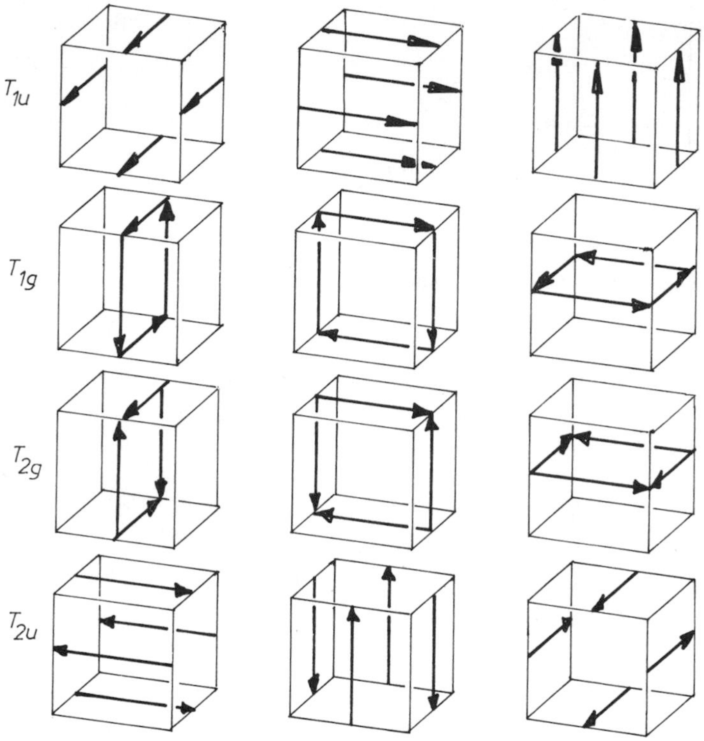

These are illustrated schematically in Fig. 10.9, where the arrows represent pπ-orbitals [these diagrams can be constructed intuitively, and are helpful for establishing the form of the linear combinations as well as for indicating their appearances].

For $2T_2$ ($\longrightarrow T_{2u}+T_{2g}$ in O_h), we have (neglecting normalization)

$$p_{T_2}\pi_1 = 3\pi_1+\pi_1+\pi_3-\pi_4-\pi_5-\pi_6-\pi_6+\pi_3-\pi_4-\pi_5$$
$$= 2\pi_1+\pi_3-\pi_4-\pi_5-\pi_6$$

$$p_{T_2}\pi_2 = 2\pi_2-\pi_3-\pi_6+\pi_5-\pi_4$$

$$p_{T_2}\pi_3 = 2\pi_3+\pi_1-\pi'_6-\pi_2-\pi'_4 \qquad p_{T_2}\pi_5 = 2\pi_5-\pi_1-\pi'_4+\pi_2-\pi'_6$$

$$p_{T_2}\pi_4 = 2\pi_4+\pi'_3-\pi_2-\pi'_5-\pi_1 \qquad p_{T_2}\pi_6 = 2\pi_6-\pi'_3-\pi_1+\pi'_5-\pi_2$$

$$p_{T_2}\pi'_1 = 2\pi'_1+\pi'_4-\pi'_5-\pi'_6-\pi'_3 \qquad p_{T_2}\pi'_4 = 2\pi'_4+\pi'_1-\pi_3-\pi'_2+\pi_5$$

$$p_{T_2}\pi'_2 = 2\pi'_2-\pi'_4-\pi'_3+\pi'_6-\pi'_5 \qquad p_{T_2}\pi'_5 = 2\pi'_5-\pi_4-\pi'_2+\pi_6-\pi'_1$$

$$p_{T_2}\pi'_3 = 2\pi'_3+\pi_4-\pi'_1-\pi_6-\pi'_2 \qquad p_{T_2}\pi'_6 = 2\pi'_6-\pi'_1-\pi_5+\pi'_2-\pi_3 .$$

As before, we select the g,u combinations :

$$\left.\begin{aligned} p_{T_2}\pi_1 - p_{T_2}\pi_2 &= \underline{\pi_1 - \pi_2 + \pi_3 - \pi_5} \\ p_{T_2}\pi'_1 - p_{T_2}\pi'_2 &= \underline{\pi'_1 - \pi'_2 + \pi'_4 - \pi'_6} \\ p_{T_3}\pi_4 - p_{T_2}\pi_6 &= \underline{\pi'_3 + \pi_4 - \pi'_5 - \pi_6} \end{aligned}\right\} \text{ parity g; } T_{2g}$$

$$\left.\begin{aligned} p_{T_2}\pi'_1 + p_{T_2}\pi'_2 &= \underline{\pi'_1 + \pi'_2 - \pi'_3 - \pi'_5} \\ p_{T_2}\pi_3 + p_{T_2}\pi_5 &= \underline{\pi_3 - \pi'_4 + \pi_5 - \pi'_6} \\ p_{T_2}\pi_1 + p_{T_2}\pi_2 &= \underline{\pi_1 + \pi_2 - \pi_4 - \pi_6} \end{aligned}\right\} \text{ parity u; } T_{2u}\,.$$

These linear combinations are also illustrated in Fig. 10.9. The code for relating the orbital numbering used here to that used in the text (p. 282) is as follows :

	x	y	z
Text	3 4 5 6	1 2 3 5	1 2 3 4
Here	1 2 4 6	5 2 1 6	3 4′ 5 6′

EXERCISE: Find the symmetry-adapted linear combinations of s-orbitals at the corners of a trigonal antiprism.

10.24 Once again, it is helpful to have a model of the tetrahedral system labelled with the orbitals. Use the same cube as in Problem 10.23, but labelled as in Fig. 10.10(a). The σ-orbital linear combinations can be constructed as follows. Consider σ_1 ; under the operations of the group (Fig. 10.10(b)) it transforms as follows:

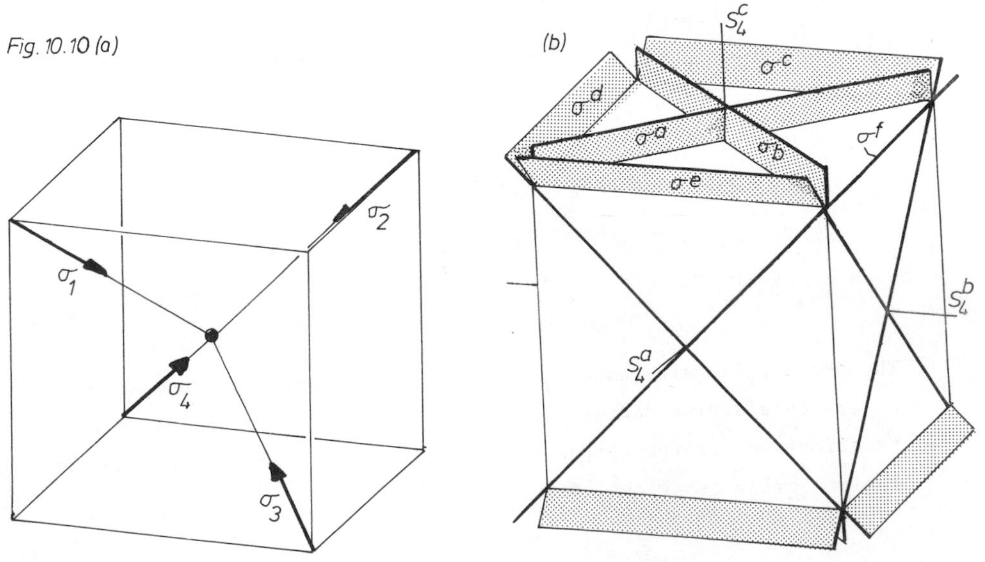

Fig. 10.10 (a) (b)

R	E	C_3^{+a}	C_3^{+b}	C_3^{+c}	C_3^{+d}	C_3^{-a}	C_3^{-b}	C_3^{-c}	C_3^{-d}	C_2^{a}	C_2^{b}	C_2^{c}
$R\sigma_1$	σ_1	σ_3	σ_2	σ_1	σ_3	σ_4	σ_4	σ_1	σ_2	σ_3	σ_4	σ_2
R	σ_d^{a}	σ_d^{b}	σ_d^{c}	σ_d^{d}	σ_d^{e}	σ_d^{f}	S_4^{a}	S_4^{b}	S_4^{c}	S_4^{-a}	S_4^{-b}	S_4^{-c}
$R\sigma_1$	σ_1	σ_2	σ_4	σ_1	σ_1	σ_3	σ_4	σ_3	σ_3	σ_2	σ_2	σ_4

Application of the projection operators to σ_6 then leads to :

$$p_{A_1}\sigma_1 = (1/24)\left\{\sigma_1+\sigma_3+\sigma_2+\ldots+\sigma_3+\sigma_2+\sigma_4\right\}$$
$$= (1/4)\left\{\sigma_1+\sigma_2+\sigma_3+\sigma_4\right\}$$

$$p_{T_2}\sigma_1 = (1/4)\left\{3\sigma_1-\sigma_2-\sigma_3-\sigma_4\right\}$$
$$p_{T_2}\sigma_2 = (1/4)\left\{3\sigma_2-\sigma_3-\sigma_4-\sigma_1\right\} \quad \text{[by symmetry]}$$
$$p_{T_2}\sigma_3 = (1/4)\left\{3\sigma_3-\sigma_4-\sigma_1-\sigma_2\right\}$$
$$p_{T_2}\sigma_4 = (1/4)\left\{3\sigma_4-\sigma_1-\sigma_2-\sigma_3\right\}.$$

Ignoring normalization, we take the following linear combinations (chosen, Fig. 10.11, so as to have the symmetries of p_x, p_y, p_z):

Fig. 10.11

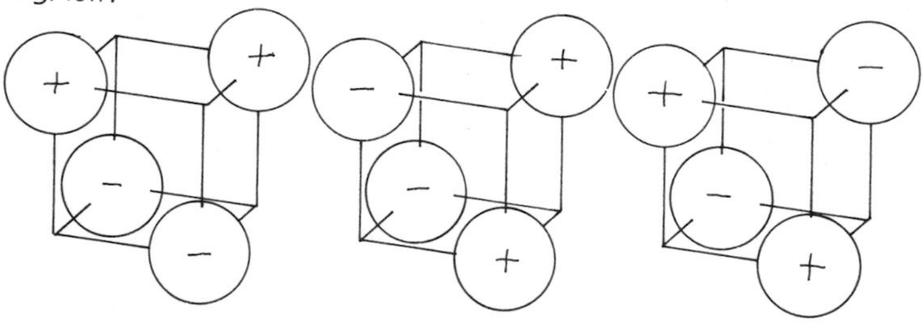

$$\left.\begin{aligned} p_{T_2}\sigma_1+p_{T_2}\sigma_2 &= \sigma_1+\sigma_2-\sigma_3-\sigma_4 \\ p_{T_2}\sigma_2+p_{T_2}\sigma_3 &= \sigma_2+\sigma_3-\sigma_4-\sigma_1 \\ p_{T_2}\sigma_3+p_{T_2}\sigma_4 &= \sigma_1+\sigma_3-\sigma_2-\sigma_4 \end{aligned}\right\} T_2 .$$

The p-orbital basis (not, at this stage, the π-orbital basis, because we are considering cartesian p-orbitals at each corner, Fig. 10.12) transforms under the operations of the group (T) as illustrated by the following behaviour of p_1 :

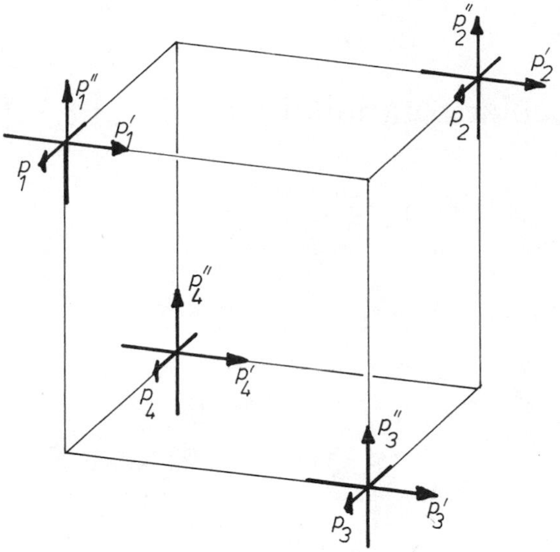

Fig. 10.12

R	E	$C_3^{\mathbf{a}}$	$C_3^{\mathbf{b}}$	$C_3^{\mathbf{c}}$	$C_3^{\mathbf{d}}$	$C_3^{-\mathbf{a}}$	$C_3^{-\mathbf{b}}$	$C_3^{-\mathbf{c}}$	$C_3^{-\mathbf{d}}$	$C_2^{\mathbf{a}}$	$C_2^{\mathbf{b}}$	$C_2^{\mathbf{c}}$
Rp_1	p_1	$-p_3''$	p_2'	p_1''	p_3'	$-p_4'$	$-p_4''$	$-p_1'$	p_2''	p_3	$-p_4$	$-p_2$
R	$\sigma_{\mathbf{d}}^{\mathbf{a}}$	$\sigma_{\mathbf{d}}^{\mathbf{b}}$	$\sigma_{\mathbf{d}}^{\mathbf{c}}$	$\sigma_{\mathbf{d}}^{\mathbf{d}}$	$\sigma_{\mathbf{d}}^{\mathbf{e}}$	$\sigma_{\mathbf{d}}^{\mathbf{f}}$	$S_4^{\mathbf{a}}$	$S_4^{\mathbf{b}}$	$S_4^{\mathbf{c}}$	$S_4^{-\mathbf{a}}$	$S_4^{-\mathbf{b}}$	$S_4^{-\mathbf{c}}$
Rp_1	$-\pi_1'$	π_2'	$-\pi_4''$	π_1	π_1''	π_3	$-\pi_4$	$-\pi_3''$	$-\pi_4'$	$-\pi_2$	π_2''	π_3' .

Since we have taken the p-orbital basis, which spans $A_1 + E + 2T_2$, there will be $A_1 + T_2$ components (corresponding to the σ-basis, as in the first part of the Problem) as well as $E + T_1 + T_2$ components (corresponding to the π-orbitals at each site). We shall construct only the E π-orbital combinations. The projection operator gives

$$p_E\pi_1 = (2/24)\Big\{2p_1 + p_3'' - p_2' - p_1'' - p_3' + p_4' + p_4'' + p_1' - p_2'' + 2p_3 - 2p_4 - 2p_2\Big\}.$$

The remaining $p_E\pi_j$ may be constructed similarly.

EXERCISE: Find the remaining E, T_1, and T_2 symmetry-adapted combinations.

11 Molecular rotational and vibrational transitions

11.1 $\mu(t) = \alpha E(t) + \frac{1}{2}\beta E^2(t)$

$$= \alpha E_0 \cos \omega t + \tfrac{1}{2}\beta E_0^2 \cos^2 \omega t$$

$$= \alpha E_0 \cos \omega t + \tfrac{1}{4}\beta E_0^2\{1+\cos 2\omega t\} .$$

Therefore, $\mu(t)$ has components oscillating at ω (of amplitude αE_0) and at 2ω (of amplitude $\frac{1}{4}\beta E_0^2$), and also possesses a static dipole (of amplitude $\frac{1}{4}\beta E_0^2$).

EXERCISE: Investigate the consequences of a response that is proportional to E^3.

11.2 The centre of mass is where $m_A R_A = m_B R_B$, with $R_A + R_B = R$. The moment of inertia is then

$$I = m_A R_A^2 + m_B R_B^2$$

$$= m_A\{m_B/(m_A + m_B)\}^2 R^2 + m_B\{m_A/(m_A + m_B)\}^2 R^2$$

$$= \{m_A m_B/(m_A + m_B)\}R^2 = \mu R^2 , \quad \mu = m_A m_B/(m_A + m_B) .$$

(a) $^1H_2$: $I = \frac{1}{2} m(^1H)R^2 = \frac{1}{2}\times 1.0078\,u \times (75.09\text{ pm})^2$

$$= \underline{4.718\times 10^{-48}\text{ kg m}^2} \quad [u=1.660\,56\times 10^{-27}\text{ kg, or use } m = m_e + m_p].$$

(b) $^2H_2$: $I = \frac{1}{2} m(^2H)R^2 = \underline{9.429\times 10^{-48}\text{ kg m}^2}$.

(c) $^1H^{35}Cl$: $I = \left\{\dfrac{m(^1H)\,m(^{35}Cl)}{m(^1H) + m(^{35}Cl)}\right\}R^2 = \left\{\dfrac{M_r(^1H)\,M_r(^{35}Cl)}{M_r(^1H) + M_r(^{35}Cl)}\right\}uR^2$

$$= 0.0796\,uR^2 = \underline{2.644\times 10^{-47}\text{ kg m}^2} .$$

(d) $^2H^{35}Cl$: $I = \left\{\dfrac{M_r(^2H)\,M_r(^{35}Cl)}{M_r(^2H) + M_r(^{35}Cl)}\right\}uR^2$

$$= 1.9044\,uR^2 = \underline{5.141\times 10^{-47}\text{ kg m}^2} .$$

EXERCISE: Find the moments of inertia of H_2O about three perpendicular axes.

11.3 The line separation is $\Delta E/hc = 2B$ [following eqn (11.3.1)]. Then, as $B = \hbar/4\pi c I_\perp$ [eqn (11.2.4)], we have

$$\tilde{\nu} = \Delta E/hc = \hbar/2\pi c I_\perp = \hbar/2\pi c \mu R^2 \qquad \text{[Problem 11.2].}$$

Therefore,

$$R = \sqrt{\{\hbar/2\pi c\mu\tilde{\nu}\}}$$

$$= \underline{162\ \text{pm}}\ [\text{when } \mu = 0.999\,86\,u = 1.6603\times10^{-27}\ \text{kg},\ \tilde{\nu} = 12.8\ \text{cm}^{-1}].$$

For $^{2}H^{127}I$, $\mu = 1.9826\,u$. As $\tilde{\nu} \propto 1/\mu$ the line separation will be $0.5037\times(12.8\ \text{cm}^{-1}) = \underline{6.45\ \text{cm}^{-1}}$.

EXERCISE: The bond length of $^{1}H^{2}H$ is 74.13_6 pm. At what wavenumbers would it absorb if its rotational transitions were active ?

11.4 Let $\boldsymbol{R}_{AB}$ be the vector from A to B, $\boldsymbol{R}_{BC}$ that from B to C, and let the centre of mass be at $\boldsymbol{R}$ from B. Its location is given by

$$-m_A(\boldsymbol{R}_{AB}+\boldsymbol{R}) - m_B\boldsymbol{R} + m_C(\boldsymbol{R}_{BC}-\boldsymbol{R}) = 0,$$

or

$$\boldsymbol{R} = (m_C\boldsymbol{R}_{BC} - m_A\boldsymbol{R}_{AB})/m.$$

It follows (since $\boldsymbol{R}_{AB}\cdot\boldsymbol{R}_{BC} = R_{AB}R_{BC}$) that

$$R^2 = \left(m_C^2R_{BC}^2 + m_A^2R_{AB}^2 - 2m_Am_CR_{AB}R_{BC}\right)\Big/m^2.$$

I' around B is given by

$$I' = m_AR_{AB}^2 + m_CR_{BC}^2.$$

Therefore, I around an axis through the centre of mass is given by

$$I = I' - mR^2$$

$$= \underline{(m_Am_C/m)(R_{AB}+R_{BC})^2 + (m_B/m)(m_AR_{AB}^2 + m_CR_{BC}^2)}.$$

The moments of inertia of the molecules can be calculated from

$m(^{16}O) = 15.9949\,u = 2.6560\times10^{-26}$ kg, $m(^{32}S) = 31.9715\,u = 5.3089\times10^{-26}$ kg,

$m(^{12}C) = 12.0000\,u = 1.9926\times10^{-26}$ kg, $m(^{34}S) = 33.9679\,u = 5.6404\times10^{-26}$ kg.

[Nuclide masses from the *Handbook of the American Institute of Physics*, D.E. Gray (ed.), McGraw Hill (1972); see also *Physical Chemistry*.]

Then,

$m(^{16}O\,^{12}C\,^{32}S) = 9.9574\times10^{-26}$ kg, $m(^{16}O\,^{12}C\,^{34}S) = 10.2890\times10^{-26}$ kg.

The moments of inertia can also be established from

$$\nu = 2Bc(J+1) \quad \text{and} \quad B = \hbar/4\pi cI,$$

so that

$$I = (J+1)\hbar/2\pi\nu.$$

This allows us to draw up the following Table:

J	1	2	3	4	mean
$I(^{16}O\,^{12}C\,^{32}S)/10^{-45}$ kg m^2	1.379 95	1.379 96	1.379 96	1.379 97	1.379 96
$I(^{16}O\,^{12}C\,^{34}S)/10^{-45}$ kg m^2	1.414 47	—	1.414 54	—	1.414 50

Comparison of the experimental and calculated values gives

$$1.379\,96\times10^{-45}\,\text{kg m}^2 = (1.4161\times10^{-26}\,\text{kg})(R_{\text{OC}}+R_{\text{CS}})^2 + (5.3150\times10^{-27}\,\text{kg})\,R_{\text{OC}}^2 + (1.0624\times10^{-26}\,\text{kg})\,R_{\text{CS}}^2\,.$$

$$1.414\,50\times10^{-45}\,\text{kg m}^2 = (1.4560\times10^{-26}\,\text{kg})(R_{\text{OC}}+R_{\text{CS}})^2 + (5.1437\times10^{-27}\,\text{kg})\,R_{\text{OC}}^2 + (1.0924\times10^{-26}\,\text{kg})\,R_{\text{CS}}^2\,.$$

These constitute two equations for the two unknown bond lengths. On solution (a tedious process analytically: it is better to proceed numerically by successive approximation) we find $R_{\text{OC}} = 116.3$ pm and $R_{\text{CS}} = 160.0$ pm.

EXERCISE: The B values of $^1H\,^{12}C\,^{14}N$ and $^2H\,^{12}C\,^{14}N$ are respectively 44 315.99 MHz and 36 207.42 MHz. What are the lengths of the bonds ?

11.5 The transition moment is

$$\langle \varepsilon J'M'|\boldsymbol{\mu}|\varepsilon JM\rangle = \langle J'M'|\boldsymbol{\mu}_0|JM\rangle\,.$$

The components of $\boldsymbol{\mu}_0$, the permanent electric dipole moment, are proportional to $\mu_0 Y_{1m}$, where μ_0 is the magnitude of the moment and $m = 0, \pm 1$. Therefore,

$$\langle \varepsilon J'M'|\mu_m|\varepsilon J'M'\rangle \propto \langle J'M'|Y_{1m}|JM\rangle\,.$$

Now proceed using group theory. $Y_{1m}|JM\rangle$ spans irreps of the full rotation group R_3 of symmetry species $\Gamma^1\times\Gamma^J = \Gamma^{J+1}+\Gamma^J+\Gamma^{J-1}$ [Clebsch-Gordan series], and so the integral vanishes unless $J' = J+1$, J, $J-1$ [because $\Gamma^{J'}\times\Gamma^1\times\Gamma^J$ contains Γ^0 only in these cases]. The parity of the integral is $(-1)^{J'+1+J}$, and so $J' = J$ is excluded. Hence $J' = J\pm1$ is allowed. Likewise, $\langle J'M'|Y_{1m}|JM\rangle$ corresponds to integrals with integrands of the form

$$e^{i(M+m-M')\phi}\,;$$

it therefore vanishes unless $M' = m+M$. Consequently, $\Delta M = 0, \pm1$ for $m = 0, \pm1$.

EXERCISE: Establish the rotational Raman selection rules (elements of α are proportional to Y_{2m}).

11.6 Tables of moments of inertia are given in *Physical chemistry* [Table 17.1], and further discussions will be found in *Microwave spectroscopy of gases*, T. M. Sugden and C. N. Kenney, Van Nostrand (1965), and in C. H. Townes and A. L. Schawlow, *Microwave spectroscopy*, McGraw-Hill (1955).

$$I_{\parallel} = 2m_B R^2(1-\cos\theta), \qquad [m_B = M_r(^{35}\text{Cl})u = 34.9688\,u]$$
$$= 2\times(34.9688\times 1.660\,56\times10^{-27}\ \text{kg})\times(204.3\ \text{pm})^2\times(1-\cos 100°\,6')$$
$$= 5.6974\times10^{-45}\ \text{kg m}^2; \qquad A = \hbar/4\pi c I_{\parallel} = 4.9133\times10^{-2}\ \text{cm}^{-1}.$$

$$I_{\perp} = m_B R^2(1-\cos\theta) + (m_A m_B/m)R^2\ (1+2\cos\theta) \quad [m_A = M_r(^{31}\text{P})u = 30.9738\,u]$$
$$= 2.8487\times10^{-45}\ \text{kg m}^2 + 7.4624\times10^{-46}\ \text{kg m}^2$$
$$= 3.5949\times10^{-45}\ \text{kg m}^2; \qquad B = \hbar/4\pi c I_{\perp} = \underline{7.7868\times10^{-2}\ \text{cm}^{-1}}.$$

(a) Pure rotational transitions will therefore be expected at

$$\tilde{\nu} = 2B(J+1), \quad J = 0,1,2,\ldots \qquad [\text{eqn (11.3.1)}]$$
$$\tilde{\nu}/\text{cm}^{-1} = \underline{0.1557,\ 0.3115,\ 0.4672,\ 0.6229,\ \ldots}\ .$$

(b) $\Delta J = \pm 2$ rotational Raman transitions are predicted to lie at

$$(\tilde{\nu}-\tilde{\nu}_0) = 4B(J-\tfrac{1}{2}), \quad J = 2,3,\ldots \quad [\text{anti-Stokes, eqn (11.3.2a)}]$$
$$(\tilde{\nu}-\tilde{\nu}_0) = -4B(J+\tfrac{3}{2}), \quad J = 0,1,\ldots \quad [\text{Stokes, eqn (11.3.2a)}],$$

the numerical values being

$$(\tilde{\nu}-\tilde{\nu}_0)/\text{cm}^{-1} = \underline{0.4672,\ 0.7787,\ 1.0902,\ 1.4016,\ \ldots} \quad [\text{anti-Stokes}]$$
$$(\tilde{\nu}-\tilde{\nu}_0)/\text{cm}^{-1} = \underline{-0.4672,\ -0.7787,\ -1.0902,\ -1.4016,\ \ldots} \quad [\text{Stokes}].$$

For $K=0 \nleftrightarrow K=0$ we also expect $\Delta J = \pm 1$ rotational Raman lines. These are predicted to lie at

$$(\tilde{\nu}-\tilde{\nu}_0)/\text{cm}^{-1} = \underline{\pm 0.1557,\ \pm 0.3115,\ \pm 0.4672,\ \pm 0.6229,\ \ldots}\ .$$

The relative populations of the $|JKM_J\rangle$ states are given by the Boltzmann factor, which in this case is

$$b(J,K) = (2J+1)\exp\left\{-[BJ(J+1)+(A-B)K^2](hc/kT)\right\}.$$

[The factor $(2J+1)$ arises from the $(2J+1)$-fold M_J degeneracy.] Taking $T = 298.15$ K (corresponding to 25 °C) results in

$$kT/hc = 207.223\ \text{cm}^{-1}.$$

Then

$$b(J,K) = (2J+1)\exp\left\{-3.758\,J(J+1)\times10^{-4} + 1.387\,K^2\times10^{-4}\right\}.$$

Many states are occupied at 25 °C. The relative populations of some of them are listed in the following Table:

J	K											$\sum_K b(J,K)$
	...	−4	−3	−2	−1	0	1	2	3	4	...	
0						1.000						1.000
1					2.998	2.998	2.998					8.994
2				4.992	4.989	4.989	4.989	4.992				24.951
3			6.977	6.972	6.969	6.969	6.969	6.972	6.977			48.807
4		8.952	8.944	8.938	8.934	8.933	8.934	8.938	8.944	8.952		80.468
5					...	10.877	...					119.81
⋮						⋮						⋮
40					...	43.73	...					3830
50					...	38.74	...					4427
60					...	30.58	...					4435
70					...	21.78	...					3950
80					...	14.10	...					3181
90					...	8.34	...					2342
100					...	4.52	...					1590

The populations (summed over all K for a given J) go through a maximum at $\underline{J \approx 60}$, and so the spectra will peak in that region.

EXERCISE: The bond angle of PH_3 is 93° 27′ and the bond length is 142.1 pm. Predict the form of its rotational and rotational Raman spectrum in a low temperature vapour at 77 K.

11.7 $B(^1H^{35}Cl) = \hbar/4\pi c I(^1H^{35}Cl)$ $[I = 2.644 \times 10^{-47}\ \text{kg m}^2$, Problem 11.2]

$= \underline{10.59\ \text{cm}^{-1}}$.

The relative populations are given by

$$b(J) = (2J+1)\exp\left\{-hcBJ(J+1)/kT\right\}$$

$$= (2J+1)\exp\left\{-0.050\,79\,J(J+1)\right\} \quad [kT/hc = 208.51\ \text{cm}^{-1}\ \text{at}\ 300\ \text{K}].$$

Draw up the following Table (b' is discussed below):

J	0	1	2	3	4	5	6	7
$b(J)$	1.000	2.710	3.687	3.805	3.259	2.397	1.540	0.873
$b'(J)$	1.000	1.807	2.212	2.175	1.811	1.307	0.829	0.465

J	8	9	10	11	12	13	14	15
$b(J)$	0.439	0.197	0.079	0.028	0.009	0.003	0.001	10^{-4}
$b'(J)$	0.232	0.103	0.041	0.015	0.005	0.001	10^{-4}	

($\sum_J b(J) = 20.026$, the rotational partition function at 300 K.) If the intensities were determined solely by the populations they would be proportional to $b(J)$, and the most intense transition would be $4 \leftarrow 3$.

If we take the J-dependence of the transition moment into account we should use

$$b'(J) = \{(J+1)/(2J+1)\}\, b(J) = (J+1)\exp\{-0.5079\, J(J+1)\},$$

which gives the entries in the Table above. The transition of maximum intensity is $3 \leftarrow 2$.

EXERCISE: Where does the maximum transition occur when $T = 500$ K? Find an analytical expression for $J_{\text{most intense}}$ at an arbitrary temperature T.

11.8 $^1H{-}C{\equiv}C{-}^1H$ behave in the same way as $^1H{-}^1H$ [the $^{12}C{\equiv}^{12}C$ component of the molecule acts simply as a link]. Therefore, the same argument applies, and we can have *ortho*- and *para*-ethyne. For the triplet of nuclear states (proton spins parallel) only odd J values are allowed; for the singlet nuclear state (proton spins antiparallel) only even J values are allowed. [Here, as elsewhere, we are supposing the molecule to be in its vibrational ground state.]

(a) Replacement of one ^{12}C by a ^{13}C destroys the symmetry of the molecule; hence all rotational states become allowed whatever the relative proton spin orientations.

(b) Replacement of both ^{12}C by ^{13}C gives rise to a double two-fermion system. The parity of the rotational states is $(-1)^J$. Following the argument in Section 11.4 shows that two nuclear state permutations must be performed: we require the parity under $p_{\text{nuc}}(^1H)$ and $p_{\text{nuc}}(^{13}C)$. Then draw up the following Table:

$2\,^1H$ states	$2\,^{13}C$ states	Overall spin parity	Allowed J †
Triplet (+)	Triplet (+)	+1	Even
Triplet (+)	Singlet (−)	−1	Odd
Singlet (−)	Triplet (+)	−1	Odd
Singlet (−)	Singlet (−)	+1	Even

† **Overall, the rotation corresponds to the relabelling of the *two* pairs of fermions. Therefore, according to the Pauli principle, the overall wavefunction must not change sign: $(-1)^2 = +1$. Therefore $9 = 1 + 10$ nuclear states are associated with J even, and $3 + 3 = 6$ with J odd. Alternate J levels have statistical weights in the ratio $10:6 = 5:3$.**

EXERCISE; Consider the three rotations of ethene, and assess which rotational levels are permissible for the various isotopic possibilities.

11.9 Consider a molecule free to undergo motion in only one dimension (parallel to the internuclear axis). Then

$$H = -(\hbar^2/2m_A)(\partial^2/\partial x_A^2) - (\hbar^2/2m_B)(\partial^2/\partial x_B^2) + V(x_A - x_B)\ .$$

This has the form of eqn (A6.3); the argument used in Appendix 6 shows that

$$H = -(\hbar^2/2m)(\partial^2/\partial X^2) - (\hbar^2/2m_B)(\partial^2/\partial x^2) + V(x) \qquad \text{[eqn (A6.4)]}.$$

with $1/\mu = 1/m_A + 1/m_B$ [eqn (A6.6)]. On writing $\psi = \Psi(X)\,\psi(x)$, the separation of variables argument leads to

$$\left.\begin{aligned} -(\hbar^2/2m)(d^2/dX^2)\,\Psi &= E_{trans}\,\Psi \\ -(\hbar^2/2\mu)(d^2/dx^2)\,\psi + V(x)\,\psi &= E_{vib}\,\psi \end{aligned}\right\} \quad E = E_{trans} + E_{vib}\,,$$

as was to be demonstrated.

The following Table lists values of μ/u for the species; in order to obtain absolute values use $u = 1.660\,56 \times 10^{-27}$ kg. Values are obtained by forming $\mu/u = M_{r,A}M_{r,B}/(M_{r,A}+M_{r,B})$; additional data will be found in Problems 11.2 and 11.3. [Tables of nuclide masses are given in the *Handbook of the American Institute of Physics*, D. E. Gray (ed.), McGraw-Hill (1972).] Values of k are obtained from $k = \mu\omega^2 = 4\pi^2\mu c^2\tilde{\nu}^2$ [$\omega = 2\pi\nu$, $\nu = c\tilde{\nu}$].

	$^1H_2$	$^1H^{19}F$	$^1H^{35}Cl$	$^1H^{81}Br$	$^1H^{127}I$
μ/u	0.5039	0.9570	0.9796	0.9954	0.9999
$k/\text{N m}^{-1}$	574.9	965.7	516.3	411.5	313.8

For the effect of replacing ^1H by ^2H, assume that k remains constant, and so form $\tilde{\nu} = (1/2\pi c)(k/\mu^*)^{\frac{1}{2}}$, with μ^* the reduced mass of the deuterated species. Draw up the following Table:

	$^2H^1H$	$^2H^{19}F$	$^2H^{35}Cl$	$^2H^{81}Br$	$^2H^{127}I$
μ^*/u	0.6167	1.8210	1.9044	1.9652	1.9826
$\tilde{\nu}/\text{cm}^{-1}$	3811	3000	2145	1885	1639

EXERCISE: In three dimensions the motion of the centre of mass separates from the internal motion, but rotations and vibrations separate only approximately. Demonstrate these features.

11.10 The result of the calculation in Problem 10.2 is

$$E_+ = E_{1s} + (j_0/a_0)\,f(s)\,,$$

$$f(s) = \left\{\frac{(1-\frac{2}{3}s^2) + (1+s)\,e^{-s}}{1 + (1+s+\frac{1}{3}s^2)\,e^{-s}}\right\}\left(\frac{e^{-s}}{s}\right) \qquad s = R/a_0$$

The minimum is at $R \approx 130$ pm $= R_e$. We require $(d^2f/dR^2)_{R=R_e} = (1/a_0^2)(d^2f/ds^2)_{R_e}$.

$$d^2f/ds^2 = \left(\frac{\frac{1}{3}(8s-2s^2-1)\,e^{-s}+4s\,e^{-2s}}{[1+(1+s+\frac{1}{3}s^2)\,e^{-s}]\,s}\right) - \left(\frac{(\frac{2}{3}s^2-\frac{4}{3}s-1)\,e^{-s}-(1+2s)\,e^{-2s}}{[1+(1+s+\frac{1}{2}s^2)\,e^{-s}]\,s^2}\right)$$

$$+\left(\frac{[(\frac{2}{3}s^2-\frac{4}{3}s-1)\,e^{-s}-(1+2s)\,e^{-2s}](1+s)\,e^{-s}}{3[1+(1+s+\frac{1}{3}s^2)\,e^{-s}]^2}\right)$$

$$+\quad f/s^2 - f'/s + \tfrac{1}{3}f'\,s(s+1)\,e^{-s}$$

$$+\left(\frac{(1+s-s^2)\,f\,e^{-s}}{3[1+(1+s+\frac{1}{3}s^2)\,e^{-s}]}\right) + \left(\frac{s^2(s+1)^2\,f\,s^{-2s}}{9[1+(1+s+\frac{1}{2}s^2)\,e^{-s}]^2}\right).$$

At $R = R_e$ the energy is at a maximum, and so $f' = 0$; hence the terms in f' may be dropped. At $R = R_e = 130$ pm, $s = 2.46$ and $f = -0.0648$, $f'' = 0.0656$. It follows that

$$k = (j_0/a_0)(d^2f/dR^2)_{R_e} = (j_0/a_0^3)\,f''$$

$$= (1556.92\ \mathrm{N\,m^{-1}})\,f'' = \underline{102\ \mathrm{N\,m^{-1}}}.$$

EXERCISE: Find an expression for the force constant of a general one-electron molecule constructed with nuclei of atomic number Z.

11.11 The transition moment is the value of the matrix element

$$-ex_{v'v} = -e\int \psi_{v'}^{*}\,x\,\psi_v\,d\tau\,.$$

The wavefunctions $\psi_v = N_v H_v(y)\,e^{-y^2/2}$ are normalized in the sense that

$$\int_{-\infty}^{\infty}\psi_v^2\,dy = 1\,, \qquad y = (m\omega/\hbar)^{\frac{1}{2}}\,x\,.$$

Therefore, in order to employ these functions, write

$$-ex_{v'v} = -e\int_{-\infty}^{\infty}\psi_{v'}(y)\,x\,\psi_v(y)\,dy$$

$$= -e\,N_v N_{v'}\,(\hbar/m\omega)^{\frac{1}{2}}\int_{-\infty}^{\infty} e^{-y^2} H_{v'}(y)\,y\,H_v(y)\,dy$$

$$-ex_{v'v} = -eN_v N_{v'}(\hbar/m\omega)^{\frac{1}{2}}$$

$$\times\left\{v\int_{-\infty}^{\infty} e^{-y^2} H_{v'}(y)H_{v-1}(y)\,dy + \tfrac{1}{2}\int_{-\infty}^{\infty} e^{-y^2} H_{v'}(y)H_{v+1}(y)\,dy\right\}.$$

The first integral vanishes unless $v' = v-1$ [orthogonality, Table 3.1]; the second vanishes unless $v' = v+1$. Therefore, the only non-vanishing transition elements are

$$\mu_{v-1,v} = -eN_{v-1}N_v(\hbar/m\omega)^{\frac{1}{2}}\, v\int_{-\infty}^{\infty} e^{-y^2} H^2_{v-1}(y)\,dy$$

$$= -e(N_v/N_{v-1})(\hbar/m\omega)^{\frac{1}{2}}\, v = \underline{-e(\hbar/2m\omega)^{\frac{1}{2}}\sqrt{v}}\,,$$

$$\mu_{v+1,v} = -\tfrac{1}{2}eN_{v+1}N_v(\hbar/m\omega)^{\frac{1}{2}}\int_{-\infty}^{\infty} e^{-y^2} H^2_{v+1}(y)\,dy$$

$$= -\tfrac{1}{2}e(N_v/N_{v+1})(\hbar/m\omega)^{\frac{1}{2}} = \underline{-e(\hbar/2m\omega)^{\frac{1}{2}}\sqrt{(v+1)}}\,.$$

[For the integrals, use

$$N_v^2\int_{-\infty}^{\infty} e^{-y^2} H_v^2(y)\,dy = 1\,,$$

the normalization condition.]

EXERCISE: Evaluate the matrix elements of x^2.

11.12 $\tilde{\nu}^{P} = \tilde{\nu} - 2(v+1)x\tilde{\nu} + \ldots - (B_{v+1}+B_v)J + (B_{v+1}-B_v)J^2 + \ldots$ [eqn (11.6.4)],

$\tilde{\nu}^{Q} = \tilde{\nu} + (B_{v+1}-B_v)J + (B_{v+1}-B_v)J^2 + \ldots$ [eqn (11.6.6)],

$\tilde{\nu}^{R} = \tilde{\nu} - 2(v+1)x\tilde{\nu} + \ldots + 2B_{v+1} + (3B_{v+1}-B_v)J + (B_{v+1}-B_v)J^2 + \ldots$ [eqn (11.6.2)].

If x is ignored these equations become

$$\tilde{\nu}^{P} - \tilde{\nu} = -(B_{v+1}+B_v)J + (B_{v+1}-B_v)J^2 + \ldots,$$

$$\tilde{\nu}^{Q} - \tilde{\nu} = (B_{v+1}-B_v)J + (B_{v+1}-B_v)J^2 + \ldots,$$

$$\tilde{\nu}^{R} - \tilde{\nu} = 2B_{v+1} + (3B_{v+1}-B_v)J + (B_{v+1}-B_v)J^2 + \ldots .$$

Then, with $B_0 = 10.4400\ \mathrm{cm^{-1}}$ and $B_1 = 10.1366\ \mathrm{cm^{-1}}$,

$$(\tilde{\nu}^{P} - \tilde{\nu})/\mathrm{cm^{-1}} = -20.5766\,J - 0.3034\,J^2\,,$$

$$(\tilde{\nu}^{Q} - \tilde{\nu})/\mathrm{cm^{-1}} = -0.3034\,J(J+1)\,,$$

$$(\tilde{\nu}^{R} - \tilde{\nu})/\mathrm{cm^{-1}} = 20.2732 + 50.3796\,J - 0.3034\,J^2.$$

The wavenumbers of the branches are plotted in Fig. 11.1.

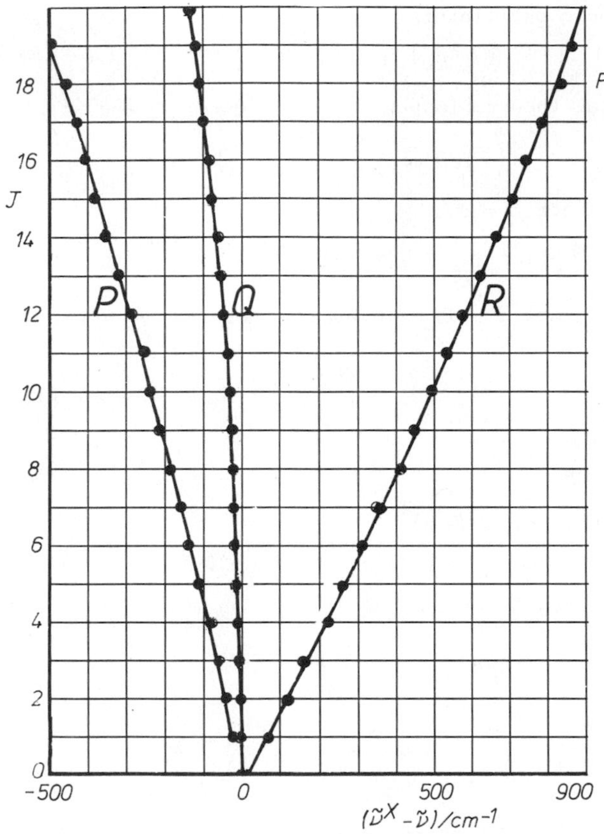

Fig. 11.1

EXERCISE: Find the location of the lines of the O- and S-branches of the Raman spectrum.

11.13 $P(R) = P$ for $R_e - \frac{1}{2}\delta R \leqslant R \leqslant R_e + \frac{1}{2}\delta R$, 0 elsewhere.

$$\int_{R_e - \frac{1}{2}\delta R}^{R_e + \frac{1}{2}\delta R} P \, dR = P\delta R = 1; \text{ therefore, } P = 1/\delta R.$$

The expectation value of $1/R^2$ is therefore

$$\langle 1/R^2 \rangle = \int_{-\infty}^{\infty} (1/R^2)\, P(R)\, dR = (1/\delta R) \int_{R_e - \frac{1}{2}\delta R}^{R_e + \frac{1}{2}\delta R} (1/R^2)\, dR$$

$$= \left\{ 1/(R_e - \tfrac{1}{2}\delta R)(R_e + \tfrac{1}{2}\delta R) \right\}$$

$$= \left\{ 1/R_e^2[1 - (\delta R/2R_e)^2] \right\} \approx \underline{(1/R_e^2)\{1 + (\delta R/2R_e)^2\}} \text{ for } \delta R^2 \ll R_e^2.$$

For a harmonic oscillator,

$$\langle (R-R_e)^2 \rangle = (2/k)\langle V \rangle \qquad [V = \tfrac{1}{2}k(R-R_e)^2].$$

By the virial theorem [Appendix 5], $\langle V \rangle = \langle T \rangle$, and so

$$\langle V \rangle_v = \tfrac{1}{2}E_v = \tfrac{1}{2}\hbar\omega(v+\tfrac{1}{2}).$$

Therefore, since $\omega = (k/\mu)^{\frac{1}{2}}$,

$$\langle (R-R_e)^2 \rangle = (\hbar^2/k\mu)^{\frac{1}{2}}(v+\tfrac{1}{2}).$$

On the basis that $P = 1/\delta R$ over the range $R_e - \frac{1}{2}\delta R \leqslant R \leqslant R_e + \frac{1}{2}\delta R$,

$$\langle (R-R_e)^2 \rangle \approx (1/\delta R)\int_{R_e-\frac{1}{2}\delta R}^{R_e+\frac{1}{2}\delta R}(R-R_e)^2\,\mathrm{d}R = (1/12)\,\delta R^2.$$

It follows that

$$\delta R^2 \approx 12(v+\tfrac{1}{2})(\hbar^2/k\mu)^{\frac{1}{2}}.$$

Consequently, $\langle 1/R^2 \rangle \approx (1/R_e^2)\left\{1+3(v+\tfrac{1}{2})(\hbar^2/k\mu)^{\frac{1}{2}}/R_e^2\right\}$

$$B_v \approx (\hbar/4\pi c\mu R_e^2) + (3\hbar/4\pi c\mu)(\hbar^2/k\mu)^{\frac{1}{2}}(v+\tfrac{1}{2})(1/R_e^4)$$

$$\approx B_e + (12\pi c B_e^2/\omega)(v+\tfrac{1}{2}) \qquad [B_e = \hbar/4\pi c\mu R_e^2]$$

$$\approx B_e + (6B_e^2/\tilde{\nu})(v+\tfrac{1}{2}) \qquad [\omega^2 = \mu/k, \quad \tilde{\nu} = \omega/2\pi c]$$

$$\underline{B_v \approx (1+\gamma_v)B_e}, \qquad \underline{\gamma_v = 6(v+\tfrac{1}{2})B_e/\tilde{\nu}}.$$

EXERCISE: Estimate $\langle 1/R^3 \rangle$ on the same basis.

11.14 $\omega_i = \sqrt{\kappa_i}$ [eqn (11.7.12)]

$$\kappa_1 = k/m(\mathrm{O}), \quad \kappa_2 = km(\mathrm{CO_2})/m(\mathrm{C})\,m(\mathrm{O}) \qquad \text{[eqn (11.7.8)]},$$

where Q_1 and Q_2 are the symmetric and antisymmetric stretches. Then use

$$m(^{16}\mathrm{O}) = 15.9949\,u = 2.655\,60 \times 10^{-26}\ \mathrm{kg},$$

$$m(^{12}\mathrm{C}) = 12.0000\,u = 1.992\,67 \times 10^{-26}\ \mathrm{kg},$$

$$m(\mathrm{CO_2}) = 43.9898\,u = 7.304\,77 \times 10^{-26}\ \mathrm{kg}.$$

Then from $\omega_i = 2\pi c\tilde{\nu}_i$,

$$\left.\begin{aligned} Q_1 &: \tilde{\nu}_1 = 1300\ \mathrm{cm^{-1}}, \quad k = 1592\ \mathrm{N\ m^{-1}} \\ Q_2 &: \tilde{\nu}_2 = 2349\ \mathrm{cm^{-1}}, \quad k = 1418\ \mathrm{N\ m^{-1}} \end{aligned}\right\} \text{mean} \quad \underline{k = 1505\ \mathrm{N\ m^{-1}}}.$$

EXERCISE: The three fundamental vibrational wavenumbers of CS_2 are 657.98 $\mathrm{cm^{-1}}$, 396.7 $\mathrm{cm^{-1}}$, and 1532.5 $\mathrm{cm^{-1}}$. What is the stretching force constant?

11.15 Consider the system depicted in Fig. 11.2. Under the operations of the group (C_{2v}) they transform as follows :

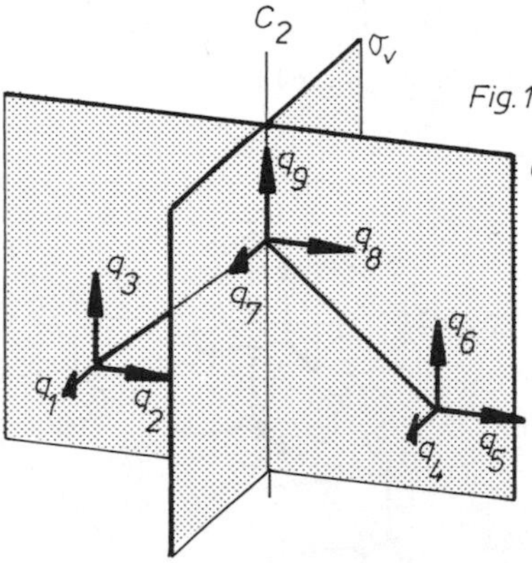

Fig. 11.2

	q_1	q_2	q_3	q_4	q_5	q_6	q_7	q_8	q_9	χ
E	q_1	q_2	q_3	q_4	q_5	q_6	q_7	q_8	q_9	9
C_2	$-q_4$	$-q_5$	q_6	$-q_1$	$-q_2$	q_3	$-q_7$	$-q_8$	q_9	-1
σ_v	q_4	$-q_5$	q_6	q_1	$-q_2$	q_3	q_7	$-q_8$	q_9	1
σ'_v	$-q_1$	q_2	q_3	$-q_4$	q_5	q_6	$-q_7$	q_8	q_9	3

[The characters, final column, is given by the net number of displacements left unchanged by the operation.] The representation decomposes as follows [eqn (7.8.4)] :

$$a(A_1) = \tfrac{1}{4}\{9-1+1+3\} = 3\,, \quad a(A_2) = \tfrac{1}{4}\{9-1-1-3\} = 1\,,$$

$$a(B_1) = \tfrac{1}{4}\{9+1+1-3\} = 2\,, \quad a(B_2) = \tfrac{1}{4}\{9+1-1+3\} = 3\,.$$

That is, the basis spans $3A_1 + A_2 + 2B_1 + 3B_2$. From the C_{2v} character table (Appendix 10), translations span $B_2 + B_2 + A_1$ (for x, y, z respectively) and rotations span $B_2 + B_1 + A_2$ (for R_x, R_y, R_z respectively) Consequently the vibrations span $\underline{2A_1 + B_2}$.

(a) Infrared-active transitions are those of the same symmetry species as the electric dipole moment, which spans $B_1 + B_2 + A_1$ (for μ_x, μ_y, μ_z respectively). Therefore, <u>all three modes are infrared active</u> (A_1 is z-polarized, B_2 y-polarized, and B_1 x-polarized).

(b) Raman-active transitions are those of the same symmetry species as the polarizability, which transforms as the quadratic forms x^2, xy, etc. These span $2A_1 + A_2 + B_1 + B_2$ [character table, Appendix 10], and so <u>all three modes are Raman active</u>.

EXERCISE: Establish the symmetry species and activities of the vibrations of H_2O_2.

11.16 The eighteen displacements of the D_{2h} ethene molecules are shown in Fig. 11.3. They transform as follows :

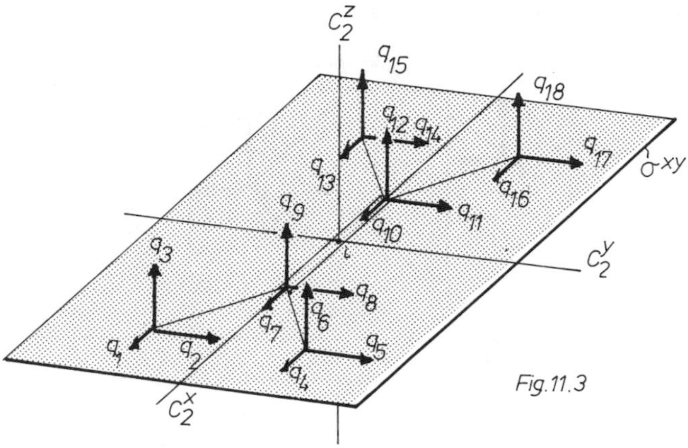

Fig.11.3

	1	2	3	4	5	6	7	8	9	10	11	12	13	14	15	16	17	18	χ
E	1	2	3	4	5	6	7	8	9	10	11	12	13	14	15	16	17	18	18
$C_2(z)$	-16	-17	18	-13	-14	15	-10	-11	12	-7	-8	9	-4	-5	6	-1	-2	3	0
$C_2(y)$	-13	14	-15	-16	17	-18	-10	11	-12	-7	8	-9	-1	2	-3	-4	5	-15	0
$C_2(x)$	4	-5	-6	1	-2	-3	7	-8	-9	10	-11	-12	16	-17	-18	13	-14	-15	-2
i	-16	-17	-18	-13	-14	-15	-10	-11	-12	-7	-8	-9	-4	-5	-6	-1	-2	-3	0
$\sigma(xy)$	1	2	-3	4	5	-6	6	8	-9	10	11	-12	13	14	-15	16	17	-18	6
$\sigma(xz)$	4	-5	6	1	-2	3	7	-8	9	10	-11	12	11	-17	18	13	-14	15	2
$\sigma(yz)$	-13	14	15	-16	17	18	-10	11	12	-7	8	9	-1	2	3	-4	5	6	0

The D_{2h} character table is as follows:

D_{2h}	E	$C_2(z)$	$C_2(y)$	$C_2(x)$	i	$\sigma(xy)$	$\sigma(xz)$	$\sigma(yz)$	$h = 8$
A_g	1	1	1	1	1	1	1	1	x^2, y^2, z^2
B_{1g}	1	1	-1	-1	1	1	-1	-1	R_z, xy
B_{2g}	1	-1	1	-1	1	-1	1	-1	R_y, xz
B_{3g}	1	-1	-1	1	1	-1	-1	1	R_x, yz
A_u	1	1	1	1	-1	-1	-1	-1	
B_{1u}	1	1	-1	-1	-1	-1	1	1	z
B_{2u}	1	-1	1	-1	-1	1	-1	1	y
B_{3u}	1	-1	-1	1	-1	1	1	-1	x

[For similar tables, see *Tables for group theory*, P. W. Atkins, M. S. Child, and C.S.G. Phillips, Oxford University Press (1970).]
The decomposition [eqn (7.8.4)] is

$$a(A_g) = (1/8)\{18-2+6+2\} = 3\,, \quad a(B_{1g}) = (1/8)\{18+2+6-2\} = 3\,,$$
$$a(B_{2g}) = (1/8)\{18+2-6+2\} = 2\,, \quad a(B_{3g}) = (1/8)\{18-2-6-2\} = 1\,,$$
$$a(A_u) = (1/8)\{18-2-6-2\} = 1\,, \quad a(B_{1u}) = (1/8)\{18+2-6+2\} = 2\,,$$
$$a(B_{2u}) = (1/8)\{18+2+6-2\} = 3\,, \quad a(B_{3u}) = (1/8)\{18-2+6+2\} = 3\,.$$

Consequently, the basis spans

$$3A_g + 3B_{1g} + 2B_{2g} + B_{3g} + A_u + 2B_{1u} + 3B_{2u} + 3B_{3u}\,.$$

Translations (x,y,z) span $B_{1u}+B_{2u}+B_{3u}$ and rotations (R_x, R_y, R_z) span $B_{1g}+B_{2g}+B_{3g}$. Therefore, vibrations span

$$\underline{3A_g + 2B_{1g} + B_{2g} + A_u + B_{1u} + 2B_{2u} + 2B_{3u}}\,.$$

Infrared-active vibrations are $\underline{B_{1u}, B_{2u}, B_{3u}}$ while Raman active vibrations are $\underline{A_g, B_{1g}, B_{2g}}$. Note that in this case, each vibration except A_u is either infrared or Raman active; A_u is inactive to both types of process.

EXERCISE: Establish the symmetry species and the activities of the benzene molecule.

11.17 The harmonic oscillator wavefunctions are proportional to $H_0 \propto 1$, $H_1 \propto x$ for oscillations in the x-direction and to $H_0 \propto 1$, $H_1 \propto y$ for oscillations in the y-direction. Therefore, the linear combinations $H_0(y)\,H_1(x) \pm iH_1(y)\,H_0(x)$ of the singly-excited degenerate states are proportional to $x \pm iy \propto e^{\pm i\phi}$, which are eigenfunctions of l_z, the angular momentum about the z-axis, with $m_l = \pm 1$.

EXERCISE: What can be said about the doubly excited bending modes ?

12 Molecular electronic transitions

12.1 Refer to the specification of the selection rules in eqn (12.3.1).

(a) $^2\Pi - {}^2\Pi$ *allowed* by $\Delta\Lambda = 0$.

(b) $^1\Sigma - {}^1\Sigma$ *allowed* by $\Delta\Lambda = 0$.

(c) $\Sigma - \Delta$ *forbidden* ($\Delta\Lambda = \pm 2$ not allowed).

(d) $\Sigma^+ - \Sigma^+$ *allowed* by $\Delta\Lambda = 0$ and $\Sigma^+ \leftrightarrow \Sigma^+$.

(e) $\Sigma^+ - \Sigma^-$ *forbidden* by $\Sigma^+ \not\leftrightarrow \Sigma^-$.

(f) $^1\Sigma_g^+ - {}^1\Sigma_u^+$ *allowed* by $\Delta\Lambda = 0$, $\Sigma^+ - \Sigma^+$, and $g \leftrightarrow u$.

(g) $^3\Sigma_g^+ - {}^3\Sigma_u^-$ *forbidden* by $\Sigma^- \not\leftrightarrow \Sigma^+$.

EXERCISE: Which of $^2\Pi - {}^2\Sigma$, $^3\Pi_g - {}^3\Sigma_u^+$, and $^3\Delta_g - {}^1\Pi_u$ are allowed?

12.2 Regard the carbonyl group as C_{2v} with z along the C=O bond direction. Then π^* and π each span B_1. The transition element $\langle \pi^* | \mu_q | \pi \rangle$ transforms as $B_1 \times \Gamma(\mu_q) = \Gamma(\mu_q)$. In C_{2v} $\Gamma(\mu_z) = A_1$; therefore the π^*–π transition is allowed, and is polarized along the bond. (The π^*–n transition, the more important in most cases, is forbidden.)

EXERCISE: Show that the π^*–π transition in ethene (D_{2h}) is polarized parallel to the bond direction.

12.3 The transition element $\langle {}^1A_2 | \mu_q | {}^1A_1 \rangle$ transforms as

$$A_2 \times \Gamma(\mu_q) \times A_1 = A_2 \times \Gamma(\mu_q)\,;$$

but as $\Gamma(\mu_q) = A_1, B_1, B_2$, it must vanish.

The normal coordinates of H_2O are of symmetry species $2A_1 + B_2$. Since $B_2 \times A_2 \times B_1$, the vibronic transition matrix element

$$\langle {}^1A_2, {}^1B_2 | \mu_q | {}^1A_1, {}^0B_2 \rangle$$

is of symmetry species $A_2 \times B_2 \times \Gamma(\mu_q) \times A_1 \times A_1 = B_1 \times \Gamma(\mu_q)$. It is A_1 when $\Gamma(\mu_q) = B_1$, which is so when $q = x$. Therefore, an x-polarized vibronic transition may occur.

EXERCISE: Show that the transition $^2B_2 \leftarrow {}^2B_1$ is forbidden in ClO_2 (a C_{2v} molecule), but may be vibronically allowed.

12.4 H_2CO belongs to C_{2v}, and μ_q transforms as B_2, B_2, A_1 for x, y, z respectively. The transition $^1A_2 \leftarrow {}^1A_1$ is therefore allowed only if it is vibronic $[\langle A_2|\mu_q|A_1\rangle$ does not span $A_1]$. Since the six vibrations of H_2CO span $3A_1 + B_1 + 2B_2$, possible singly excited vibronic states of the A_2 electronic state are of symmetry species $A_1 \times A_2 = A_2$, $B_1 \times A_2 = B_2$, and $B_2 \times A_2 = B_1$. These vibronic states may be stimulated from the A_1 state by y-polarized (B_2) or x-polarized (B_1) radiation.

Ethene belongs to D_{2h}, and μ_q transforms as B_{3u}, B_{2u}, B_{1u} for x, y, z respectively [the character table is given in the solution of Problem 11.16]. Therefore $B_{2u} \leftarrow A_g$ is allowed for y-polarized radiation. The vibrations of ethene span

$$3A_g + 2B_{1g} + B_{2g} + A_u + B_{1u} + 2B_{2u} + 2B_{3u}$$

[Problem 11.16], and so the possible vibronic states of the B_{2u} electronic state are

$$B_{2u} \times (A_g, B_{1g}, B_{2g}, A_u, B_{1u}, B_{2u}, B_{3u}) = B_{2u}, B_{3u}, A_u, B_{2g}, B_{3g}, A_g, B_{1g}$$

[refer to the character table, form $\chi(R)\chi'(R)$ for $\Gamma \times \Gamma'$, and identify the set of characters so produced]. Of these, B_{2u} and B_{3u} may be reached by an electric dipole transition from the A_g ground state.

EXERCISE: Assess the polarization of the $^1B_2 \leftarrow {}^1B_1$ transition in H_2CO and the $B_{1g} \leftarrow A_g$ transition in ethene

12.5 Both transitions are $g \leftarrow g$, and so are parity forbidden. [More formally: μ_q is u in a centrosymmetric complex, and so $\langle g|\mu_q|g\rangle$ must vanish.] In O_h, μ_q spans T_{1u}; therefore the upper vibronic state must be such that $\Gamma \times T_{1u}$ contains A_{1g}. Refer to the direct product tables in Appendix 11. We see that $\Gamma \times T_{1u}$ contains A_{1g} when $\Gamma = T_{1u}$. Therefore the vibration excited must give T_{1u} when multiplied by the electronic T_{1g} species. Since $T_{2u} \times T_{1g}$, $T_{1u} \times T_{1g}$, $E_u \times T_{1g}$, and $A_{1u} \times T_{1g}$ contain T_{1u}, vibrations of symmetry species T_{2u}, T_{1u}, E_u, and A_{1u} may be involved. Furthermore, since p-orbitals span T_{1u} in O_h, their admixture may account for the transition intensity because $T_{1u} \leftarrow A_g$ is allowed.

In the case $^1T_{2g} \leftarrow {}^1A_{1g}$ we require a vibrational state of species Γ such that $\Gamma \times T_{2g}$ contains T_{1u}. This is so for $\Gamma = A_{2u}$, E_u, T_{1u}, and T_{2u}, and so such vibrations may give rise to intensity. In this case also, admixture of p-orbitals (T_{1u}) may account for intensity because $T_{1u} \leftarrow A_g$ is allowed.

EXERCISE: An electron was trapped in an icosahedral cavity. Two transitions were ascribed to $T_{1u} \leftarrow A_g$ and $G_g \leftarrow A_g$. Are these transitions forbidden? If so, what vibrations would permit them to occur ?

12.6 The vibrational wavefunctions are

Lower: $\psi_v = N_v H_v(y)\, e^{-\frac{1}{2}y^2}, \quad y = (m\omega/\hbar)^{\frac{1}{2}} x \qquad [x = R - R_e],$

Upper: $\psi_v = N_v H_v(y')\, e^{-\frac{1}{2}y'^2}, \quad y' = (m\omega/\hbar)^{\frac{1}{2}} (x - \delta R).$

Then the overlap integral is

$$S_{v0} = N_v N_0 \int_{-\infty}^{\infty} H_v(y')\, H_0(y)\, e^{-\frac{1}{2}(y^2 + y'^2)}\, dy$$

$$= (1/2^v\, v!\, \pi)^{\frac{1}{2}} \int_{-\infty}^{\infty} H_v(y')\, e^{-\frac{1}{2}(y^2 + y'^2)}\, dy.$$

$$S_{00} = (1/\pi)^{\frac{1}{2}} \int_{-\infty}^{\infty} e^{-\frac{1}{2}(y^2 + y'^2)}\, dy$$

$$= (1/\pi)^{\frac{1}{2}}\, e^{-\frac{1}{4}z^2} \int_{-\infty}^{\infty} e^{-(y' + \frac{1}{2}z)^2}\, dy' \qquad [z = (m\omega/\hbar)^{\frac{1}{2}}\, \delta R]$$

$$= e^{-\frac{1}{4}z^2}.$$

$$S_{00}^2 = e^{-\frac{1}{2}z^2} = \underline{\exp\{-(m\omega/2\hbar)\,\delta R^2\}}.$$

$$S_{10} = (2/\pi)^{\frac{1}{2}} \int_{-\infty}^{\infty} y'\, e^{-\frac{1}{2}(y^2 + y'^2)}\, dy$$

$$= (2/\pi)^{\frac{1}{2}}\, e^{-\frac{1}{4}z^2} \int_{-\infty}^{\infty} y'\, e^{-(y' + \frac{1}{2}z)^2}\, dy'$$

$$= (2/\pi)^{\frac{1}{2}}\, e^{-\frac{1}{4}z^2} \int_{-\infty}^{\infty} (w - \tfrac{1}{2}z)\, e^{-w^2}\, dw \qquad [w = y' + \tfrac{1}{2}z]$$

$$= (2/\pi)^{\frac{1}{2}}\, e^{-\frac{1}{4}z^2} (-\tfrac{1}{2}z) \int_{-\infty}^{\infty} e^{-w^2}\, dw = -(z/\sqrt{2})\, e^{-\frac{1}{4}z^2}.$$

$$S_{10}^2 = \tfrac{1}{2}z^2\, e^{-\frac{1}{2}z^2} = \underline{(m\omega/2\hbar)\,\delta R^2 \exp\{-(m\omega/2\hbar)\,\delta R^2\}}.$$

$$S_{20} = (1/8\pi)^{\frac{1}{2}} \int_{-\infty}^{\infty} (4y'^2 - 2)\, e^{-\frac{1}{2}(y^2 + y'^2)}\, dy$$

$$= (1/8\pi)^{\frac{1}{2}}\, e^{-\frac{1}{4}z^2} \int_{-\infty}^{\infty} (4w^2 - 4wz + z^2 - 2)\, e^{-w^2}\, dw$$

$$= (1/8\pi)^{\frac{1}{2}}\, e^{-\frac{1}{4}z^2} \left\{4(\tfrac{1}{2}\sqrt{\pi}) - 0 + (z^2 - 2)\sqrt{\pi}\right\} = (z^2/2\sqrt{2})\, e^{-\frac{1}{4}z^2}$$

$$S_{20}^2 = \tfrac{1}{8}z^4\, e^{-\frac{1}{2}z^2} = \underline{\tfrac{1}{8}(m\omega/\hbar)^2 \delta R^4 \exp\{-(m\omega/2\hbar)\,\delta R^2\}}.$$

The intensities are therefore in the ratio

$$S_{20}^2 : S_{10}^2 : S_{00}^2 = \tfrac{1}{8}(m\omega/\hbar)^2\,\delta R^4 : \tfrac{1}{2}(m\omega/\hbar)\,\delta R^2 : 1\,.$$

$1 \leftarrow 0$ is more intense than $0 \leftarrow 0$ when $\delta R^2 > 2\hbar/m\omega$, and $1 \leftarrow 0$ is more intense than $2 \leftarrow 0$ when $\delta R^2 < 4\hbar/m\omega$. Consequently, $1 \leftarrow 0$ is the most intense of the three when $\underline{\sqrt{(2\hbar/m\omega)} < \delta R < \sqrt{(4\hbar/m\omega)}}$.

EXERCISE: Evaluate S_{30}^2. Go on to consider the case where the force constants are different in the two electronic states.

12.7 $\ln(I_l/I_0) = -\alpha C l$ [eqn (12.6.1)]. Since

$I_l/I_0 = 0.16$, $C = 0.05$ mol dm^{-3}, and $l = 1$ mm,

$$\alpha = -(\ln 0.16)/(0.05\ \text{mol dm}^{-3}) \times (0.1\ \text{cm}) = \underline{3.7 \times 10^2\ \text{mol}^{-1}\ \text{dm}^3\ \text{cm}^{-1}}.$$

$$A \approx c\alpha\Delta\tilde{\nu} \approx c \times (3.7 \times 10^2\ \text{mol}^{-1}\ \text{dm}^3\ \text{cm}^{-1}) \times (4000\ \text{cm}^{-1})$$

$$\approx 4.4 \times 10^{16}\ \text{dm}^3\ \text{cm}^{-1}\ \text{mol}^{-1}\text{s}^{-1} = \underline{4.4 \times 10^{15}\ \text{m}^2\ \text{mol}^{-1}\ \text{s}^{-1}}.$$

$$f = 6.257\,30 \times 10^{-19}\ (A/\text{m}^2\ \text{mol}^{-1}\ \text{s}^{-1}) \qquad \text{[eqn (A17.8)]}$$

$$= \underline{2.8 \times 10^{-3}}.$$

$$|\mu_{fi}|^2 = (3e^2\hbar/4\pi m_e\nu)f \qquad \text{[eqn (12.6.5)]}$$

$$= \{7.0946 \times 10^{-43}/(\nu/\text{Hz})\}\,f\ \text{C}^2\ \text{m}^2$$

$$= \{2.3665 \times 10^{-53}/(\tilde{\nu}/\text{cm}^{-1})^{\frac{1}{2}}\}\,f\ \text{C}^2\text{m}^2$$

$$|\mu_{fi}| = \{4.8647 \times 10^{-27}/(\tilde{\nu}/\text{cm}^{-1})^{\frac{1}{2}}\}\,f^{\frac{1}{2}}\ \text{C}^2\text{m}^2$$

$$= 1458.4\,D\,\{f/(\tilde{\nu}/\text{cm}^{-1})\}^{\frac{1}{2}}$$

$$= \underline{0.39\,D} \text{ when } f = 2.8 \times 10^{-3},\ \lambda = 256\ \text{nm}\ (\widehat{=}\ 3.91 \times 10^4\ \text{cm}^{-1}).$$

EXERCISE: Another compound gave a reduction of intensity to 0.011 of the incident value in the same case. Its absorption was roughly constant from 300 nm to 400 nm. Estimate α, A, f, and μ.

12.8 $\mu_{v+1,v} = -e\langle v+1|x|v\rangle = \underline{-e(v+1)^{\frac{1}{2}}(\hbar/2m\omega)^{\frac{1}{2}}}$ [Problem 5.15].

$$f_{v+1,v} = (4\pi m_e\nu_{v+1,v}/3e^2\hbar)|\mu_{v+1,v}|^2 \qquad \text{[eqn (A17.9)]}$$

$$= (2m_e\omega/3e^2\hbar)(e^2\hbar/2m\omega)(v+1) = \tfrac{1}{3}(m_e/m)(v+1).$$

Therefore, if $m = m_e$ and $v = 0$, $f_{1,0} = \frac{1}{3}$. (Note that if the particles were two electrons, then since m is the reduced mass of the oscillator, $m = \frac{1}{2}m_e$ and $f_{1,0} = \frac{2}{3}$. The condition $m = m_e$ would arise if one particle were an electron and the other an infinitely massive object—such as might be approximated by a nucleus.)

In the general case,

$$A = (\pi\nu_{fi}/3\varepsilon_0\hbar c)\,L|\mu_{fi}|^2 \qquad \text{[eqn (A17.7)]}$$
$$= (\omega/6\,\varepsilon_0\hbar c)\,L(e^2\hbar/2m\omega)(v+1) = \underline{(e^2L/12\,\varepsilon_0 mc)\,(v+1)}\,.$$

EXERCISE: Calculate the oscillator strength for an isotropic three-dimensional harmonic oscillator.

12.9 $\langle n+1|x|n\rangle = (2/L)\int_0^L x\sin[(n+1)\pi x/L]\sin[n\pi x/L]\,dx$
$$= -(8L/\pi^2)\,n\,(n+1)/(2n+1)^2.$$

[Use $\int x\sin ax\sin bx\,dx = \frac{1}{2}\{(x/c')\sin c'x - (x/c)\sin cx + (1/c'^2)\cos c'x - (1/c^2)\cos cx\}$, where $c = a+b$, $c' = a-b$.] Assume that all the electrons except one, together with the nuclei, constitute a uniform positive background,

$$\langle n+1|\mu|n\rangle = -e\langle n+1|x|n\rangle = (8eL/\pi^2)\{n(n+1)/(2n+1)^2\}.$$

$$f = (4\pi m_e/3\hbar e^2)[(n+1)^2-n^2](h/8m_eL^2)(8eL/\pi^2)^2\{n(n+1)/(2n+1)^2\}^2$$
$$= (64/3\pi^2)\{n^2(n+1)^2/(2n+1)^3\}.$$

$$A = (\pi/3\,\varepsilon_0\hbar c)((2n+1)\,h/8m_eL^2)\,L_A(8eL/\pi^2)^2\{n(n+1)/(2n+1)^2\}^2$$
$$= (16/3\pi^2)(e^2L_A/m_e c\,\varepsilon_0)\{n^2(n+1)^2/(2n+1)^3\},\quad [L_A\text{: Avogadro's constant}].$$

For a linear polyene, $L = (N-1)R_{CC}$ and $n = \frac{1}{2}N$ [Problem 3.14]. Therefore,

$$\mu = \underline{(2/\pi^2)\,eR_{CC}\{N(N-1)(N+2)/(N+1)^2\}}\,,$$
$$f = \underline{(4/3\pi^2)\,N^2(N+2)^2/(N+1)^3}\,,$$
$$A = \underline{(e^2L_A/3\pi^2 m_e c\,\varepsilon_0)\{N^2(N+2)^2/(N+1)^3\}}.$$

Taking $N = 20$, $R_{CC} = 140$ pm [Problem 3.14] gives

$$\mu = 3.84\,eR_{CC} = 8.62\times10^{-29}\ \text{C m} \mathrel{\hat{=}} \underline{26\,D}\,,$$
$$f = \underline{2.82}\,,\quad A = \underline{4.51\times10^{18}\ \text{m}^2\ \text{mol}^{-1}\ \text{s}^{-1}}\,.$$

EXERCISE: What are the transition dipoles and the oscillator strengths for the $n+2 \leftarrow n$ and $n+3 \leftarrow n$ transitions of a particle in a square well?

12.10 $H_{so} = \xi_1\,\boldsymbol{l}_1\cdot\boldsymbol{s}_1 + \xi_2\,\boldsymbol{l}_2\cdot\boldsymbol{s}_2$ [eqn (12.9.2)]

$$= \xi_1 l_{1z}s_{1z} + \tfrac{1}{2}\xi_1(l_1^+s_1^- + l_1^-s_1^+) + \xi_2 l_{2z}s_{2z} + \tfrac{1}{2}(l_2^+s_2^- + l_2^-s_2^+)\,.$$

Consider the effect of $l_{1z}s_{1z}$:

$$l_{1z}s_{1z}\,\phi(1)\phi(2)(1/\sqrt{2})(\alpha_1\beta_2-\beta_1\alpha_2)=\phi_z(1)\,\phi(2)(1/\sqrt{2})\{\tfrac{1}{2}\hbar\alpha_1\beta_2+\tfrac{1}{2}\hbar\beta_1\alpha_2\}$$
$$=\tfrac{1}{2}\hbar\phi_z(1)\,\phi(2)\,T_0\,;\quad T_0=(1/\sqrt{2})(\alpha_1\beta_2+\beta_1\alpha_2);\quad \phi_z\equiv l_z\phi\,.$$

$$l_1^+s_1^-\phi(1)\,\phi(2)(1/\sqrt{2})(\alpha_1\beta_2-\beta_1\alpha_2)=\phi_+(1)\,\phi(2)(1/\sqrt{2})(0-\hbar\alpha_1\alpha_2)$$
$$=-(\hbar/\sqrt{2})\phi_+(1)\,\phi(2)\,T_{+1};\quad T_{+1}=\alpha_1\alpha_2\,,\ \phi_+\equiv l^+\phi\,.$$

$$l^-s_1^+\phi(1)\,\phi(2)(1/2)^{\frac{1}{2}}(\alpha_1\beta_2-\beta_1\alpha_2)=\phi(1)\,\phi(2)(1/\sqrt{2})(\hbar\beta_1\beta_2-0)$$
$$=(\hbar/\sqrt{2})\phi(1)\,\phi(2)\,T_{-1}\,;\quad T_{-1}=\beta_1\beta_2\,,\ \phi_-\equiv l^-\phi\,.$$

Likewise for the effect of $\boldsymbol{l}_2\cdot\boldsymbol{s}_2$. Then, taking $\xi_1=\xi_2=\xi$ [for simplicity],

$$H_{so}|S\rangle=\tfrac{1}{2}\hbar\xi\,\{\phi_z(1)\,\phi(2)-\phi(1)\,\phi_z(2)\}\,T_0$$
$$-(\hbar\xi/2\sqrt{2})\{\phi_+(1)\,\phi(2)-\phi(1)\,\phi_+(2)\}\,T_{+1}$$
$$+(\hbar\xi/2\sqrt{2})\{\phi_-(1)\,\phi(2)-\phi(1)\,\phi_-(2)\}\,T_{-1}\,.$$

Then, with

$$|T_0\rangle=(1/\sqrt{2})\{\phi_z(1)\,\phi(2)-\phi(1)\,\phi_z(2)\}\,T_0$$
$$|T_\pm\rangle=(1/\sqrt{2})\{\phi_\pm(1)\,\phi(2)-\phi(1)\,\phi_\pm(2)\}\,T_{-1}$$

we have

$$H_{so}|S\rangle=\underline{(\hbar\xi/\sqrt{2})|T_0\rangle-(\hbar\xi/2)|T_+\rangle+(\hbar\xi/2)|T_-\rangle}\,.$$

EXERCISE: Investigate the case when $\xi_1\neq\xi_2$. Go on to specialize to $\xi_1=\xi$, $\xi_2=0$.

12.11 $P_2(t)=(2V/\Omega)^2\sin^2(\tfrac{1}{2}\Omega t)$, $\Omega^2=\omega_{21}^2+4V^2$ [eqn (8.5.10)]. $\hbar\omega_{21}=\hbar J$. For state 2 taken as T_0, $\hbar V=\hbar\xi/\sqrt{2}$ [preceding Problem solution]; for state 2 corresponding to $T_\pm$, $\hbar V=\hbar\xi/2$. Therefore,

$$P(T_0)=\underline{\{2\xi^2/(J^2+2\xi^2)\}\sin^2\{\tfrac{1}{2}(J^2+2\xi^2)^{\frac{1}{2}}t\}}\,,$$

$$P(T_\pm)=\underline{\{\xi^2/(J^2+\xi^2)\}\sin^2\{\tfrac{1}{2}(J^2+\xi^2)^{\frac{1}{2}}t\}}\,.$$

$$P(T)=\underline{P(T_0)+P(T_+)+P(T_-)}\,.$$

For a range of initial times, $0\leqslant t_0\leqslant T$; the time t then corresponds to the duration since initiation, which is $t-t_0$ for a given member of the system. Therefore, since t_0 ranges from 0 to T, the average population is:

$$\bar{P}(T_0) = \left\{2\xi^2/(J^2+2\xi^2)\right\}(1/T)\int_0^T \sin^2\left\{\tfrac{1}{2}(J^2+2\xi^2)^{\frac{1}{2}}(t-t_0)\right\}\mathrm{d}t_0$$

$$= \left\{2\xi^2/(J^2+2\xi^2)\right\}(1/T)\int_{t-T}^{t}\sin^2(a\tau)\mathrm{d}\tau \quad [\tau=t-t_0,\ a=\tfrac{1}{2}(J^2+2\xi^2)^{\frac{1}{2}}]$$

$$= \left\{\xi^2/(J^2+2\xi^2)\right\}\left\{1-\gamma/2aT\right\},$$

$$\gamma = [1-\cos(2at)]\sin(2at)+\sin(2aT)\sin(2at).$$

If T is long in the sense $2aT \gg 1$,

$$\bar{P}(T_0) \approx \xi^2/(J^2+2\xi^2).$$

Likewise,

$$\bar{P}(T_\pm) \approx \tfrac{1}{2}\xi^2/(J^2+\xi^2).$$

Overall, therefore,

$$\bar{P}(T) \approx \xi^2/(J^2+2\xi^2)+\xi^2/(J^2+\xi^2) = \underline{(2J^2+3\xi^2)\xi^2/(J^2+\xi^2)(J^2+2\xi^2)}.$$

When $J^2 \gg \xi^2$, $\bar{P}(T) \approx 2\xi^2/J^2$.

EXERCISE: Suppose a magnetic field is present. How does $\bar{P}(T)$ depend on it ?

12.12 Determine which states are mixed by rotations. In D_{6h} rotations transform as A_{2g} and E_{1g}. Therefore, $B_{1u}\times\{A_{2g}, E_{1g}\} = \{B_{2u}, E_{2u}\}$ and $B_{2u}\times\{A_{2g}, E_{1g}\} = \{B_{1u}, E_{2u}\}$. Therefore $\underline{^1B_{2u}\text{ and }^1E_{2u}}$ may by mixed into $^3B_{1u}$ and $\underline{^1B_{1u}\text{ and }^1E_{2u}}$ may be mixed into $^3B_{2u}$.

EXERCISE: What triplet states may be mixed into the ^1E state of NH_3. What states would be mixed if the molecule were planar in the ^1E excited state ?

12.13 In D_{2h} rotations span B_{3g}, B_{2g}, B_{1g} (for R_x, R_y, R_z respectively). Since $B_{1u}\times\{B_{3g}, B_{2g}, B_{1g}\} = \{B_{2u}, B_{3u}, A_{1u}\}$ [see *Tables for group theory*, P. W. Atkins, M. S. Child, and C. S. G. Phillips, OUP (1970)], the $^3B_{1u}$ term is contaminated by $^1B_{2u}$, $^1B_{3u}$, and $^1A_{1u}$ terms. In D_{2h} the electric dipole moment components transform as B_{3u}, B_{2u}, B_{1u} (for x, y, z respectively). The terms having allowed transitions to the $^1A_{1g}$ ground state are therefore

$$A_{1g}\times\{B_{3u}, B_{2u}, B_{1u}\} = \{B_{3u}, B_{2u}, B_{1u}\}.$$

Therefore, the phosphorescent radiation may occur *via*

$\underline{B_{2u}\rightarrow A_{1g}\ (y\text{-polarized})}$ and $\underline{B_{3u}\rightarrow A_{1g}\ (x\text{-polarized})}$.

EXERCISE: In an aromatic molecule of D_{5d} symmetry the lowest triplet term was identified as $^3A_{1u}$. What is the polarization of its phosphorescence ?

12.14 $H\Psi = \xi\Psi$, $\Psi = a\psi + \sum_n b_n \phi_n$; $H(\text{bath})\,\phi_n = E_n \phi_n$. ψ should be interpreted as $\psi_{\mathbf{system}} f_{\mathbf{bath}}$ and ϕ as $\phi_{\mathbf{bath}} g_{\mathbf{system}}$, with $H_{\mathbf{bath}} f_{\mathbf{bath}} = 0$ and $H_{\mathbf{system}} g_{\mathbf{system}} = 0$ (so that in each case no energy resides in the relevant component). Then

$$H\Psi = aE_{\mathbf{system}}\psi + \sum_n b_n E_n \phi_n + H'a\psi + \sum_n b_n H'\phi_n$$
$$= \mathcal{E} a\psi + \sum_n b_n \mathcal{E}\phi_n .$$

Multiply by (a) ψ^* and integrate, (b) $\phi^*_{n'}$ and integrate (follow that by setting $n' \to n$):

$$(E_{\mathbf{system}} - \mathcal{E})\, a + V\sum_n b_n = 0 \qquad [\langle\psi|\phi_n\rangle = 0,\ \langle\psi|H'|\psi\rangle = 0]$$
$$Va + (E_n - \mathcal{E})\, b_n = 0 \qquad [\langle\phi_n|\psi\rangle = 0,\ \langle\phi_n|H'|\phi_{n'}\rangle = 0].$$

Then, $b_n = \{V/(\mathcal{E} - E_n)\}\, a$.

Substitute this expression for b_n back into the first equation of the pair:

$$(E_{\mathbf{system}} - \mathcal{E}) + V^2 \sum_n \{1/(\mathcal{E} - E_n)\} = 0 .$$

Then, with $\mathcal{E} - E_n = \gamma\varepsilon - n\varepsilon$ and $\rho = 1/\varepsilon$,

$$E_{\mathbf{system}} - \mathcal{E} = -(V^2/\varepsilon)\sum_n \{1/(\gamma - n)\}$$
$$= \underline{(V^2/\varepsilon)\,\pi\cot(\pi\gamma)} .$$

Since Ψ is normalized to unity, $a^2 + \sum_n b_n^2 = 1$; consequently [from above]

$$a^2 + a^2V^2\sum_n \{1/(E_n - \mathcal{E}_n)^2\} = 1 .$$

Since

$$\sum_{n=-\infty}^{\infty} (\gamma - n)^{-2} = \pi^2 \operatorname{cosec}^2 \pi\gamma ,$$

[*Handbook of mathematical functions*]

$$a^2 = \{1 + \pi^2\rho^2V^2 \operatorname{cosec}^2(\pi\gamma)\}^{-1}$$
$$= \{1 + \pi^2\rho^2V^2 + \pi^2\rho^2V^2\cot^2(\pi\gamma)\}^{-1}$$
$$= \{1 + \pi^2\rho^2V^2 + \pi^2\rho^2V^2[(E_{\mathbf{system}} - \mathcal{E})/\pi\rho V^2]^2\}^{-1}$$
$$= \underline{V^2/\{(E_{\mathbf{system}} - \mathcal{E})^2 + V^2 + (\pi V^2\rho)^2\}} .$$

EXERCISE: Find an expression for $\sum_n b_n^4$.

13 The electronic properties of molecules

13.1 $\langle \mu_z \rangle = \alpha E$ [eqn (13.1.7), $\mu_{0z} = 0$; $\alpha_{zz} = \alpha$]

$= 4\pi\varepsilon_0 \alpha'$ [eqn (13.1.16); $\alpha = 4\pi\varepsilon_0 \alpha'$]

$= (1.112\,65 \times 10^{-10}\ \mathrm{J^{-1}\ C^2\ m^{-1}}) \times (10.5 \times 10^{-30}\ \mathrm{m^3}) \times (10^4\ \mathrm{V\ m^{-1}})$

$= (1.17 \times 10^{-39}\ \mathrm{J^{-1}\ C^2\ m^2}) \times (10^4\ \mathrm{V\ m^{-1}})$

$= 1.17 \times 10^{-35}\ \mathrm{C\ m} \,\hat{=}\, \underline{3.5 \times 10^{-6}\ \mathrm{D}}$.

$\mathcal{E}(E) - \mathcal{E}_0 = -\frac{1}{2}\alpha E^2 = -5.85 \times 10^{-32}\ \mathrm{J} \,\hat{=}\, \underline{-3.52 \times 10^{-5}\ \mathrm{kJ\ mol^{-1}}}$.

EXERCISE: Calculate the dipole moment induced by a singly charged ion at a distance of (a) 0.1 nm, (b) 1.0 nm from a tetrachloromethane molecule.

13.2 $\alpha_{xx} = 2\sum_n{}' \{\mu_{x,0n}\mu_{x,n0}/\Delta_{n0}\}$ [eqn (13.1.14) with $z \to x$].

$\mu_{x,0n} = \langle 0|-ex + \frac{1}{2}eL|n\rangle = -e\langle 0|x|n\rangle$ [$\langle 0|L|n\rangle = \langle 0|n\rangle L = 0$, $n \neq 0$].

$$\langle 0|x|n\rangle = \begin{cases} -(8/\pi^2)Ln/(n^2-1), & n \text{ even} \\ 0 & n \text{ odd} \end{cases}$$ [Problem 8.5].

$\Delta_{n0} = (n^2-1)(h^2/8mL^2)$.

$\alpha_{xx} = 2e^2(8L/\pi^2)^2(8mL^2/h^2)\sum_n{}' \{n^2/(n^2-1)^3\}$

$= 2(8/\pi^2)^2 a \sum_n{}' \{n^2/(n^2-1)^3\}$, $a = (eL)^2/(h^2/8mL^2)$.

$\sum_n{}' \{n^2/(n^2-1)^3\} = 0.1542$ [Problem 8.5].

Therefore, $\alpha_{xx} = 0.2026a = \underline{0.2026\, e^2L^2/(h^2/8mL^2)}$.

For $m = m_e$, $\alpha_{xx} = 8.633\ (L/\mathrm{pm})^4 \times 10^{-50}\ \mathrm{J^{-2}\ C^2\ m^2}$,

$\alpha'_{xx} = \alpha_{xx}/4\pi\varepsilon_0 = 7.759 \times 10^{-34}\ (L/\mathrm{pm})^4\ \mathrm{cm^3}$.

With $L = 150$ pm, $\alpha'_{xx} = \underline{0.393 \times 10^{-24}\ \mathrm{cm^3}}$.

EXERCISE: Calculate the polarizability volume of a rectangular, three-dimensional box of sides X, Y, Z, and the mean polarizability volume, and relate α' to $V = XYZ$.

13.3 $\alpha_{xx} \approx (2/\Delta)\{\langle 0|\mu_x^2|0\rangle - \langle 0|\mu_x|0\rangle^2\} = (2e^2/\Delta)\{\langle x^2\rangle - \langle x\rangle^2\}$.

$\langle x^2\rangle = \frac{1}{3}L^2\{1-(3/2\pi^2)\}$ [Problem 3.11 with $n = 1$],

$\langle x\rangle^2 = \frac{1}{4}L^2$,

$\langle x^2\rangle - \langle x\rangle^2 = \{(1/12)-(1/2\pi^2)\}L^2 = 0.032\,67\ L^2$.

$\alpha_{xx} \approx 0.065\,34\ (e^2 L^2/\Delta)$.

Take $\Delta = \lambda(h^2/8mL^2)$ with $\lambda = 0.3225$ [Problem 8.9]; then

$\alpha_{xx} \approx 0.2026\ \{e^2 L^2/(h^2/8mL^2)\}$, as above.

EXERCISE: The positive charge is located at a distance R from the left-hand side ($x=0$) of the box treated above. Calculate α_{xx} on the basis of the closure approximation with $\lambda = 0.3225$. How does α_{xx} depend on R? Why?

13.4 $\alpha_{zz} = 2 \sum'_{nlm_l} \{|\langle nlm_l|\mu_z|1s\rangle|^2/(E_n - E_1)\}$

$= 2 \sum'_{n} \{|\langle np_z|\mu_z|1s\rangle|^2/(E_n - E_1)\}$ [only p_z-orbitals contribute]

$E_n - E_1 = -hcR_H\{(1/n^2) - 1\} = hcR_H\{1 - (1/n^2)\}$.

$\langle np_z|\mu_z|1s\rangle = -e\langle np_z|z|1s\rangle$.

These integrals are very difficult to evaluate for general n, but they are tabulated in various places: for instance, p. 154 of *Intermediate quantum mechanics*, H. A. Bethe, Benjamin (1964), gives

$$\langle n', l+m, m_l|z|nlm_l\rangle = \left\{\frac{(l+1)^2 - m_l^2}{(2l+3)(2l+1)}\right\}^{\frac{1}{2}} \langle n', l+1|r|nl\rangle .$$

Since in our case $l=0$, $n=1$, $m_l=0$,

$\langle np_z|z|1s\rangle = (1/\sqrt{3})\langle np|r|1s\rangle$ [n' has been replaced by n].

This result is also listed on p. 132 of *The theory of atomic spectra*, E. U. Condon and G. H. Shortley, Cambridge University Press (1963), who go on to give the values of the radial integrals. In particular their Table 3.5 gives

$$\langle np|r|1s\rangle/a_0 = \{2^8 n^7 (n-1)^{2n-5}/(n+1)^{2n+5}\}^{\frac{1}{2}}.$$

Therefore

$$\alpha_{zz} = (2^9 e^2 a_0^2/3hcR_H) \sum_{n=2}^{\infty} \{n^9(n-1)^{2n-5}/(n+1)^{2n+5}(n^2-1)\}$$

$$= (2^9 e^2 a_0^2/3hcR_H) \sum_{n=2}^{\infty} \{n^9(n-1)^{2n-6}/(n+1)^{2n+6}\}$$

$$= (2^9 e^2 a_0^2/3hcR_H)\{0.0087 + 0.0012 + 0.0004 + \ldots\}$$

$$= (2^9 e^2 a_0^2/3hcR_H) \times 0.0106 = 5.966 \times 10^{-41}\ \mathrm{J^{-1}\,C^2\,m^2}$$

$$\alpha'_{zz} = \alpha_{zz}/4\pi\varepsilon_0 = \underline{5.36 \times 10^{-25}\ \mathrm{cm}^3}.$$

EXERCISE: Calculate the polarizability of one-electron ions with atomic number Z.

13.5 Take as an unnormalized trial function $\psi = \psi_{1s} + a\psi_{2p_z}$. The hamiltonian is $H = H_0 + ezE$. The Rayleigh ratio is therefore

$$\mathcal{E} = \langle \psi_{1s} + a\psi_{2p_z} | H_0 + ezE | \psi_{1s} + a\psi_{2p_z}\rangle / \langle \psi_{1s} + a\psi_{2p_z} | \psi_{1s} + a\psi_{2p_z}\rangle$$

$$= \left\langle \psi_{1s} + a\psi_{2p_z} \middle| \left\{ \mathcal{E}_{1s} |\psi_{1s}\rangle + a\,\mathcal{E}_{2p_z} |\psi_{2p_z}\rangle + ezE |\psi_{1s} + a\psi_{2p_z}\rangle \right\}\right\rangle \Big/ (1+a^2)$$

$$= \left\{ \mathcal{E}_{1s} + a^2\,\mathcal{E}_{2p_z} + 2eaE\langle \psi_{1s}|z|\psi_{2p_z}\rangle \right\} \Big/ (1+a^2).$$

[$\langle\psi_{1s}|\psi_{2p}\rangle = 0$, $\langle\psi_{1s}|z|\psi_{1s}\rangle = \langle\psi_{2p}|z|\psi_{2p}\rangle = 0$.] Write $\langle\psi_{1s}|z|\psi_{2p}\rangle = z_{12}$. Then

$$(\partial\mathcal{E}/\partial a) = (2a\mathcal{E}_{2p} + 2ez_{12}E)/(1+a^2) - 2a(\mathcal{E}_{1s} + a^2\mathcal{E}_{2p} + 2eaEz_{12})/(1+a^2)^2$$

$$= \left\{ 2a(\mathcal{E}_{2p} - \mathcal{E}_{1s}) - 2ez_{12}Ea^2 \right\} \Big/ (1+a^2)^2.$$

The optimum value of a is given by $d\mathcal{E}/da = 0$, which is satisfied by $(a^2-1)\,eEz_{12} + a(\mathcal{E}_{1s} - \mathcal{E}_{2p}) = 0$,

$$a = \left\{ (\mathcal{E}_{2p} - \mathcal{E}_{1s}) \pm (\mathcal{E}_{2p} - \mathcal{E}_{1s})\sqrt{\{1 + 4e^2z_{12}^2E^2/(\mathcal{E}_{2p} - \mathcal{E}_{1s})^2\}} \right\} \Big/ 2ez_{12}E.$$

Since $4e^2z_{12}^2E^2/(\mathcal{E}_{2p} - \mathcal{E}_{1s})^2 \ll 1$ for all reasonable fields,

$$a = \left\{ (\mathcal{E}_{2p} - \mathcal{E}_{1s}) - (\mathcal{E}_{2p} - \mathcal{E}_{1s})[1 + 2e^2z_{12}^2E^2/(\mathcal{E}_{2p} - \mathcal{E}_{1s})^2 + \dots] \right\} \Big/ (2ez_{12}E)$$

[take the negative root, because $a = 0$ when $E = 0$]. Then

$$a \approx -ez_{12}E/(\mathcal{E}_{2p} - \mathcal{E}_{1s}).$$

The expectation value of μ_z in the presence of the field is

$$\langle\mu_z\rangle = \langle\psi_{1s} + a\psi_{2p_z}|\mu_z|\psi_{1s} + a\psi_{2p_z}\rangle / \langle\psi_{1s} + a\psi_{2p_z}|\mu_z|\psi_{1s} + a\psi_{2p_z}\rangle$$

$$= 2a\langle\psi_{1s}|\mu_z|\psi_{2p_z}\rangle/(1+a^2) \approx 2a\langle\psi_{1s}|\mu_z|\psi_{2p_z}\rangle \qquad [a^2 \ll 1].$$

$$\approx -2eaz_{12} = 2e^2z_{12}^2E/(\mathcal{E}_{2p} - \mathcal{E}_{1s}).$$

It follows that

$$\alpha_{zz} = 2e^2z_{12}^2/(\mathcal{E}_{2p} - \mathcal{E}_{1s}) \qquad [\langle\mu_z\rangle = \alpha_{zz}E].$$

Now, $z_{12} = (2^7\sqrt{2/3^5})\,a_0$ [Problem 13.4, or use wavefunctions in Tables 4.1 and 4.2] and $\mathcal{E}_{2p} - \mathcal{E}_{1s} = -hcR_H(\frac{1}{4} - 1) = \frac{3}{4}hcR_H$. Therefore,

$$\alpha_{zz} = (2^{17}/3^{11})(e^2/hcR_H)\,a_0^2 = 2.44 \times 10^{-41}\,\mathrm{J^{-1}\,C^2\,m^2}.$$

$$\alpha'_{zz} = \alpha_{zz}/4\pi\varepsilon_0 = \underline{2.19 \times 10^{-25}\ \mathrm{cm^3}}.$$

The experimental value is 6.6×10^{-25} cm^3. [The exact theoretical expression will be found on p. 204 of *The theory of electric and magnetic susceptibilities*, J. H. van Vleck, Oxford University Press (1932).]

EXERCISE: Investigate the improvement to the value of α_{zz} brought about by including a $3p_z$-orbital in the trial function as well as the $2p_z$-orbital. Also investigate the '$1p_z$-orbital' trial function $(1+az)\psi_{1s}$, as suggested in the hint.

13.6 $\beta = -(d^3\mathcal{E}/dE^3)_0$ [eqn (13.1.8c)]

$$\mathcal{E}^{(3)} = \sum_{mn}{}' \left\{H^{(1)}_{0m}\, H^{(1)}_{mn}\, H^{(1)}_{n0}/(E_m - E_0)(E_n - E_0)\right\} - H^{(1)}_{00} \sum_{n}{}' \left\{H^{(1)}_{0n}\, H^{(1)}_{n0}/(E_n - E_0)^2\right\}.$$

For $H^{(1)} = -\mu_z E$, since $H^{(1)}_{00} = -\mu_{0z}E = 0$ for a non-polar species,

$$\mathcal{E}^{(3)} = -\sum_{mn}{}' \left\{\mu_{z,0m}\,\mu_{z,mn}\mu_{z,n0}\, E^3/\Delta_{m0}\,\Delta_{n0}\right\}$$

$$(d^3\mathcal{E}/dE^3)_0 = -6\sum_{mn}{}' \left\{\mu_{z,0m}\,\mu_{z,mn}\mu_{z,n0}/\Delta_{m0}\,\Delta_{n0}\right\}.$$

Therefore, $\underline{\beta_{zzz} = 6\sum_{mn}{}' \left\{\mu_{z,0m}\,\mu_{z,mn}\,\mu_{z,n0}/\Delta_{m0}\,\Delta_{n0}\right\}}$.

EXERCISE: Deduce an expression for the general component $\beta_{qq'q''}$ of the first hyperpolarizability.

13.7 The component β_{xyz} transforms as xyz. In T_d the product xyz spans A_1; all other cubic combinations span other symmetry species. Since β_{xyz} must be invariant under symmetry operations of the group, only β_{xyz} (and its cyclic transformations) is non-zero.

In order to demonstrate explicitly that xyz spans A_1, consider a representative selection of symmetry operations, (i.e. one from each class). Draw up the following Table:

	x	y	z	xyz	χ
E	x	y	z	xyz	1
C_3	$-y$	z	$-x$	xyz	1
C_2	$-x$	$-y$	z	xyz	1
S_4	$-y$	x	$-z$	xyz	1
σ_d	$-y$	$-x$	z	xyz	1

The characters are all 1 for the basis xyz; hence xyz spans A_1.

Since β_{xxz}, β_{yyz}, and β_{zzz} are each zero, the mean β is zero.

EXERCISE: Determine what components of the polarizability may be non-zero for a molecule belonging to the point group O. What is the mean value β?

13.8 $$\beta_{xxx} = -6e^3 \sum_{mn}{}' \{x_{0m}x_{mn}x_{n0}/\Delta_{m0}\Delta_{n0}\}$$ [Problem 13.6, $z \to x$]

$$= -6e^3 \sum_{v'v''}{}' \{x_{vv'}x_{v'v''}x_{v''v}/\Delta_{v'v}\Delta_{v''v}\}$$ [v: oscillator quantum number].

$x_{v''v}$ vanishes unless $v'' = v \pm 1$; $x_{vv'}$ vanishes unless $v' = v \pm 1$; $x_{v'v''}$ vanishes unless $v' = v'' \pm 1$. Therefore the product of matrix elements vanishes [it consists of terms such as $x_{v,v+1}x_{v+1}$, $v \times x_{vv} = 0$; three steps up or down can never return you to the same place]. Therefore, $\underline{\beta_{xxx} = 0}$.

EXERCISE: Evaluate β_{zzz} for a hydrogen atom.

13.9 $$[H, x^2] = -(\hbar^2/2m_e)[(d^2/dx^2), x^2]$$ $[[V, x^2] = 0]$

$$= -(\hbar^2/2m_e)\{(d^2/dx^2)x^2 - x^2(d^2/dx^2)\}$$

$$= -(\hbar^2/2m_e)\{2 + 4x(d/dx) + x^2(d^2/dx^2) - x^2(d^2/dx^2)\}$$

$$= -(\hbar^2/m_e) - 2(\hbar^2/m_e)x(d/dx)$$

$$= -(\hbar^2/m_e) - 2i(\hbar/m_e)xp.$$

$$\langle m|[H, x^2]|n\rangle = (E_m - E_n)\langle m|x^2|n\rangle = \hbar\omega_{mn}(x^2)_{mn}$$

$$= \langle m| -(\hbar^2/m_e) - 2i(\hbar/m_e)xp|n\rangle$$

$$= -(\hbar^2/m_e)\delta_{mn} - 2i(\hbar/m_e)\sum_f x_{mf}p_{fn}$$

$$= -(\hbar^2/m_e)\delta_{mn} - 2i(\hbar/m_e)(im_e)\sum_f x_{mf}\omega_{fn}x_{fn}$$ [eqn (A18.3)]

$$= -(\hbar^2/m_e)\delta_{mn} + 2\hbar\sum_f x_{mf}x_{fn}\omega_{fn}.$$

Therefore

$$\sum_f x_{mf}x_{fn}\omega_{fn} = (\hbar/2m_e)\delta_{mn} + \tfrac{1}{2}\omega_{mn}(x^2)_{mn}.$$

EXERCISE: Devise a sum rule based on $[H, x^3]$.

13.10 $\alpha \approx (2/3\,\Delta)\, e^2 \langle r^2 \rangle$ [eqn (13.1.18)].

This is the mean value. The individual components for $q = x, y, z$, are

$$\alpha_{qq} \approx (2/\Delta)\, e^2 \langle q^2 \rangle .$$

Use

$$\psi_{2p_z} = (Z^{*5}/32\,\pi\, a_0^5)^{\frac{1}{2}}\, r \cos\theta\, e^{-Z^* r/2a_0}$$ [Section 9.10 of text],

$$\psi_{2s} = (Z^{*5}/96\,\pi\, a_0^5)^{\frac{1}{2}}\, r\, e^{-Z^* r/2a_0},$$

$$\psi_{1s} = (Z^{*3}/\pi\, a_0^3)^{\frac{1}{2}}\, e^{-Z^* r/a_0} .$$

$$Z^*_{2p} = Z^*_{2s} = 3.25, \quad Z^*_{1s} = 5.70$$ [Table 9.1].

$$\langle z^2 \rangle_{2p_z} = (Z^{*5}/32\,\pi\, a_0^5) \int_0^{2\pi} d\phi \int_0^{\pi} \sin\theta\, d\theta \int_0^{\infty} r^2 \left\{ r^4 \cos^4\theta\, e^{-Z^* r/a_0} \right\} dr$$

$$= (Z^{*5}/32\,\pi\, a_0^5)(2\pi)(2/5)(6!\,a_0^7/Z^{*7}) = 18\, a_0^2/Z^{*2}_{2p} .$$

$$\langle x^2 \rangle_{2p_z} = (Z^{*5}/32\pi\, a_0^5) \int_0^{2\pi} d\phi \int_0^{\pi} \sin\theta\, d\theta \int_0^{\infty} r^2 \left\{ r^4 \cos^2\phi\, \sin^2\theta\, \cos^2\theta\, e^{-Z^* r/a_0} \right\} dr$$

$$= (Z^{*5}/32\,\pi\, a_0^5)(\pi)(4/15)(6!\, a_0^7/Z^{*7}) = 6\, a_0^2/Z^{*2}_{2p} .$$

$$\langle y^2 \rangle_{2p_z} = \langle x^2 \rangle_{2p_z} = 6 a_0^2/Z^{*2}_{2p} .$$

Consequently, (a) $\alpha_{\parallel} = \alpha_{zz} = (2e^2/\Delta) \langle z^2 \rangle_{2p_z} = \underline{36\, e^2\, a_0^2/Z^{*2}_{2p_z}\,\Delta}$

(b) $\alpha_{\perp} = \alpha_{xx} = \alpha_{yy} = (2\,e^2/\Delta) \langle x^2 \rangle_{2p_z} = \underline{12\, e^2\, a_0^2/Z^{*2}_{2p}\,\Delta}$.

Note that $\alpha_{\parallel}/\alpha_{\perp} = 3$. With $Z^*_{2p} = 3.25$ (Table 9.1) and $\Delta \,\hat{=}\, 11.264$ eV $\hat{=}\, 1.805 \times 10^{-18}$ J,

(a) $\alpha_{\parallel} \approx 4.5 \times 10^{-41}\ J^{-2}\ C^2\ m^2$, $\alpha'_{\parallel} \approx \underline{4.1 \times 10^{-25}\ cm^3}$,

(b) $\alpha_{\perp} \approx 1.5 \times 10^{-41}\ J^{-1}\ C^2\ m^2$, $\alpha'_{\perp} \approx \underline{1.4 \times 10^{-25}\ cm^3}$.

In the case of a 2s-electron,

$$\langle z^2 \rangle_{2s} = (Z^{*5}/96\pi\, a_0^5) \int_0^{2\pi} d\phi \int_0^{\pi} \sin\theta\, d\theta \int_0^{\infty} r^2 \left\{ r^4 \cos^2\theta\, e^{-Z^* r/a_0} \right\} dr$$

$$= (Z^{*5}/96\pi a_0^5)(2\pi)(2/3)(6!\,a_0^7/Z^{*7}) = 10\, a_0^2/Z^{*2}$$

$$= \langle x^2 \rangle_{2s} = \langle y^2 \rangle_{2s} .$$

Consequently,

$$\alpha_{zz}(2s) = \alpha_{yy}(2s) = \alpha_{xx}(2s) = \underline{20\, e^2\, a_0^2/\Delta' Z^{*2}_{2s}} .$$

For a 1s-electron,

$$\langle z^2\rangle_{1s} = (Z^{*3}/\pi a_0^3)\int_0^{2\pi} d\phi \int_0^{\pi} \sin\theta\, d\theta \int_0^{\infty} r^2 \left\{r^2 \cos^2\theta\, e^{-2Z^*r/a_0}\right\} dr$$
$$= (Z^{*3}/\pi a_0^3)(2\pi)(2/3)(4!\, a_0^5/2^5 Z^{*5}) = a_0^2/Z^{*2}\,.$$

Consequently,

$$\alpha_{zz}(1s) = \alpha_{yy}(1s) = \alpha_{xx}(1s) = \underline{2e^2 a_0^2/Z_{1s}^{*2}\Delta''}\,.$$

For a carbon atom in its valence configuration $1s^2\, 2s\, 2p_x\, 2p_y\, 2p_z$, the overall polarizability is

$$\alpha_{zz} = 2\alpha_{zz}(1s) + \alpha_{zz}(2s) + 2\alpha_{zz}(2p_x) + \alpha_{zz}(2p_z) \qquad [\alpha_{zz}(2p_x) = \alpha_{zz}(2p_y)]$$
$$= \left\{(4/Z_{1s}^{*2}\Delta'') + (20/Z_{2s}^{*2}\Delta') + (60/Z_{2p}^{*2}\Delta)\right\} e^2 a_0^2\,.$$

Take $Z_{2s}^* = Z_{2p}^* = 3.25$, $Z_{1s}^* = 5.70$ [Table 9.1]

$$\alpha_{zz} = (0.123/\Delta'' + 1.893/\Delta' + 5.680/\Delta)\, e^2 a_0^2\,.$$

Then, with

$$\Delta \approx 1.805\times 10^{-18}\ \mathrm{J}, \quad \Delta' \approx \Delta + (75\,000\ \mathrm{cm^{-1}})hc = 3.29\times 10^{-18}\ \mathrm{J},$$

and $\Delta'' \mathrel{\hat=} 13.6\ \mathrm{eV}\times 5.7^2$ $[I(\mathrm{H}) \mathrel{\hat=} 13.6\ \mathrm{eV},\ I \propto Z^2] \mathrel{\hat=} 8.78\times 10^{-18}\ \mathrm{J}$:

$$\alpha_{zz} \approx \left\{(1.40 + 57.5 + 315)\times 10^{16}\ \mathrm{J^{-1}}\right\} e^2 a_0^2 = \underline{2.69\times 10^{-40}\ \mathrm{J^{-1}\,C^2\,m^{-1}}},$$
$$\alpha'_{zz} = \alpha_{zz}/4\pi\varepsilon_0 = \underline{2.41\times 10^{-24}\ \mathrm{cm^3}}\,.$$

The experimental value for CH_4 is $2.60\times 10^{-24}\ \mathrm{cm^3}$.

EXERCISE: Find the contribution to the polarizability components of an electron in a d_{z^2}-orbital.

13.11 $\alpha = (\hbar^2 e^2/m_e)\sum_n{}' \left\{f_{n0}/\Delta_{n0}^2\right\}$ [eqn (13.1.21)]

$\approx (\hbar^2 e^2/m_e)(f/\Delta^2)$ [one transition dominating]

$\approx (e^2/4\pi^2 m_e c^2)\lambda^2 f$ $[\Delta = hc/\lambda]$.

$$\alpha' = \alpha/4\pi\varepsilon_0 = (e^2/16\pi^3 \varepsilon_0 m_e c^2)\, \lambda^2 f$$
$$= (7.138\times 10^{-17}\ \mathrm{m})\lambda^2 f = (7.138\times 10^{-29}\ \mathrm{cm^3})(\lambda/\mathrm{nm})^2 f.$$

For $\lambda = 160\ \mathrm{nm}$ and $f = 0.3$, $\alpha' \approx \underline{5\times 10^{-25}\ \mathrm{cm^3}}$, which is an order of magnitude smaller than the experimental value.

EXERCISE: Find an expression for α' in terms of the integrated absorption coefficient of a band.

13.12 $n_r \approx 1 + (2\pi L\rho/M_m)\,\alpha'(\omega)$ [eqn (13.3.11)],

$\approx 1 + (2\pi\rho/m_e)\,\alpha'(\omega)$ $[M_m/L = m_e]$.

$$\alpha'(\omega) \approx -e^2/m_e\omega^2 \qquad [\text{eqn } (13.3.10),\ N_e = 1].$$

Therefore,

$$n_r \approx 1-(2\pi\rho/m_e)(e^2/m_e\omega^2) = \underline{1-(2\pi\rho e^2/m_e^2\,\omega^2)}.$$

EXERCISE: Find an expression for the refractive index of a dense swarm of free electrons, and evaluate it for a density corresponding to a pressure of 1 atm.

13.13
$$\left.\begin{aligned} n_r(\omega) &\approx 1 + (L\rho/3\hbar\varepsilon_0 M_m)\,C(\omega) \\ C(\omega) &= \sum_n \left\{\omega_{n0}|\mu_{0n}|^2/(\omega_{n0}^2-\omega^2)\right\} \end{aligned}\right\} \quad [\text{eqn } (13.3.11)]\ .$$

Evaluate $C(\omega)$ numerically, drawing on the information in the solution of Problem 13.4.

$$C(\omega) = \sum_{nlm_l} \left\{\omega_{n,1s}\left|\mu_{1s,nlm_l}\right|^2\Big/(\omega_{n,1s}^2-\omega^2)\right\} \qquad [0 \mathrel{\hat{=}} 1s]$$

$$= \sum_{nlm_l} \left\{\omega_{n,1}\left|\mu_{z;1s,nlm_l}\right|^2\Big/(\omega_{n,1}^2-\omega^2)\right\} \qquad [\mu_x^2=\mu_y^2=\mu_z^2]$$

$$= 3e^2\sum_{n} \left\{\omega_{n,1}\left|z_{np_z,1s}\right|^2\Big/(\omega_{n,1}^2-\omega^2)\right\} \qquad [\text{only } np_z\text{-orbitals contribute}]$$

$$= (3e^2R_H/2\pi c)\sum_n \left\{\frac{[1-(1/n^2)]\left|z_{np_z,1s}\right|^2}{[1-(1/n^2)]^2R_H^2-(1/\lambda^2)}\right\} \qquad \left[\hbar\omega_{n,1}=hcR_H\left(1-\frac{1}{n^2}\right)\right]$$

$$= (2^7/\pi)(e^2a_0^2R_H/c)\sum_n \left\{\frac{[1-(1/n^2)]\,n^7\,(n-1)^{2n-5}/(n+1)^{2n+5}}{[1-(1/n^2)]^2R_H^2-(1/\lambda^2)}\right\}$$

$$= (2^7/\pi)(e^2a_0^2/R_H\,c)\,D(\lambda)\ ,$$

$$D(\lambda) = \sum_n \left\{\frac{[1-(1/n^2)]\,n^7\,(n-1)^{2n-5}/(n+1)^{2n+5}}{[1-(1/n^2)]^2-(1/\lambda R_H)^2}\right\}$$

$$= \sum_n \left\{\frac{n^9(n-1)^{2n-4}/(n+1)^{2n+4}}{(n^2-1)^2-\gamma^2n^4}\right\}, \qquad \gamma = 1/\lambda R_H\ .$$

Since $\gamma = 1/(590\text{ nm})\times(1.097\times10^5\text{ cm}^{-1}) = 0.155$, numerical evaluation of the sum (up to $n\approx 20$) leads to

$$D(590\text{ nm}) = 0.0536\ .$$

Therefore,

$$C = (8.903\times10^{-73}\text{ C}^2\text{ m}^2\text{ s})\,D = 4.772\times10^{-74}\text{ C}^2\text{ m}^2\text{ s}\ .$$

Consequently,

$$n_r \approx 1+(N/3\hbar\varepsilon_0)C \qquad [\rho = Nm_H/V\ ,\ m_H = M_m(H)/L]$$

$$\approx 1+(N/\text{atoms m}^{-3})\times(1.703\times10^{-29}).$$

When $N \approx 5$ atoms m^{-3},

$$n_r - 1 \approx \underline{8.5 \times 10^{-29}}\,.$$

For a gas of atoms at 1 atm pressure and 25 °C,

$$N = p/kT = 2.46 \times 10^{25}\ m^{-3}\,,$$

and then $n_r \approx 1.000\,42$.

EXERCISE: Find an expression for the refractive index of a gas of one-electron ions of atomic number Z.

13.14 $(n_r^2 - 1)/(n_r^2 + 2) = R\rho/M_m$ [eqn (13.3.14)].

This rearranges to

$$n_r^2 = (2x+1)/(1-x)\,,\quad x = R\rho/M_m\,.$$

For gases, $\rho/M_m = p/RT = 40.9\ \mathrm{mol\,m^{-3}}$ at $p = 1$ atm, $T = 298$ K.

For water, $\rho/M_m = (1.00 \times 10^3\ \mathrm{kg\,m^{-3}})/(18.02\ \mathrm{g\,mol^{-1}}) = 5.55 \times 10^4\ \mathrm{mol\,m^{-3}}$.

For ethanoic acid, $\rho/M_m = 1.74 \times 10^4\ \mathrm{mol\,m^{-3}}$.

(a) $R(CH_3COOH)/\mathrm{cm^3\,mol^{-1}} = 3 \times 1.65 + 1.20 + 3.34 + 1.41 + 1.85 = 12.75.$
$x = (12.75 \times 10^{-6}\ \mathrm{m^3\,mol^{-1}}) \times (1.74 \times 10^4\ \mathrm{mol\,m^{-3}}) = 0.222.$
$n_r^2 = 1.855$; $\underline{n_r = 1.36}$ [experimental value : 1.3718].

(b) $R(CH_3{\cdot}CH_3)/\mathrm{cm^3\,mol^{-1}} = 6 \times 1.65 + 1.20 = 11.10$; $x = 4.54 \times 10^{-4}$.
$n_r^2 = 1.0014$; $\underline{n_r = 1.0007}$.

(c) $R(CH_2{:}CH_2)/\mathrm{cm^3\,mol^{-1}} = 4 \times 1.65 + 2.79 = 9.39$; $x = 3.84 \times 10^{-4}$.
$n_r^2 = 1.0012$; $\underline{n_r = 1.0006}$ [experimental value: 1.000 696 at 20 °C].

(d) $R(H_2O)/\mathrm{cm^3\,mol^{-1}} = 2 \times 1.85 = 3.70$; $x = 0.205$.
$n_r^2 = 1.775$; $\underline{n_r = 1.332}$ [experimental value: 1.3325].

EXERCISE : Calculate the refractive index of (a) CH_3CHO, (b) CH_3CH_2COOH, (c) $(COOH)_2$.

13.15 $\mu_A = \mu_A\hat{\boldsymbol{i}}$, $\boldsymbol{\mu}_B = \mu_B\hat{\boldsymbol{i}}$ [$\hat{\boldsymbol{i}}$ is the unit vector along x].

$\mu_A = -ex_1 + \frac{1}{2}eL$, $\mu_B = -ex_2 + \frac{1}{2}eL'$ [charge $+e$ at centres of boxes].

(a) $H^{(1)} = (1/4\pi\varepsilon_0 R^3)\{\mu_A\mu_B - 3\mu_A\mu_B\} = -2\mu_A\mu_B/4\pi\varepsilon_0 R^3$,

(b) $H^{(1)} = (1/4\pi\varepsilon_0 R^3)\{\mu_A\mu_B - 0\} = +\mu_A\mu_B/4\pi\varepsilon_0 R^3$.

$$\mu_A\mu_B = e^2\{x_1 x_2 + \tfrac{1}{4}LL' - \tfrac{1}{2}(x_1 L' + x_2 L)\}\,.$$

For the ground state set $|0\rangle \to |1_A 1_B\rangle$, then

$$\langle n_A n_B | H^{(1)} | 0 \rangle = \langle n_A n_B | \mu_A \mu_B | 1_A 1_B \rangle K \quad \begin{cases} \text{(a)} & K = -1/2\pi\varepsilon_0 R^3 \\ \text{(b)} & K = 1/4\pi\varepsilon_0 R^3 \end{cases}\,.$$

$$= e^2 \langle n_A n_B \rangle | x_1 x_2 | 1_A 1_B \rangle - \tfrac{1}{2} e^2 \langle n_A n_B | (x_1 L' + x_2 L) | 1_A 1_B \rangle$$

[LL' has no off-diagonal elements]

$$= e^2 x_{n_A 1} x_{n_B 1} - \tfrac{1}{2} e^2 \left(x_{n_A 1} L' \delta_{n_B 1} + x_{n_B 1} L \delta_{n_A 1} \right).$$

$$\mathcal{E}_{n_A n_B} = (h^2/8m_e)\left\{ (n_A/L)^2 + (n_B/L')^2 \right\}.$$

$$\mathcal{E}^{(2)} = K^2 e^4 \sum_{n_A n_B}{}' \left\{ x_{1n_A} x_{1n_B} - \tfrac{1}{2}\left(x_{1n_A} L' \delta_{n_B 1} + x_{1n_B} L \delta_{n_A 1} \right) \right\} \Big/ \left(\mathcal{E}_{11} - \mathcal{E}_{n_A n_B} \right)$$

$$= K^2 e^4 \left\{ \sum_{n_A n_B}{}' \left[x_{1n_A}^2 x_{1n_B}^2 \Big/ \left(\mathcal{E}_{11} - \mathcal{E}_{n_A n_B} \right) \right] + \tfrac{1}{4} L'^2 \sum_{n_A}{}' \left[x_{1n_A}^2 \Big/ \left(\mathcal{E}_1 - \mathcal{E}_{n_A} \right) \right] \right.$$
$$+ \tfrac{1}{4} L^2 \sum_{n_B}{}' \left[x_{1n_B}^2 \Big/ \left(\mathcal{E}_1 - \mathcal{E}_{n_B} \right) \right]$$
$$- L' \sum_{n_A}{}' \left[x_{1n_A}^2 x_{11} \Big/ \left(\mathcal{E}_1 - \mathcal{E}_{n_A} \right) \right]$$
$$\left. - L \sum_{n_B}{}' \left[x_{1n_B}^2 x_{11} \Big/ \left(\mathcal{E}_1 - \mathcal{E}_{n_B} \right) \right] \right\}.$$

[The primes exclude the $n_A = n_B = 1$ terms (but not $n_A = 1$, $n_B \neq 1$ and vice versa).]

$$x_{11}^{A} = \tfrac{1}{2} L, \quad x_{11}^{B} = \tfrac{1}{2} L$$

$$x_{1n_A} = \begin{cases} -(8/\pi^2)\, L\, \{ n_A/(n_A^2 - 1) \} & n_A \text{ even [Problem 8.5]} \\ 0 & n_A \text{ odd,} \end{cases}$$

$$x_{1n_B} = \begin{cases} -(8/\pi^2)\, L'\, \{ n_B/(n_B^2 - 1) \} & n_B \text{ even} \\ 0 & n_B \text{ odd.} \end{cases}$$

Be careful to allow for $n_A = 1$, $n_B \neq 1$ and $n_B = 1$, $n_A \neq 1$ in the first term. Then note that

$$\sum_{n_A n_B}{}' \left\{ x_{1n_A} x_{1n_B} \Big/ \left(\mathcal{E}_{11} - \mathcal{E}_{n_A n_B} \right) \right\}$$
$$= \sum_{n_A}^{\text{even}} \left\{ \frac{x_{1n_A}^2 x_{11}^2}{\left(\mathcal{E}_1 - \mathcal{E}_{n_A} \right)} \right\} + \sum_{n_B}^{\text{even}} \left\{ \frac{x_{11}^2 x_{1n_B}^2}{\mathcal{E}_1 - \mathcal{E}_{n_B}} \right\} + \sum_{n_A n_B}^{\text{both even}} \left\{ \frac{x_{1n_A}^2 x_{1n_B}^2}{\mathcal{E}_{11} - \mathcal{E}_{n_A n_B}} \right\}$$
$$= \tfrac{1}{4} L'^2 \sum_{n_A}^{\text{even}} \left\{ \frac{x_{1n_A}^2}{\mathcal{E}_1 - \mathcal{E}_{n_A}} \right\} + \tfrac{1}{4} L^2 \sum_{n_B}^{\text{even}} \left\{ \frac{x_{1n_B}^2}{\mathcal{E}_1 - \mathcal{E}_{n_B}} \right\} + \sum_{n_A n_B}^{\text{both even}} \left\{ \frac{x_{1n_A}^2 x_{1n_B}^2}{\mathcal{E}_{11} - \mathcal{E}_{n_A n_B}} \right\}.$$

But since $-L'\sum_{n_A}\left\{\dfrac{x_{1n_A}^2 x_{11}^B}{\mathcal{E}_1-\mathcal{E}_{n_A}}\right\} = -\tfrac{1}{2}L'^2\sum_{n_A}^{\text{even}}\left\{\dfrac{x_{1n_A}^2}{\mathcal{E}_1-\mathcal{E}_{n_A}}\right\}$

the terms involving L'^2 (and L^2) cancel, and we are left with

$$\mathcal{E}^{(2)} = K^2e^4\sum_{n_An_B}\left\{x_{1n_A}^2 x_{1n_B}^2\Big/\left(\mathcal{E}_{11}-\mathcal{E}_{n_An_B}\right)\right\} \qquad (n_A, n_B \text{ both even})$$

$$= \left\{\frac{K^2e^4(8/\pi^2)^4L^2L'^2}{h^2/8m_e}\right\}\sum_{n_An_B}\left\{\frac{[n_A^2n_B^2/(n_A^2-1)^2(n_B^2-1)^2]}{[(1-n_A^2)/L^2]+[(1-n_B^2)/L'^2]}\right\}$$

$$= (2^{15}K^2e^4m_e/\pi^8h^2)(LL')^4\sum_{n_An_B}\left\{\frac{[n_An_B/(n_A^2-1)^2(n_B^2-1)^2]}{(1-n_A^2)L'^2+(1-n_B^2)L^2}\right\}.$$

For simplicity, set $L'=L$; then

$$\mathcal{E}^{(2)} = -(2^{15}K^2e^4m_e/\pi^8h^2)L^6\sum_{n_An_B}\left\{n_A^2n_B^2/(n_A^2-1)^2(n_B^2-1)^2(n_A^2+n_B^2-2)\right\}.$$

The sum has the value 0.037 88 (over n_A, n_B each running to 30). Hence

$$\mathcal{E}^{(2)} = \underline{-0.1308(K^2e^4m_e/h^2)L^6}$$

$$= \begin{cases}\text{(a)} & -(5.779\times10^{-20}\text{ J})\times(L/R)^6\\ \text{(b)} & -(1.445\times10^{-20}\text{ J})\times(L/R)^6.\end{cases}$$

Therefore,

$$\mathcal{E}^{(2)}/\text{kJ mol}^{-1} = \begin{cases}\text{(a)} & \underline{-34.80(L/R)^6}\\ \text{(b)} & \underline{-\ 8.70(L/R)^6}.\end{cases}$$

For $L/R\approx1$, $-\mathcal{E}^{(2)}/\text{kJ mol}^{-1}\approx$ (a) 34.8, (b) −8.7.

EXERCISE: Evaluate the dispersion energy between electrons confined to two cubic boxes.

13.16 Under closure, starting at the point

$$-\mathcal{E}^{(2)}/K^2e^4 = -\sum_{n_An_B}^{\text{even}}\left\{x_{1n_A}^2x_{1n_B}^2/(\mathcal{E}_{11}-\mathcal{E}_{n_An_B})\right\}$$

$$\approx (\Delta_A+\Delta_B)^{-1}\sum_{n_An_B}^{\text{even}}x_{1n_A}^2x_{1n_B}^2 \qquad (\Delta_A=\mathcal{E}_{n_A}-\mathcal{E}_1;\ \Delta_B=\mathcal{E}_{n_B}-\mathcal{E}_1)$$

$$\approx (\Delta_A+\Delta_B)^{-1}\left\{\left(\sum_{n_A}x_{1n_A}^2\right)-x_{11}^2\right\}\left\{\left(\sum_{n_B}x_{1n_B}^2\right)-x_{11}^2\right\}$$

$$\approx (\Delta_A+\Delta_B)^{-1}\left\{(x^2)_{11}-(x_{11})^2\right\}\left\{(x^2)_{11}-(x_{11})^2\right\}.$$

We have $x_{11} = \frac{1}{2}L$ and $(x^2)_{11} = \frac{1}{3}L^2(1 - 3/2\pi^2)$ [Problem 13.3].
Consequently,

$$-\mathcal{E}^{(2)}/K^2e^4 \approx (\Delta_A + \Delta_B)^{-1}\left\{(1/12) - (1/2\pi^2)\right\}^2 L^2 L'^2$$
$$\approx 1.068 \times 10^{-3}\, L^2 L'^2/(\Delta_A + \Delta_B)\,.$$

For $L' = L$ and $\Delta_A = \Delta_B = \Delta$,

$$-\mathcal{E}^{(2)} \approx 5.34 \times 10^{-4}(L^4 K^2 e^4/\Delta)$$

Set $\Delta = \lambda(h^2/8m_e L^2)$; then

$$-\mathcal{E}^{(2)} \approx 4.27 \times 10^{-3}\, m_e e^4 K^2 L^6/\lambda h^2\,.$$

Compare this with

$$-\mathcal{E}^{(2)} = 0.1308(m_e e^4 K^2 L^6/h^2) \qquad \text{[Problem 13.15]}\,.$$

It follows that we should choose $\lambda = 0.0326$.

EXERCISE: Find the appropriate value of λ for the dispersion interaction treated in the *Exercise* to Problem 13.15.

13.17 $\mathcal{E}^{(2)} \approx -\frac{3}{2}\eta\,[I_A I_B/(I_A + I_B)](\alpha'_A\alpha'_B/R^6)$.

$I_A = I_B \approx 13.6$ eV $\hat{=}$ 1312 kJ mol^{-1}; $\alpha'_A = \alpha'_B = 6.6 \times 10^{-25}$ cm^3 [Problem 13.5].

Consequently,

$$\mathcal{E}^{(2)} \approx -\frac{3}{4}\eta\, I\alpha'^2/R^6 = \underline{-(4.29 \times 10^{-4}\ \text{kJ mol}^{-1}) \times \{\eta/(R/\text{nm})^6\}}\,.$$

EXERCISE: Evaluate the dispersion energy directly on the basis of eqn (13.4.8) and the matrix elements listed in the solution to Problem 13.4.

13.18 Take as a trial function $\psi = \psi_{1s} + a\,\psi_{2p_z}$ for each atom, so that the overall trial function is

$$\psi = (\psi_{A,1s} + a_A\psi_{A,2p_z})(\psi_{B,1s} + a_B\psi_{B,2p_z})\,.$$

The denominator of the Rayleigh ratio is therefore

$$\int\psi^2\,d\tau = \int(\psi_{A,1s} + a_A\psi_{A,2p_z})^2(\psi_{B,1s} + a_B\psi_{B,2p_z})^2\,d\tau_A\,d\tau_B$$
$$= (1 + a_A^2)(1 + a_B^2) \qquad \text{[basis functions are orthonormal]}.$$

The hamiltonian is

$$H = H_A + H_B + H^{(1)}\,, \qquad H_A\psi_{A,nl} = \mathcal{E}_n\psi_{A,nl}\,.$$

The numerator of the Rayleigh ratio is therefore

$$\int \psi H \psi \, d\tau = \int \left(\psi_{A,1s} + a_A \psi_{A,2p_z}\right)\left(\psi_{B,1s} + a_B \psi_{B,2p_z}\right)$$
$$\times \left\{\left(\mathcal{E}_1 \psi_{A,1s} + a_A \mathcal{E}_2 \psi_{A,2p_z}\right)\left(\psi_{B,1s} + a_B \psi_{B,2p_z}\right) + \left(\psi_{A,1s} + a_A \psi_{A,2p_z}\right)\left(\mathcal{E}_1 \psi_{B,1s} + a_B \mathcal{E}_2 \psi_{B,2p_z}\right) + H^{(1)}\left(\psi_{A,1s} + a_A \psi_{A,2p_z}\right)\left(\psi_{B,1s} + a_B \psi_{B,2p_z}\right)\right\}$$
$$= \left(\mathcal{E}_1 + a_A^2 \mathcal{E}_2\right)\left(1 + a_B^2\right) + \left(\mathcal{E}_1 + a_B^2 \mathcal{E}_2\right)\left(1 + a_A^2\right)$$
$$+ \int \left(\psi_{A,1s}\psi_{B,1s} + a_A \psi_{A,2p_z}\psi_{B,1s} + a_B \psi_{A,1s}\psi_{B,2p_z} + a_A a_B \psi_{A,2p_z}\psi_{B,2p_z}\right)$$
$$\times H^{(1)}\left(\psi_{A,1s}\psi_{B,1s} + a_A \psi_{A,2p_z}\psi_{B,1s} + a_B \psi_{A,1s}\psi_{B,2p_z} + a_A a_B \psi_{A,2p_z}\psi_{B,2p_z}\right) d\tau_A \, d\tau_B .$$

Only the $z_A z_B$ components of $H^{(1)}$ contributes to the integral (because only it has non-vanishing matrix elements between 1s and $2p_z$), and so we take

$$H^{(1)} = -2(1/4\pi\varepsilon_0 R^3)\, \mu_{Az}\, \mu_{Bz} .$$

Then the only surviving terms are

$$a_A a_B \left\{\int \psi_{A,1s}\psi_{B,1s} H^{(1)} \psi_{A,2p_z}\psi_{B,2p_z} d\tau_A \, d\tau_B + \int \psi_{A,2p_z}\psi_{B,2p_z} H^{(1)} \psi_{A,1s}\psi_{B,1s} d\tau_A \, d\tau_B\right\}$$
$$= -(e^2/\pi\varepsilon_0 R^3)\, a_A a_B\, z_{A;1s,2p_z} z_{B;1s,2p_z}$$
$$= -a_A a_B KZ, \quad K = e^2/\pi\varepsilon_0 R^3, \quad z = z_{A;1s,2p_z} z_{B;1s,2p_z} .$$

The Rayleigh ratio is then

$$\mathcal{E} = (\mathcal{E}_1 + a_A^2 \mathcal{E}_2)/(1 + a_A^2) + (\mathcal{E}_1 + a_B^2 \mathcal{E}_2)/(1 + a_B^2)$$
$$- a_A a_B KZ/(1 + a_A^2)(1 + a_B^2) .$$

The optimum values of $a_A a_B$ are those for which $d\mathcal{E}/da_A = d\mathcal{E}/da_B = 0$.

$$d\mathcal{E}/da_A = 2 a_A \mathcal{E}_2/(1 + a_A^2) - 2a_A(\mathcal{E}_1 + a_A^2 \mathcal{E}_2)/(1 + a_A^2)^2$$
$$- a_B KZ/(1 + a_A^2)(1 + a_B^2) + 2a_A^2 a_B KZ/(1 + a_A^2)^2(1 + a_B^2) = 0 .$$

Likewise for $d\mathcal{E}/da_B$. Therefore we must solve

$$2(\mathcal{E}_2-\mathcal{E}_1)\,a_A + 2(\mathcal{E}_2-\mathcal{E}_1)\,a_B^2 a_A - a_B KZ + a_B a_A^2 KZ = 0$$

$$2(\mathcal{E}_2-\mathcal{E}_1)\,a_B + 2(\mathcal{E}_2-\mathcal{E}_1)\,a_A^2 a_B - a_A KZ + a_A a_B^2 KZ = 0 \quad .$$

These solve to $a_A a_B = KZ/2\Delta$, with $\Delta = \mathcal{E}_2 - \mathcal{E}_1 = \frac{3}{4}hcR_H$. Then, since $a_A^2 = a_B^2$ by symmetry, we also have

$$a_A = \pm\sqrt{(KZ/2\Delta)}\ , \quad a_B = \pm\sqrt{(KZ/2\Delta)}\,.$$

It follows that, setting $\gamma = KZ/2\Delta$,

$$\mathcal{E} = \left\{\frac{2\mathcal{E}_1 + 2\gamma\mathcal{E}_2}{1+\gamma}\right\} - \left\{\frac{\gamma KZ}{(1+\gamma)^2}\right\}.$$

The question now arises as to the identity of the dispersion energy. If we were to identify it as $\mathcal{E}_2 - 2\mathcal{E}_1$ we would find

$$\mathcal{E} - 2\,\mathcal{E}_1 = [2\gamma/(1+\gamma)^2]\,\Delta$$

which is (a) positive, (b) proportional to R^{-3} (when $\gamma \ll 1$). The alternative procedure (for which I can give no justification) is to identify the first term of $\mathcal{E}$ as the change of energies of the atoms themselves, and to identify the second term (the one proportional to γKZ) as the energy due to their interaction. This division of the global energy, and then identifying a component as the term of interest, is plainly artificial, but I know of no discussion of the point. It may be that, noting that the first term is proportional to $1/R^3$ (for $\gamma \ll 1$) while the second term is proportional to $1/R^6$ (because $\gamma K \propto 1/R^6$), the second term dominates the first at sufficiently small distances. Nevertheless, there is obviously some difficulty even with this view, because $\mathcal{E} - 2\mathcal{E}_1$ is always positive. Perhaps when a more extended trial function is used the second term increases its contribution more quickly than the first, and what we are seeing is the beginnings of the emergence of an attractive dispersion interaction. In any event, let us set

$$\mathcal{E}_{\mathbf{disp}} = \underline{-\gamma KZ/(1+\gamma)^2}\,.$$

We have $\gamma = KZ/2\Delta = (2^{16}/3^{11})(e^2 a_0^2/\pi\varepsilon_0 R^3 hcR_H)$

$$= 4.386 \times 10^{-4}/(R/\mathrm{nm})^3\,,$$

$$KZ = (2^{15}/3^{10})e^2 a_0^2/\pi\varepsilon_0 R^3 = (1.3953\times 10^{-21}\ \mathrm{J})/(R/\mathrm{nm})^3$$

$$\hat{=}\ (0.840\ \mathrm{kJ\ mol^{-1}})/(R/\mathrm{nm})^3.$$

Then

$$\gamma KZ = \{3.684 \times 10^{-4}/(R/\mathrm{nm})^6\}\ \mathrm{kJ\ mol^{-1}}\,.$$

$$\mathcal{E}_{\mathrm{disp}}/\mathrm{kJ\ mol^{-1}} = -\frac{[3.684\times10^{-4}/(R/\mathrm{nm})^6]}{\{1+[4.386\times10^{-4}/(R/\mathrm{nm})^3]\}^2} \approx \underline{-3.684\times10^{-4}/(R/\mathrm{nm})^6},$$

so long as $4.386\times10^{-4}/(R/\mathrm{nm})^3 \ll 1$, or $R \gg 0.08$ nm. The value of $\mathcal{E}_{\mathrm{disp}}$ calculated here is consistent with that in Problem 13.17.

EXERCISE: Calculate the dispersion energy on the basis that the trial function (a) also includes a 3p-orbital component, (b) includes a '1p-orbital' component.

13.19 Upper orbital: $\psi = \mathrm{p}_z\cos\eta + \mathrm{d}_{xy}\sin\eta$. Lower orbital: pure p_x.

$$l_y\,\mathrm{p}_x = (\hbar/\mathrm{i})\left\{z(\partial/\partial x) - x(\partial/\partial z)\right\} xf(r)$$
$$= (\hbar/\mathrm{i})\,zf(r) = -\mathrm{i}\hbar\,\mathrm{p}_z.$$

$$\langle\psi|m_y|\mathrm{p}_x\rangle = \gamma_{\mathrm{e}}\langle\mathrm{p}_z|l_y|\mathrm{p}_z\rangle\cos\eta = -\mathrm{i}\hbar\gamma_{\mathrm{e}}\cos\eta$$
$$= \mathrm{i}\mu_{\mathrm{B}}\cos\eta \qquad [\mu_{\mathrm{B}} = -\hbar\gamma_{\mathrm{e}}]$$

$$\langle\mathrm{p}_x|\mu_y|\psi\rangle = -e\langle\mathrm{p}_x|y|\mathrm{d}_{xy}\rangle\sin\eta$$
$$= -ea_0\left\{2^6(6\zeta_{\mathrm{p}}^5\zeta_{\mathrm{d}}^7)^{\frac{1}{2}}/(\zeta_{\mathrm{p}}+\zeta_{\mathrm{d}})^7\right\}\sin\eta \qquad [\textit{Example}, \text{ p. 372}].$$

Therefore, from eqn. (13.6.18),

$$R = -ea_0\,\mu_{\mathrm{B}}\left\{2^5(6\zeta_{\mathrm{p}}^5\zeta_{\mathrm{d}}^7)^{\frac{1}{2}}/(\zeta_{\mathrm{p}}+\zeta_{\mathrm{d}})^7\right\}\sin 2\eta$$
$$= -(6.16\times10^{-51}\ \mathrm{C^2\,m^3\,s^{-1}})\,A\sin 2\eta, \qquad A = (\zeta_{\mathrm{p}}^5\zeta_{\mathrm{d}}^7)^{\frac{1}{2}}/(\zeta_{\mathrm{p}}+\zeta_{\mathrm{d}})^7\,a_0.$$

For a carbon atom, $\zeta_{\mathrm{p}} = 1.95/a_0$, $\zeta_{\mathrm{d}} = 0.33/a_0$ [Section 9.10]. Therefore, $A = 3.42\times10^{-4}$. Consequently,

$$R = \underline{-(2.108\times10^{-54}\ \mathrm{C^2\,m^3\,s^{-1}})\sin 2\eta}.$$

The angle of optical rotation is given by

$$\theta \approx -3.0\times10^3\,(C/\mathrm{mol\,dm^{-3}})(l/\mathrm{dm})\{\omega^2/(\omega_{k0}^2-\omega^2)\}\,A\sin 2\eta \qquad [\textit{Example}, \text{ p.372}]$$
$$\approx -1.03(C/\mathrm{mol\ dm^{-3}})(l/\mathrm{dm})\{\omega^2/(\omega_{k0}^2-\omega^2)\}\sin 2\eta.$$

For $\lambda = 590$ nm ($\omega = 2\pi c/\lambda = 3.19\times10^{15}\ \mathrm{s^{-1}}$) and $\lambda_{k0} \approx 200$ nm ($\omega_{k0} = 9.42\times10^{15}\ \mathrm{s^{-1}}$), for $C = 1$ mol dm^{-1} and $l = 1$ dm,

$$\underline{\theta \approx -0.13\sin 2\eta}.$$

This is the angle in radians; it corresponds to $\underline{-7.6^\circ\sin 2\eta}$.

EXERCISE: Calculate the rotational strength of the $\psi \leftarrow 3\mathrm{p}_x$ transition, where $\psi = (3\mathrm{p}_z)\cos\eta + (3\mathrm{d}_{xy})\sin\eta$. Evaluate it for the sulphur atom, and estimate the angle of rotation.

13.20 $\psi_0 = |0_X 0_Y 0_Z\rangle$, the unperturbed ground state. The perturbed state is

$$\psi = |0_X 0_Y 0_Z\rangle + \sum_{v_X v_Y v_Z} a_{v_X v_Y v_Z} |v_X v_Y v_Z\rangle ,$$

$$a_{v_X v_Y v_Z} = \langle v_X v_Y v_Z | H^{(1)} | 0_X 0_Y 0_Z\rangle / (E_0 - E_{v_X v_Y v_Z})$$

The matrix element $\langle v_X v_Y v_Z | H^{(1)} | 0_X 0_Y 0_Z\rangle = A\langle v_X v_Y v_Z | xyz | 0_X 0_Y 0_Z\rangle$ vanishes unless $v_X = v_Y = v_Z = 1$. Therefore,

$$\psi = |0_X 0_Y 0_Z\rangle + a_{111}|1_X 1_Y 1_Z\rangle = |000\rangle + a_{111}|111\rangle \quad \text{for brevity.}$$

$$a_{111} = A\langle 111|xyz|000\rangle/(E_0 - E_{111}) = -\frac{A(\hbar^3/8m_e^3\omega_X\omega_Y\omega_Z)^{\frac{1}{2}}}{\hbar(\omega_X+\omega_Y+\omega_Z)} \quad \text{[Problem 5.15].}$$

The upper state contributing to R_{n0} is also perturbed. Suppose we are considering the $1_Z \leftarrow 0_Z$ transition, then the upper state, nominally $|001\rangle$, is in fact

$$\psi_n = |0_X 0_Y 1_Z\rangle + \sum_{v_X v_Y v_Z} a_{v_X v_Y v_Z} |v_X v_Y v_Z\rangle$$

$$= |001\rangle + a_{110}|110\rangle + a_{112}|112\rangle$$

$$a_{110} = -A(\hbar^3/8m_e^3\omega_X\omega_Y\omega_Z)^{\frac{1}{2}}/\hbar(\omega_X+\omega_Y-\omega_Z) \qquad \text{[Problem 5.15]}$$

$$a_{112} = -A(\hbar^3/8m_e^3\omega_X\omega_Y\omega_Z)^{\frac{1}{2}}(\surd 2)/\hbar(\omega_X+\omega_Y+\omega_Z).$$

Therefore,

$$\psi_n = |001\rangle - (\hbar^3/4m_e^3\omega_X\omega_Y\omega_Z)^{\frac{1}{2}} A\left\{\left(\frac{|112\rangle}{\omega_X+\omega_Y+\omega_Z}\right) + \left(\frac{|110\rangle}{\omega_X+\omega_Y-\omega_Z}\right)\right\}.$$

The electric dipole transition moment is therefore

$$\mu_{z,0n} = -e\left\{\langle 000| + a_{111}\langle 111|\right\} z \left\{|001\rangle + a_{112}|112\rangle + a_{110}|110\rangle\right\}$$

$$= -e\left\{\langle 000|z|001\rangle + a_{111}a_{112}\langle 111|112\rangle + a_{111}a_{110}\langle 111|110\rangle\right\}$$

$$= -e\langle 000|z|001\rangle \qquad \text{[to first-order in } A\text{]}$$

$$= -e(\hbar/2m_e\omega_Z)^{\frac{1}{2}}.$$

Now consider $m_{z,n0} = \gamma_e l_{z,n0}$. Note that, since $l_z = xp_y - yp_x$,

$$\langle v_X v_Y v_Z | l_z | 000\rangle = \langle v_X v_Y v_Z | xp_y | 000\rangle - \langle v_X v_Y v_Z | yp_x | 000\rangle$$

$$= \begin{cases} \langle 110|xp_y|000\rangle - \langle 110|yp_x|000\rangle & \text{for } v_X v_Y v_Z = 110 \\ 0 \quad \text{for all other values.} \end{cases}$$

From the matrix elements listed in Problem 5.15,

$$\langle 110|l_z|000\rangle = \mathrm{i}(\hbar/2m_\mathrm{e}\omega_X)^{\frac{1}{2}}(\hbar m_\mathrm{e}\omega_Y/2)^{\frac{1}{2}} - \mathrm{i}(\hbar/2m_\mathrm{e}\omega_Y)^{\frac{1}{2}}(\hbar m_\mathrm{e}\omega_X/2)^{\frac{1}{2}}$$
$$= \tfrac{1}{2}\mathrm{i}\hbar\left\{(\omega_Y/\omega_X)^{\frac{1}{2}} - (\omega_X/\omega_Y)^{\frac{1}{2}}\right\}.$$

Likewise:

$$\langle v_X v_Y v_Z|l_z|111\rangle = \begin{cases} \langle 221|l_z|111\rangle & \text{for } v_X v_Y v_Z = 221 \\ \langle 201|l_z|111\rangle & \text{for } v_X v_Y v_Z = 201 \\ \langle 021|l_z|111\rangle & \text{for } v_X v_Y v_Z = 021 \\ \langle 001|l_z|111\rangle & \text{for } v_X v_Y v_Z = 001 \\ 0 & \text{for all other values.} \end{cases}$$

We shall need only the last:

$$\langle 001|l_z|111\rangle = \langle 001|xp_y|111\rangle - \langle 001|yp_x|111\rangle$$
$$= -\mathrm{i}(\hbar/2m_\mathrm{e}\omega_X)^{\frac{1}{2}}(\hbar m_\mathrm{e}\omega_Y/2)^{\frac{1}{2}} + \mathrm{i}(\hbar/2m_\mathrm{e}\omega_Y)^{\frac{1}{2}}(\hbar m_\mathrm{e}\omega_X/2)^{\frac{1}{2}}$$
$$= \tfrac{1}{2}\mathrm{i}\hbar\left\{(\omega_X/\omega_Y)^{\frac{1}{2}} - (\omega_Y/\omega_X)^{\frac{1}{2}}\right\}.$$

The magnetic dipole transition moment is therefore

$$m_{z,n0} = \gamma_\mathrm{e}\left\{\langle 001| + a_{112}\langle 112| + a_{110}\langle 110|\right\} l_z \left\{|000\rangle + a_{111}|111\rangle\right\}$$
$$= \gamma_\mathrm{e} a_{110}\langle 110|l_z|000\rangle + \gamma_\mathrm{e} a_{111}\langle 001|l_z|111\rangle$$
$$= \tfrac{1}{2}\mathrm{i}\hbar\gamma_\mathrm{e} a_{110}\left\{(\omega_Y/\omega_X)^{\frac{1}{2}} - (\omega_X/\omega_Y)^{\frac{1}{2}}\right\}$$
$$- \tfrac{1}{2}\mathrm{i}\hbar\gamma_\mathrm{e} a_{111}\left\{(\omega_Y/\omega_X)^{\frac{1}{2}} - (\omega_X/\omega_Y)^{\frac{1}{2}}\right\}$$
$$= \tfrac{1}{2}\mathrm{i}\hbar\gamma_\mathrm{e}(a_{110} - a_{111})\left\{(\omega_Y/\omega_X)^{\frac{1}{2}} - (\omega_X/\omega_Y)^{\frac{1}{2}}\right\}.$$

Since

$$a_{110} - a_{111} = -(\hbar^3/8m_\mathrm{e}^3\omega_X\omega_Y\omega_Z)^{\frac{1}{2}}(A/\hbar)\left\{\frac{1}{\omega_X+\omega_Y-\omega_Z} - \frac{1}{\omega_X+\omega_Y+\omega_Z}\right\}$$
$$= -A(\hbar/2m_\mathrm{e}^3)^{\frac{1}{2}}\left\{\frac{(\omega_Z/\omega_X\omega_Y)^{\frac{1}{2}}}{(\omega_X+\omega_Y)^2 - \omega_Z^2}\right\},$$

it follows that

$$R_{n0}^{(z)} = \mathrm{im}(\mu_{0n}^z m_{n0}^z)$$
$$= e(\hbar/2m_\mathrm{e}\omega_Z)^{\frac{1}{2}}(-\tfrac{1}{2}\mu_\mathrm{B})A(\hbar/2m_\mathrm{e}^3)^{\frac{1}{2}}\left\{\frac{(\omega_Z/\omega_X\omega_Y)^{\frac{1}{2}}\{(\omega_Y/\omega_X)^{\frac{1}{2}} - (\omega_X/\omega_Y)^{\frac{1}{2}}\}}{(\omega_X+\omega_Y)^2 - \omega_Z^2}\right\}$$
$$= (\hbar A\mu_\mathrm{B}/4m_\mathrm{e}^2)\left\{\frac{(\omega_X - \omega_Y)/\omega_X\omega_Y}{(\omega_X+\omega_Y)^2 - \omega_Z^2}\right\}.$$

Since the perturbation and the ground state are symmetric under cyclic interchange of x, y, z, the contribution from the x- and y-components may be written down by cyclic interchange of X, Y, Z:

$$\underline{R_{n0} = (\hbar A \mu_{\mathrm{B}}/4m_{\mathrm{e}}^2)\, F(\omega_X, \omega_Y, \omega_Z)}\,,$$

$$F(\omega_X,\omega_Y,\omega_Z) = \left(\frac{(\omega_X-\omega_Y)/\omega_X\omega_Y}{[(\omega_X+\omega_Y)^2-\omega_Z^2]}\right) + \left(\frac{(\omega_Z-\omega_X)/\omega_X\omega_Z}{[(\omega_X+\omega_Z)^2-\omega_Y^2]}\right) + \left(\frac{(\omega_Y-\omega_Z)/\omega_Z\omega_Y}{[(\omega_Z+\omega_Y)^2-\omega_X^2]}\right)$$

Note that if any two frequencies are the same (i.e. the set of oscillators has cylindrical symmetry), then $F=0$ and consequently $R_{n0}=0$.

EXERCISE: Evaluate $R_{n'n}$ for a general initial state of the oscillators.

14 The magnetic properties of molecules

14.1 $\chi = 6.3001\times10^{-3}\{S(S+1)/M_r(T/\mathrm{K})\}$ [eqn (14.1.10)]. For hydrogen atoms at 298 K, $S=\frac{1}{2}$, $M_r = 1.008$, and $T/\mathrm{K} = 298$; then $\underline{\chi = 1.57\times10^{-5}}$.

EXERCISE: Calculate the spin contribution to χ for the ground state of a nitrogen atom at 298 K.

14.2 Under the influence of the perturbation $H^{(1)} = -\gamma_e l_z B$ the non-degenerate, real, groundstate ψ_0 changes to $\psi = \psi_0 + a\psi_1$, with $a = -\langle\psi_1|H^{(1)}|\psi_0\rangle/\Delta$ [eqn (8.1.22)]. In this case $a = \gamma_e B l_{z,10}/\Delta$, which is imaginary. Therefore ψ is complex. It follows that

$$\langle\psi|l_q|\psi\rangle = \langle\psi|l_q|\psi\rangle^* \quad \text{[hermiticity]}$$

$$\langle\psi|l_q|\psi\rangle = \langle\psi_0|l_q|\psi_0\rangle + \langle\psi_0|l_q|\psi_1\rangle a + a^*\langle\psi_1|l_q|\psi_0\rangle + O(a^2)$$

$$\langle\psi|l_q|\psi\rangle^* = \langle\psi_0|l_q|\psi_0\rangle^* + \langle\psi_0|l_q|\psi_1\rangle^* a^* + a\langle\psi_1|l_q|\psi_0\rangle^* + O(a^2)$$

$$= -\langle\psi_0|l_q|\psi_0\rangle - \langle\psi_0|l_q|\psi_1\rangle a^* - a\langle\psi_1|l_q|\psi_0\rangle + O(a^2).$$

On comparing the two expressions, noting that $\langle\psi_0|l_q|\psi_0\rangle = 0$, we are left with

$$\langle\psi_0|l_q|\psi_1\rangle a + a^*\langle\psi_1|l_q|\psi_0\rangle = \langle\psi_0|l_q|\psi_1\rangle a + a^*\langle\psi_1|l_q|\psi_0\rangle,$$

which does not require a value of zero. Hence $\langle\psi|l_q|\psi\rangle$ need not disappear.

EXERCISE: Show that the expectation value of linear momentum is zero when the state is real but may be non-zero in the presence of an appropriate perturbation.

14.3 $\langle S_z^2\rangle = \sum_{M_S} P_{M_S}\langle M_S|S_z^2|M_S\rangle$, $P_{M_S} = 1/(2S+1)$

$$= \{1/(2S+1)\}\hbar^2\sum_{M_S=-S}^{S} M_S^2 = 2\{\hbar^2/(2S+1)\}\sum_{M_S=0}^{S} M_S^2$$

$$= \underline{\tfrac{1}{3}S(S+1)\hbar^2}.$$

$$\langle S_x S_z\rangle = \sum_{M_S} P_{M_S}\langle M_S|S_x S_z|M_S\rangle = \sum_{M_S} P_{M_S}\langle M_S|S_x|M_S\rangle\hbar M_S$$

$$= \underline{0} \qquad [S_x \text{ is off-diagonal}].$$

$$\langle S_z^4\rangle = \sum_{M_S} P_{M_S}\langle M_S|S_z^4|M_S\rangle = \{\hbar^4/(2S+1)\}\sum_{M_S=-S}^{S} M_S^4$$

$$= 2\{\hbar^4/(2S+1)\}\sum_{M_S=0}^{S} M_S^4$$

$$= \underline{(1/15)\hbar^4\{(S+1)/(2S+1)\}\{6(S+1)^4-15(S+1)^3+10(S+1)^2-1\}}.$$

[See *Handbook of mathematical functions*, M. Ambramowitz and I.A. Stegun, Dover (1965), eqn (23.1.4) and Table 23.1.]

EXERCISE: Evaluate S_z^6 for the same system.

14.4 Since $\langle S_x^2\rangle = \langle S_y^2\rangle = \langle S_z^2\rangle$, it follows that

$$\langle S_z^2\rangle = \tfrac{1}{3}\langle S_x^2 + S_y^2 + S_z^2\rangle = \tfrac{1}{3}\langle S^2\rangle.$$

But $\langle S^2\rangle = \hbar^2 S(S+1)$ for each state. Therefore, $\langle S_z^2\rangle = \underline{\tfrac{1}{3}S(S+1)\hbar^2}$.

EXERCISE: Evaluate $\langle S_z^4\rangle$ in this way.

14.5 See Fig. 14.1. Since $V_x = -z$, $V_y = 0$, and $V_z = x$, we have

Fig. 14.1

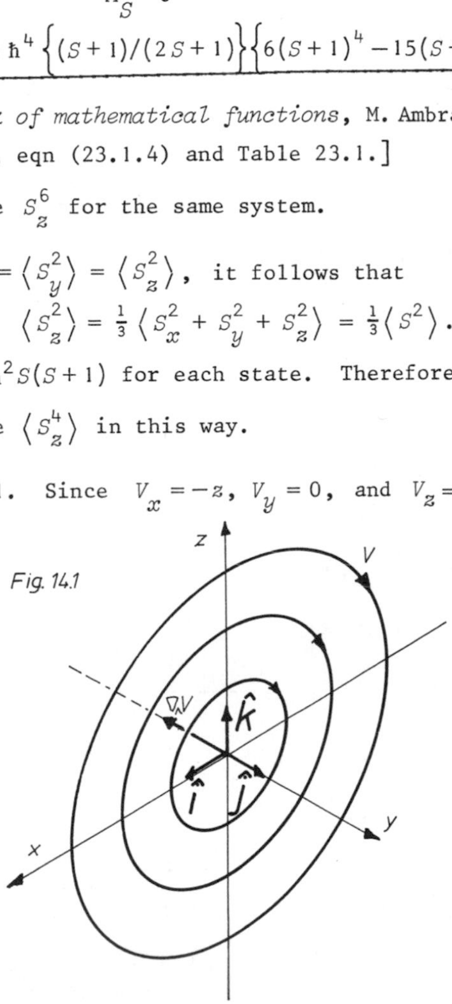

$$\nabla\cdot V = (\partial/\partial x)(-z) + (\partial/\partial y)0 + (\partial/\partial z)x = \underline{0}.$$

$$\nabla\wedge V = \begin{vmatrix} \hat{i} & \hat{j} & \hat{k} \\ (\partial/\partial x) & (\partial/\partial y) & (\partial/\partial z) \\ z & 0 & x \end{vmatrix} = \hat{i}[(\partial x/\partial y) - (\partial 0/\partial x)]$$

$$- \hat{j}[(\partial x/\partial x) + (\partial z/\partial z)] + \hat{k}[(\partial 0/\partial x) + (\partial z/\partial y)]$$

$$= \underline{-2\hat{j}}.$$

EXERCISE: Sketch the form of $\boldsymbol{V}=x^2\hat{\boldsymbol{k}}-z^2\hat{\boldsymbol{i}}$, and calculate its divergence and curl.

14.6 $\nabla\wedge\boldsymbol{A}=\frac{1}{2}\nabla\wedge(\boldsymbol{B}\wedge\boldsymbol{r})$

$$=\tfrac{1}{2}\left\{\boldsymbol{B}(\nabla\cdot\boldsymbol{r})-(\nabla\cdot\boldsymbol{B})\boldsymbol{r}+(\boldsymbol{r}\cdot\nabla)\boldsymbol{B}-(\boldsymbol{B}\cdot\nabla)\boldsymbol{r}\right\} \qquad \text{[eqn (A20.7f)]}$$

$$=\tfrac{1}{2}\left\{3\boldsymbol{B}-0+0-(\boldsymbol{B}\cdot\nabla)\boldsymbol{r}\right\}.$$

$$\boldsymbol{B}\cdot\nabla=(B_x\hat{\boldsymbol{i}}+B_y\hat{\boldsymbol{j}}+B_z\hat{\boldsymbol{k}})\cdot\left\{\hat{\boldsymbol{i}}(\partial/\partial x)+\hat{\boldsymbol{j}}(\partial/\partial y)+\hat{\boldsymbol{k}}(\partial/\partial z)\right\}$$

$$=B_x(\partial/\partial x)+B_y(\partial/\partial y)+B_z(\partial/\partial z).$$

$$(\boldsymbol{B}\cdot\nabla)\boldsymbol{r}=\left\{B_x(\partial/\partial x)+B_y(\partial/\partial y)+B_z(\partial/\partial z)\right\}\{x\hat{\boldsymbol{i}}+y\hat{\boldsymbol{j}}+z\hat{\boldsymbol{k}}\}$$

$$=B_x\hat{\boldsymbol{i}}+B_y\hat{\boldsymbol{j}}+B_z\hat{\boldsymbol{k}}=\boldsymbol{B}.$$

Consequently, $\nabla\wedge\boldsymbol{A}=\frac{1}{2}\{3\boldsymbol{B}-\boldsymbol{B}\}=\boldsymbol{B}$.

(a) $\boldsymbol{B}=B\hat{\boldsymbol{i}}$;

$$\boldsymbol{A}=\tfrac{1}{2}B\hat{\boldsymbol{i}}\wedge\boldsymbol{r}=\tfrac{1}{2}\begin{vmatrix}\hat{\boldsymbol{i}} & \hat{\boldsymbol{j}} & \hat{\boldsymbol{k}}\\ B & 0 & 0\\ x & y & z\end{vmatrix}=\underline{-\tfrac{1}{2}B(-z\hat{\boldsymbol{j}}+y\hat{\boldsymbol{k}})}.$$

(b) $\boldsymbol{B}=B(\hat{\boldsymbol{i}}+\hat{\boldsymbol{j}}+\hat{\boldsymbol{k}})/\sqrt{3}$;

$$\boldsymbol{A}=(B/2\sqrt{3})(\hat{\boldsymbol{i}}+\hat{\boldsymbol{j}}+\hat{\boldsymbol{k}})\wedge\boldsymbol{r}$$

$$=(B/2\sqrt{3})\begin{vmatrix}\hat{\boldsymbol{i}} & \hat{\boldsymbol{j}} & \hat{\boldsymbol{k}}\\ 1 & 1 & 1\\ x & y & z\end{vmatrix}=\underline{(B/2\sqrt{3})\{(z-y)\hat{\boldsymbol{i}}-(z-x)\hat{\boldsymbol{j}}+(y-x)\hat{\boldsymbol{k}}\}}.$$

For a uniform field,

$$A^2=\{\tfrac{1}{2}(\boldsymbol{B}\wedge\boldsymbol{r})\}\cdot\{\tfrac{1}{2}(\boldsymbol{B}\wedge\boldsymbol{r})\}$$

$$=\tfrac{1}{4}\{(\boldsymbol{B}\cdot\boldsymbol{B})(\boldsymbol{r}\cdot\boldsymbol{r})-(\boldsymbol{B}\cdot\boldsymbol{r})(\boldsymbol{r}\cdot\boldsymbol{B})\}=\underline{\tfrac{1}{4}\{B^2r^2-(\boldsymbol{B}\cdot\boldsymbol{r})^2\}}.$$

(a) $\boldsymbol{B}\cdot\boldsymbol{r}=Bx$; $\quad A^2=\frac{1}{4}B^2(r^2-x^2)=\underline{\frac{1}{4}B^2(x^2+y^2)}$.

(b) $\boldsymbol{B}\cdot\boldsymbol{r}=(B/\sqrt{3})(x+y+z)$; $\quad A^2=\underline{\frac{1}{4}B^2\{r^2-\frac{1}{3}(x+y+z)^2\}}$.

EXERCISE: Find the expressions for the vector potentials representing uniform fields directed towards the corners of a regular tetrahedron. Evaluate $\boldsymbol{B}$ for $\boldsymbol{A}=B(x^2\hat{\boldsymbol{k}}-z^2\hat{\boldsymbol{i}})$.

14.7 $\boldsymbol{A}=\frac{1}{2}B\boldsymbol{V}+\lambda\boldsymbol{V}'$ [eqn (14.2.10)].

$H^{(0)}=p^2/2m_e\rightarrow H=(\boldsymbol{p}+e\boldsymbol{A})^2/2m_e$ [rule on p. 383 of text].

$$H = (p^2/2m_e) + (e/2m_e)(\boldsymbol{p}\cdot\boldsymbol{A} + \boldsymbol{A}\cdot\boldsymbol{p}) + (e^2/2m_e)A^2 .$$

$$\boldsymbol{p}\cdot\boldsymbol{A} = (\hbar/\mathrm{i})\nabla\cdot\boldsymbol{A} = (\hbar/\mathrm{i})(\nabla\cdot\boldsymbol{A}) + (\hbar/\mathrm{i})\boldsymbol{A}\cdot\nabla$$

$$= \lambda(\hbar/\mathrm{i})(\nabla\cdot\boldsymbol{V}') + \boldsymbol{A}\cdot\boldsymbol{p} \qquad [(\nabla\cdot V) = 0] .$$

$$A^2 = \tfrac{1}{4}BV^2 + \lambda^2 V'^2 + \lambda B\boldsymbol{V}\cdot\boldsymbol{V}' .$$

$$\boldsymbol{V} = -y\hat{\boldsymbol{i}} + x\hat{\boldsymbol{j}}, \quad V' = x\hat{\boldsymbol{i}} + y\hat{\boldsymbol{j}}; \quad (\nabla\cdot\boldsymbol{V}') = 2 ;$$

$$V^2 = x^2 + y^2, \quad V'^2 = x^2 + y^2, \quad \boldsymbol{V}\cdot\boldsymbol{V}' = -yx + xy = 0 .$$

Therefore,

$$H = H^{(0)} + (e/m_e)\boldsymbol{A}\cdot\boldsymbol{p} + \lambda(e\hbar/2\mathrm{i}m_e)(\nabla\cdot\boldsymbol{V}') + (e^2/2m_e)A^2$$

$$\boldsymbol{A}\cdot\boldsymbol{p} = \tfrac{1}{2}B\boldsymbol{V}\cdot\boldsymbol{p} + \lambda\boldsymbol{V}'\cdot\boldsymbol{p}$$

$$= \tfrac{1}{2}B(-yp_x + xp_y) + \lambda(xp_x + yp_y)$$

$$= \tfrac{1}{2}Bl_z + \lambda(xp_x + yp_y),$$

$$A^2 = \tfrac{1}{4}B(x^2+y^2) + \lambda^2(x^2+y^2) .$$

$$H = \underline{H^{(0)} + (e/2m_e)Bl_z + (\lambda e/m_e)(xp_x + yp_y)}$$

$$\underline{-\mathrm{i}\lambda(e\hbar/m_e) + (e^2/8m_e)B^2(x^2+y^2) + (e^2\lambda^2/2m_e)(x^2+y^2)} .$$

On choosing $\lambda^2 = -\tfrac{1}{4}B^2$, so that $\lambda = \tfrac{1}{2}\mathrm{i}B$,

$$H = \underline{H^{(0)} + (e/2m_e)Bl_z + \mathrm{i}(eB/2m_e)(xp_x + yp_y) + e\hbar B/2m_e},$$

which is linear in B, and so $H^{(2)}$ is absent.

EXERCISE: Find the form of the hamiltonian for a general uniform field in three dimensions, and its form when $H^{(2)}$ is absent. Find an expression for the magnetic susceptibility using the transformed hamiltonian. Can $H^{(1)}$ be caused to disappear instead?

14.8 $\chi = -(e^2\mu_0 L\rho^{\ominus}/6m_e M_m)\langle r^2\rangle$ [eqn (14.2.23)].

$$\psi_{1s} = (Z^{*3}/\pi a_0^3)^{\frac{1}{2}} e^{-Z^*r/a_0} \qquad [\text{Section 9.10}],$$

$$\psi_{2s} = (Z^{*5}/96\pi a_0^5)^{\frac{1}{2}} r\,\mathrm{e}^{-Z^*r/a_0} .$$

$$\langle r^2\rangle_{1s} = (Z^{*3}/\pi a_0^3)\int_0^{2\pi}\mathrm{d}\phi\int_0^{\pi}\sin\theta\,\mathrm{d}\theta\int_0^{\infty} r^2\left\{r^2\,\mathrm{e}^{-2Z^*r/a_0}\right\}\mathrm{d}r$$

$$= (Z^{*3}/\pi a_0^3)(2\pi)(2)\left\{4!/(2Z^*/a_0)^5\right\} = 3a_0^2/Z^{*2} .$$

$$\langle r^2\rangle_{2s} = (Z^{*5}/96\,\pi a_0^5)(2\pi)(2)\int_0^\infty r^6\,e^{-Z^*r/a_0}\,dr$$

$$= (Z^{*5}/96\,\pi a_0^5)(4\pi)\left\{6!/(Z^*/a_0)^7\right\} = 30\,a_0^2/Z^{*2}\,.$$

Therefore, in each case we write

$$\chi = -(e^2\mu_0 L\rho^{\ominus}/6m_e M_m)\,Ka_0^2/Z^{*2}\,, \quad K = 3 \text{ for 1s, 30 for 2s}$$

$$= -(9.953\times 10^{-9}/M_r)\,K/Z^{*2} \qquad [M_m = M_r \text{ g mol}^{-1},\ \rho^{\ominus} = 1\text{ kg m}^{-3}].$$

(i) For the hydrogen atom, take $M_r = 1.008$, $Z^* = 1$, $K = 3$;
$\chi(H) = -2.96\times 10^{-8}$.

(ii) For the carbon atom, take $M_r = 12.00$, $Z^*(1s) = 5.70$, $Z^*(2s) = 3.25$ [Table 9.1];

$$\chi(C,1s) = -7.66\times 10^{-11};\quad \chi(C,2s) = -2.36\times 10^{-9}.$$

EXERCISE: Use the true hydrogenic 2s-orbitals to calculate $\chi(H,2s)$ and compare it with the Slater orbital result. Does the orthogonalization of the Slater H1s and H2s improve the agreement ?

14.9 $\chi_{\parallel} = -(e^2\mu_0 L\rho^{\ominus}/4m_e M_m)\langle x^2 + y^2\rangle$. [eqns (14.2.20) and (14.2.22)].

$\chi_{\perp} = -(e^2\mu_0 L\rho^{\ominus}/4m_e M_m)\langle x^2 + z^2\rangle$.

$x^2 + y^2 = r^2\sin^2\theta(\cos^2\phi + \sin^2\phi = r^2\sin^2\theta$;

$\psi_{2p_z} = (Z^{*5}/32\,\pi a_0^5)^{\frac{1}{2}}\,r\cos\theta\,e^{-Z^*r/2a_0}$ [Section 9.10].

(a)

$$\langle x^2 + y^2\rangle = (Z^*/32\pi a_0^5)\int_0^{2\pi} d\phi\int_0^{\pi}\sin\theta\,d\theta\int_0^\infty r^2\left\{r^4\sin^2\theta\cos^2\theta\,e^{-Z^*r/a_0}\right\}dr$$

$$= (Z^{*5}/32\pi a_0^5)(2\pi)(4/15)\left\{6!/(Z^*/a_0)^7\right\} = 12\,a_0^2/Z^{*2}.$$

(b)

$$\langle x^2 + z^2\rangle = \langle r^2(\sin^2\theta\cos^2\phi + \cos^2\theta)\rangle$$

$$= (Z^{*5}/32\pi a_0^5)\int_0^{2\pi} d\phi\int_0^{\pi}\sin\theta\,d\theta\int_0^\infty r^2\left\{r^4(\sin^2\theta\cos^2\phi + \cos^2\theta)\cos^2\theta \times e^{-Z^*r/a_0}\right\}dr$$

$$= (Z^*/32\pi a_0^5)\left\{(\pi)(4/15) + (2\pi)(2/5)\right\}\left\{6!/(Z^*/a_0)^7\right\}$$

$$= 24a_0^2/Z^{*2}.$$

Consequently,

(a) $\chi_{\parallel} = -(3e^2 a_0^2 \mu_0 L\rho^{\ominus}/m_e M_m)(1/Z^{*2})$

$= \underline{-1.791 \times 10^{-7}/M_r Z^{*2}}$.

(b) $\chi_{\perp} = -(6e^2 a_0^2 \mu_0 L\rho^{\ominus}/m_e M_m)(1/Z^{*2})$

$= \underline{-3.583 \times 10^{-7}/M_r Z^{*2}}$.

$\chi = \frac{1}{3}(\chi_{\parallel} + 2\chi_{\perp}) = \frac{5}{3}\chi_{\parallel}$ [$\chi_{\perp} = 2\chi_{\parallel}$]

$= \underline{-2.985 \times 10^{-7}/M_r Z^{*2}}$.

For the carbon atom, $Z^* = 3.25$ [Table 9.1]; consequently

$\underline{\chi_{\parallel} = -1.70 \times 10^{-8}/M_r}$; $\underline{\chi_{\perp} = -3.39 \times 10^{-8}/M_r}$; $\underline{\chi = -2.83 \times 10^{-8}/M_r}$,

Where M_r is the appropriate RMM of the species.

EXERCISE: Evaluate $\chi_{\parallel}$ and $\chi_{\perp}$ for Slater $3d_{z^2}$-orbitals.

14.10 $\rho = -e\Psi^*\Psi$

$\partial\rho/\partial t = -e\dot{\Psi}^*\Psi - e\Psi^*\dot{\Psi} = -e(\Psi\dot{\Psi}^* + \Psi^*\dot{\Psi})$

$= (e/i\hbar)(\Psi H\Psi^* - \Psi^* H\Psi)$ [$H\Psi = i\hbar\dot{\Psi}$]

$= (ie\hbar/2m_e)(\Psi\nabla^2\Psi^* - \Psi^*\nabla^2\Psi)$ [$H = (-\hbar^2/2m_e)\nabla^2 + V$] .

$\nabla\cdot\boldsymbol{j} = \nabla\cdot\left\{(-e/2m_e)[\Psi^* p\Psi - \Psi p\Psi^*]\right\}$ [eqn (14.3.1), $p^* = -p$]

$= -i(e\hbar/2m_e)\nabla\cdot\left\{\Psi\nabla\Psi^* - \Psi^*\nabla\Psi\right\}$ [$p = (\hbar/i)\nabla$]

$= -i(e\hbar/2m_e)\left\{(\nabla\Psi)\cdot\nabla\Psi^* + \Psi\nabla^2\Psi^* - (\nabla\Psi^*)\cdot\nabla\Psi - \Psi^*\nabla^2\Psi\right\}$

$= -i(e\hbar/2m_e)\left\{\Psi\nabla^2\Psi^* - \Psi^*\nabla^2\Psi\right\} = -\partial\rho/\partial t$.

EXERCISE: Evaluate $\partial\rho/\partial t$ and $\nabla\cdot\boldsymbol{j}$ for one-dimensional free motion of an electron.

14.11 $\psi_{3d} = Nr^2 e^{-Z^* r/3a_0} Y_{22}(\theta,\phi)$, $N = (8/5\cdot 3^9)^{\frac{1}{2}}(Z^*/a_0)^{7/2}$ [Section 9.10]

$Y_{22} = \frac{1}{4}(15/2\pi)^{\frac{1}{2}} \sin^2\theta\, e^{2i\phi}$ [Table 4.1].

Write

$\psi_{3d} = f(r,\theta)e^{2i\phi}$, $f(r,\theta) = (1/162)(Z^{*7}/\pi a_0^7)^{\frac{1}{2}} \sin^2\theta\, r^2 e^{-Z^* r/3a_0}$.

$\boldsymbol{j}_0 = -(e/2m_e)(\hbar/i)\left\{\psi^*\nabla\psi - \psi\nabla\psi^*\right\}$ [eqn (14.3.1)]

$= -(e/2m_e)(\hbar/i)\left\{f e^{-2i\phi}\nabla f e^{2i\phi} - f e^{2i\phi}\nabla f e^{-2i\phi}\right\}$

$= -2(e\hbar/m_e) f^2 (\nabla\phi)$.

$(\nabla\phi) = (-y\hat{\boldsymbol{i}} + x\hat{\boldsymbol{j}})/(x^2 + y^2) = \boldsymbol{V}/(x^2 + y^2)$ [eqn (14.3.4)].

$$x^2+y^2 = r^2\sin^2\theta\cos^2\phi + r^2\sin^2\theta\sin^2\phi = r^2\sin^2\theta .$$

$$\boldsymbol{j}_0 = -(2e\hbar/m_e)\,f^2\boldsymbol{V}/r^2\sin^2\theta$$
$$= -(e\hbar/m_e)(Z^{*2}/13\,122\pi a_0^7)\,r^2\sin^2\theta\,e^{-2Z^*r/3a_0}\,\boldsymbol{V} .$$

$$(e\hbar/m_e)(1/13\,122\pi a_0^7) = 3.872\times10^{44}\ \mathrm{A\,m^{-5}} ;$$

$$\boldsymbol{j}_0 = -(3.872\times10^{44}\ \mathrm{A\,m^{-5}})\,Z^{*7}r^2\sin^2\theta\,e^{-2Z^*r/3a_0}\,\boldsymbol{V}$$
$$= -(3.872\times10^{44}\ \mathrm{A\,m^{-5}})\,Z^{*7}r^2\sin^2\theta(x\hat{\boldsymbol{j}}-y\hat{\boldsymbol{i}})e^{-2Z^*r/3a_0}$$
$$= -(3.872\times10^{44}\ \mathrm{A\,m^{-5}})\,Z^{*7}r^3\sin^3\theta(-\hat{\boldsymbol{i}}\sin\phi+\hat{\boldsymbol{j}}\cos\phi)\,e^{-2Z^*r/3a_0}$$
$$= -(3.872\times10^{44}\ \mathrm{A\,m^{-2}})\,Z^{*7}(r/\mathrm{m})^3\sin^3\theta\,(-\hat{\boldsymbol{i}}\sin\phi+\hat{\boldsymbol{j}}\cos\phi)\,e^{-2Z^*r/3a_0} .$$

Consequently,

$$-\boldsymbol{j}_0/\mathrm{A\,m^{-2}} = \underline{(3.872\times10^{44})\,Z^{*7}(r/\mathrm{m})^3\sin^3\theta(-\hat{\boldsymbol{i}}\sin\phi+\hat{\boldsymbol{j}}\cos\phi)e^{-2Z^*r/3a_0}} .$$

When $Z^*=1$,

$$-\boldsymbol{j}_0/\mathrm{A\,m^{-2}} = 3.872\times10^{44}(r/\mathrm{m})^3\sin^3\theta\,(-\hat{\boldsymbol{i}}\sin\phi+\hat{\boldsymbol{j}}\cos\phi)\,e^{-2r/3a_0}$$
$$= 5.738\times10^{13}\,s^3\sin^3\theta\,(-\hat{\boldsymbol{i}}\sin\phi+\hat{\boldsymbol{j}}\cos\phi)\,e^{-2s/3} ,\qquad s=r/a_0 .$$

The magnitude of $-\hat{\boldsymbol{i}}\sin\phi+\hat{\boldsymbol{j}}\cos\phi$ is 1; therefore the current density can be represented by circular contours, their magnitudes denoting

$$|(\boldsymbol{j}_0/\mathrm{A\,m^{-2}})| = \underline{5.738\times10^{13}\,s^3\sin^3\theta\,e^{-2s/3}} .$$

When $\theta=90°$ [in the equatorial plane]

$$|(\boldsymbol{j}_0/\mathrm{A\,m^{-2}})| = \underline{5.738\times10^{13}\,s^3\,e^{-2s/3}} .$$

This function is depicted in Fig. 14.2.

EXERCISE: Evaluate the current density for an electron with $l=2$, $m_l=+1$.

Fig.14.2

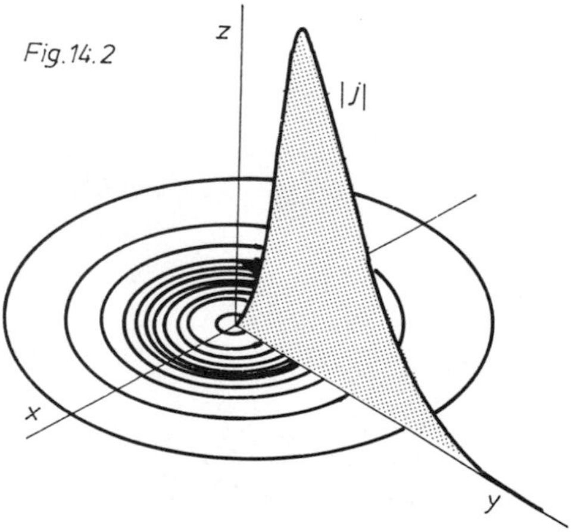

14.12 $$\boldsymbol{j}^{\mathrm{d}} = -(e^2B/2\pi m_{\mathrm{e}}a_0^3)(-y\hat{\boldsymbol{i}} + x\hat{\boldsymbol{j}})\,\mathrm{e}^{-2r/a_0} \qquad \text{[eqn (14.4.2)]}$$
$$= -(e^2B/2\pi m_{\mathrm{e}}a_0^2)(-\hat{\boldsymbol{i}}\sin\phi + \hat{\boldsymbol{j}}\cos\phi)\sin\theta\, r\,\mathrm{e}^{-2r/a_0}$$
$$= -(e^2B/2\pi m_{\mathrm{e}}a_0^2)(-\hat{\boldsymbol{i}}\sin\phi + \hat{\boldsymbol{j}}\cos\phi)\sin\theta\, s\,\mathrm{e}^{-2s}, \qquad s = r/a_0.$$

The heights 0, a_0, $2a_0$ correspond to the following values or r, θ for horizontal distances σa_0 from the nucleus (Fig. 14.3):

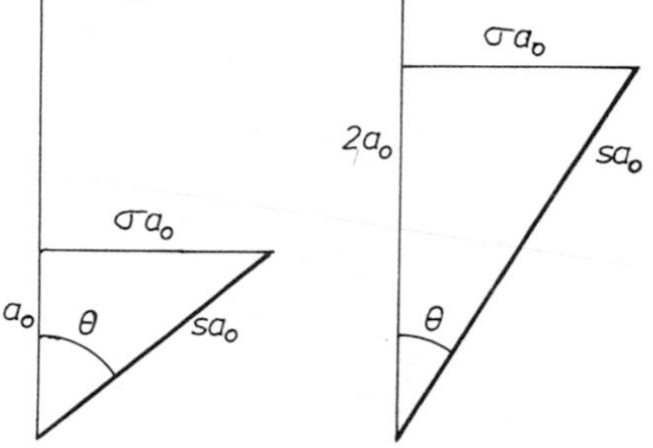

Fig. 14.3

(a) 0: $s = \sigma$, $\theta = 90°$,

(b) a_0: $s\sin\theta = \sigma$, $s = \sqrt{(1+\sigma^2)}$,

(c) $2a_0$: $s\sin\theta = \sigma$, $s = \sqrt{(4+\sigma^2)}$.

Write
$$j^{\ominus} = -(e^2B/2\pi m_{\mathrm{e}}a_0^2), \text{ then}$$
$$\boldsymbol{j}^{\mathrm{d}}/j^{\ominus} = (-\hat{\boldsymbol{i}}\sin\phi + \hat{\boldsymbol{j}}\cos\phi)\sin\theta\, s\,\mathrm{e}^{-2s},$$

which correspond to circles denoting magnitudes

(a) $j^{\mathrm{d}}/j^{\ominus} = \underline{\sigma\mathrm{e}^{-2\sigma}}$,

(b) $j^{\mathrm{d}}/j^{\ominus} = \underline{\sigma\exp\{-2\sqrt{(1+\sigma^2)}\}}$,

(c) $j^{\mathrm{d}}/j^{\ominus} = \underline{\sigma\exp\{-2\sqrt{(4+\sigma^2)}\}}$.

These are plotted in Fig. 14.4.

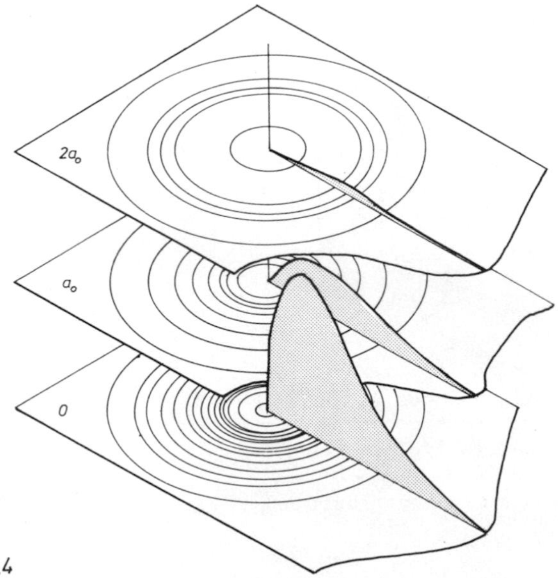

Fig. 14.4

EXERCISE: Calculate and plot the current densities in the same three planes for an electron in a hydrogenic 2s-orbital.

14.13 (a) $\psi(3d_{xy}) = xyf(r) = \sin^2\theta \sin\phi \cos\phi r^2 f(r) \propto \sin^2\theta \sin 2\phi r^2 f(r)$.

(b) $\psi(3d_{x^2-y^2}) = (x^2-y^2) f(r) = \sin^2\theta (\cos^2\theta - \sin^2\phi) r^2 f(r)$

$$\propto \sin^2\theta \cos 2\phi \, r^2 f(r).$$

In each case $f(r) \propto e^{-Z^* r/3a}$. Then use

$$j^d = -(e^2B/2m_e)\psi_0^2 V, \quad V = -y\hat{i} + x\hat{j} = r(-\hat{i}\sin\phi + \hat{j}\cos\phi)\sin\theta .$$

(a) $j^d \propto \sin^4\theta \sin^2 2\phi \, r^5 e^{-2Z^* r/3a_0}(-\hat{i}\sin\phi + \hat{j}\cos\phi)\sin\theta$

$$\propto \sin^5\theta(-\hat{i}\sin\phi + \hat{j}\cos\phi)\sin^2 2\phi \, s^5 e^{-2Z^* s/3a_0}, \qquad s = r/a_0 .$$

In the equatorial plane $\theta = 90°$, and for $Z^* = 1$,

$$j^d \propto \underline{(-\hat{i}\sin\phi + \hat{j}\cos\phi)\sin^2 2\phi \, s^5 e^{-2s/3}} .$$

This is sketched in Fig. 14.5(a).

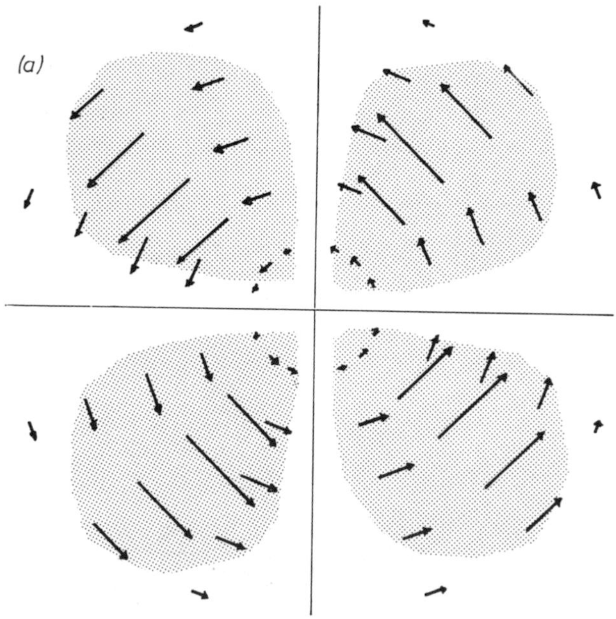

Fig. 14.5

(b) $j^d \propto \underline{(-\hat{i}\sin\phi + \hat{j}\cos\phi)\cos^2 2\phi \, s^5 e^{-2s/3}}$, [for $\theta = 90°$, $Z^* = 1$].

This is sketched in Fig. 14.5(b).

For the paramagnetic contributions, note that

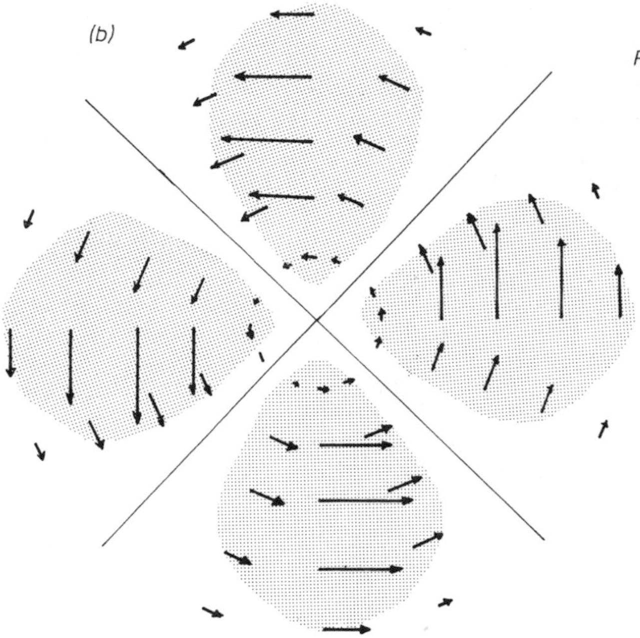

Fig. 14.5

$$l_z\psi(3d_{xy}) = (\hbar/i)\{x(\partial/\partial y)-y(\partial/\partial x)\}\, xyf(r)/r = (\hbar/i)(x^2-y^2)\,f(r)/r$$
$$= (\hbar/i)\,\psi\,(3d_{x^2-y^2}).$$

Therefore, the magnetic field induces mixing between d_{xy} and $d_{x^2-y^2}$. By the same argument that led to eqn (14.5.4), we expect j^{p} to be proportional to the current density for $\Lambda=\pm 2$ orbitals. These were calculated in Problem 14.11, and so the paramagnetic current densities are the same as those in Fig. 14.2, but of decreased magnitude (by a factor of $2\mu_{\mathrm{B}}B/\Delta$).

EXERCISE: Calculate the diamagnetic and paramagnetic current densities for non-degenerate $f_3 \pm f_{-3}$ orbitals.

14.14 Take the following hydrogenic orbitals [Tables 4.1, 4.2; $s=r/a_0$]:

$$\psi_{\mathbf{2s}} = R_{20}Y_{00} = (Z^3/8\pi a_0^3)^{\frac{1}{2}}\,(1-Zs)\,e^{-Zs}$$
$$\psi_{\mathbf{3p}} = R_{31}Y_{10} = (1/27)(2Z^5/\pi a_0^3)^{\frac{1}{2}}\,s\,(1-\tfrac{1}{3}Zs)\,e^{-Zs/3}\cos\theta.$$

When the field is along z there is no paramagnetic contribution. The diamagnetic current densities are given by

$$\boldsymbol{j}^{\mathbf{d}} = -(e^2B/2m_{\mathbf{e}})\,\psi_0^2\,\boldsymbol{V} \qquad \text{[eqn (14.4.1)]}$$

$$\boldsymbol{j}^{\mathrm{d}}(2\mathrm{s}) = -(Z^3e^2B/16\,\pi m_{\mathrm{e}}a_0^3)(1-Zs)^2\,\boldsymbol{V}\mathrm{e}^{-2Zs}$$
$$= -(Z^3e^2B/16\,\pi m_{\mathrm{e}}a_0^2)\,s\,(1-Zs)^2\,(-\hat{\boldsymbol{i}}\sin\phi+\hat{\boldsymbol{j}}\cos\phi)\mathrm{e}^{-2Zs}\,\sin\theta,$$

which correspond to circular contours denoting magnitudes

$$\boldsymbol{j}^{\mathrm{d}}(2\mathrm{s})/\boldsymbol{j}^{\theta} = s\,(1-Zs)^2\,\mathrm{e}^{-2Zs}\sin\theta\,,\quad \boldsymbol{j}^{\theta} = Z^3e^2B/16\,\pi m_{\mathrm{e}}a_0^2\,.$$

When $\theta = 90°$ and $Z = 1$,

$$\boldsymbol{j}^{\mathrm{d}}(2\mathrm{s})/\boldsymbol{j}^{\theta} = \underline{s(1-s)^2\,\mathrm{e}^{-2s}},$$

which is sketched in Fig. 14.6(a)

$$\boldsymbol{j}^{\mathrm{d}}(3\mathrm{p}_z) = -\,(2Z^5e^2B/729\,\pi a_0^2)\,s^3\,(1-\tfrac{1}{3}Zs)^2\,\mathrm{e}^{-2Zs/3}\cos^2\theta\sin\theta$$
$$\times\,(-\hat{\boldsymbol{i}}\sin\phi+\hat{\boldsymbol{j}}\cos\phi).$$

This also corresponds to circular contours denoting magnitudes

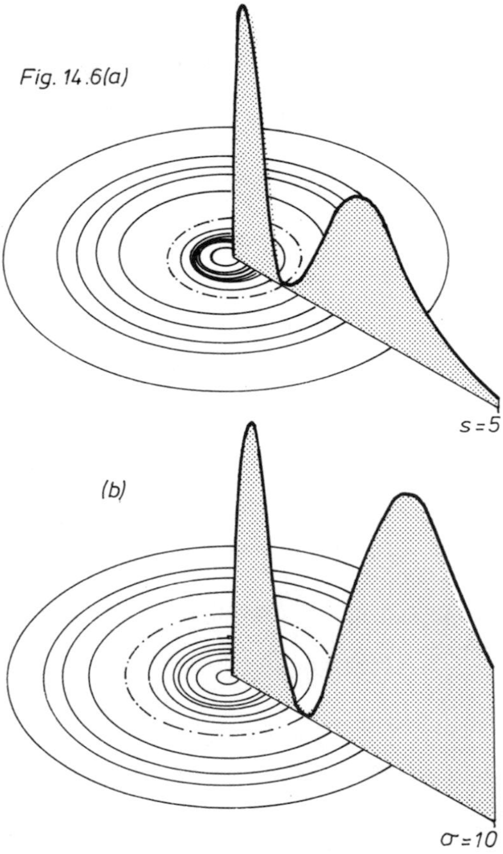

$$j^{d}(3p_z)/j^{\ominus} = s^3(1-\tfrac{1}{3}Zs)^2\cos^2\theta\,\sin^2\theta\,e^{-2Zs/3},\quad j^{\ominus} = 2Z^5e^2B/729\,\pi a_0^2.$$

The magnitudes are zero in the equatorial plane ($\sin\theta = 0$). For a plane parallel to the equatorial plane but at a height ha_0 above it, we have $s\sin\theta = \sigma$, as in Fig. 14.3 (problem 14.12), $\cos\theta = h/s$, and $s^2 = \sigma^2 + h^2$. Then, with $Z = 1$,

$$j^{d}(3p_z)/j^{\ominus} \quad \underline{h^2\sigma\,(1-\tfrac{1}{3}[\sigma^2+h^2]^{\frac{1}{2}})^2\exp\{-\tfrac{2}{3}[\sigma^2+h^2]^{\frac{1}{2}}\}}\,.$$

This is sketched in Fig. 14.6(b) for $h = 1$, when

$$j^{d}(3p_z)/j^{\ominus} = \sigma\{1-\tfrac{1}{3}\surd(1+\sigma^2)\}^2\exp\{-\tfrac{2}{3}\surd(1+\sigma^2)\}\,.$$

EXERCISE: Evaluate $\boldsymbol{j}(4p_z)$ and $\boldsymbol{j}(3s)$ for hydrogenic orbitals, the magnetic field being applied in the z-direction.

14.15 $\psi = (\psi_N - \psi_O)/\surd 2$;

$$\psi_{2p} = (Z^{*5}/32\,\pi a_0^5)^{\frac{1}{2}}\,r\cos\theta\,e^{-Z^*r/2a_0},\quad Z_N^* = 3.90,\ Z_O^* = 4.55\ [\text{Table 9.1}].$$

Let the vector potential be centred on a point a fraction λ of the bond (of length R) from N, Fig. 14.7. Then for a uniform field along x, $\boldsymbol{A} = \tfrac{1}{2}\boldsymbol{B}\wedge\boldsymbol{r} = \tfrac{1}{2}B(-z\hat{\boldsymbol{j}} + y\hat{\boldsymbol{k}})$. We need to express $\cos\theta$ and r (on both N and O) in terms of the same y and z coordinates. (Note in passing that there is no 'natural' origin for A: different values of λ correspond to different choices of gauge. In due course we shall see that the diamagnetic current density depends on the choice of gauge. The total current density, however, is independent of gauge.)

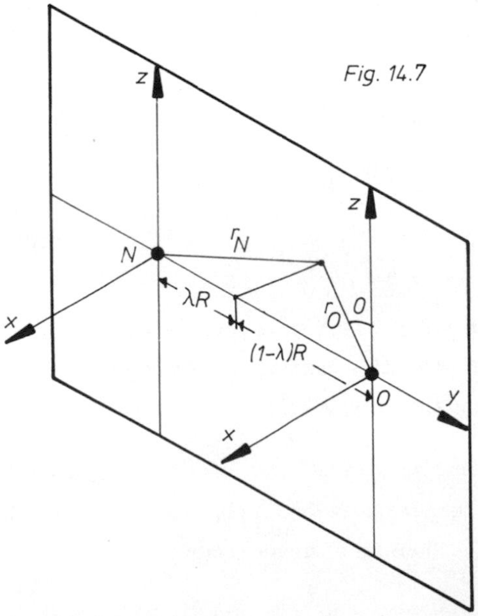

Fig. 14.7

On N: $r_N^2 = y_N^2 + z_N^2 = (y+\lambda R)^2 + z^2$,

$\cos\theta_N = z_N / r_N = z/\{(y+\lambda R)^2 + z^2\}^{\frac{1}{2}}$.

On O: $r_O^2 = y_O^2 + z_O^2 = [y-(1-\lambda)R]^2 + z^2$,

$\cos\theta_O = z_O / r_O = z/\{[y-(1-\lambda)R]^2 + z^2\}^{\frac{1}{2}}$.

Consequently,

$$\psi_{N2p_z} = (Z_N^{*5}/32\pi a_0^5)^{\frac{1}{2}} z \exp\left\{-Z_N^*[(y+\lambda R)^2 + z^2]^{\frac{1}{2}}/2a_0\right\}$$

$$\psi_{O2p_z} = (Z_O^{*5}/32\pi a_0^5)^{\frac{1}{2}} z \exp\left\{-Z_O^*\{[y-(1-\lambda)R]^2 + z^2\}^{\frac{1}{2}}/2a_0\right\}.$$

The diamagnetic current density in the yz-plane is therefore

$$\boldsymbol{j}^{\mathrm{d}} = -(e^2B/2m_e)\psi_0^2 V \qquad [\text{eqn } (14.4.1),\ \boldsymbol{V} = -z\hat{\boldsymbol{j}} + y\hat{\boldsymbol{k}}]$$

$$= -(e^2B/4m_e)(\psi_N^2 + \psi_O^2 - 2\psi_N\psi_O)(-z\hat{\boldsymbol{j}} + y\hat{\boldsymbol{k}})$$

$$= -j^{\ominus}\left\{Z_N^{*5}\,e^{-2f_N} + Z_O^{*5}\,e^{-2f_O} - 2(Z_N^* Z_O^*)^{5/2}\,e^{-(f_N+f_O)}\right\} \times z^2(-z\hat{\boldsymbol{j}} + y\hat{\boldsymbol{k}}),$$

with $j^{\ominus} = e^2B/128\pi m_e a_0^5$

$$f_N = (Z_N^*/2a_0)\left\{[y+\lambda R]^2 + z^2\right\}^{\frac{1}{2}}$$

$$f_O = (Z_O^*/2a_0)\left\{[y-(1-\lambda)R]^2 + z^2\right\}^{\frac{1}{2}}.$$

Now write

$$\zeta = z/a_0,\quad \eta = y/a_0,\quad s = R/a_0,\quad \gamma = (Z_N^*/Z_O^*)^{5/2},\quad j^{\dagger} = (Z_N^* Z_O^*)^{5/2} j^{\ominus}$$

Then

$$-\boldsymbol{j}^{\mathrm{d}}/j^{\dagger} = \left\{\gamma e^{-2f_N} + (1/\gamma)e^{-2f_O} - 2e^{-(f_O+f_N)}\right\}\zeta^2(-\zeta\hat{\boldsymbol{j}} + \eta\hat{\boldsymbol{k}}),$$

$$f_N = \tfrac{1}{2}Z_N^*\left\{(\eta+\lambda s)^2 + \zeta^2\right\}^{\frac{1}{2}},$$

$$f_O = \tfrac{1}{2}Z_O^*\left\{[\eta-(1-\lambda)s]^2 + \zeta^2\right\}^{\frac{1}{2}}.$$

Take $R = 115$ pm (so that $s = 2.17$). The current densities should be plotted for (a) $\lambda = 0$, (b) $\lambda = \frac{1}{2}$, (c) $\lambda = 1.0$.

EXERCISE: Use a computer graphics system to plot the current densities. Calculate and plot the current density for a plane parallel to the one considered above, but offset from it by a distance la_0.

14.16 The current density around the axis is that corresponding to an electron in a degenerate π-orbital, and corresponds to the j_0 of eqn (14.3.5). When the degeneracy is removed, we arrive at the arrangement that led to eqn (14.5.4): $\boldsymbol{j}^{\mathrm{p}} = (\mu_{\mathrm{B}}B/\Delta)\,\boldsymbol{j}_0$. Therefore, the field required is of the order of $B = \Delta/\mu_{\mathrm{B}}$ ['of the order of' because first-order perturbation theory fails when $\mu_{\mathrm{B}}B \approx \Delta$]. Since 1 eV corresponds to 1.602×10^{-19} J,

$$B \approx (1.602\times10^{-19}\ \mathrm{J})/(9.274\times10^{-24}\ \mathrm{J\ T^{-1}}) = 1.727\times10^{4}\ \mathrm{T}.$$

Hence a field of the order of <u>17 kT</u> or 17 megagauss, is required.

EXERCISE: Typical laboratory fields are of the order of 1 T. What current density would then be induced?

14.17 $\mathcal{E}^{(1)} = -\int \boldsymbol{A}_{\mathrm{nuc}}\cdot\boldsymbol{j}\,\mathrm{d}\tau$ [eqn (14.7.14)].

The calculation in Problem 14.15 needs to be augmented to include the current density in all planes, not simply the plane containing the two nuclei. This involves replacing r_{N} and r_{O} by

$$r_{\mathrm{N}}^2 = x_{\mathrm{N}}^2 + y_{\mathrm{N}}^2 + z_{\mathrm{N}}^2 = x^2 + [y+\lambda R]^2 + z^2$$
$$r_{\mathrm{O}}^2 = x_{\mathrm{O}}^2 + y_{\mathrm{O}}^2 + z_{\mathrm{O}}^2 = x^2 + [y-(1-\lambda)R]^2 + z^2 .$$

The implication then is that

$$-j^{\mathrm{d}}/j^{\dagger} = \left\{\gamma\,\mathrm{e}^{-2f_{\mathrm{N}}} + (1/\gamma)\,\mathrm{e}^{-2f_{\mathrm{O}}} - 2\,\mathrm{e}^{-(f_{\mathrm{O}}+f_{\mathrm{N}})}\right\}\zeta^2(-\zeta\hat{\boldsymbol{j}} + \eta\hat{\boldsymbol{k}})$$

with
$$f_{\mathrm{N}} = \tfrac{1}{2}Z_{\mathrm{N}}^{*}\left\{\chi^2 + (\eta+\lambda s)^2 + \zeta^2\right\}^{\frac{1}{2}} \qquad \chi = x/a_0$$
$$f_{\mathrm{O}} = \tfrac{1}{2}Z_{\mathrm{O}}^{*}\left\{\chi^2 + [\eta-(1-\lambda)s]^2 + \zeta^2\right\}^{\frac{1}{2}} .$$

Since $\boldsymbol{A}_{\mathrm{nuc}} = (\gamma_{\mathrm{N}}\mu_0/4\pi r_{\mathrm{N}}^3)\,\boldsymbol{I}\wedge\boldsymbol{r}_{\mathrm{N}}$,

$$\boldsymbol{I}\wedge\boldsymbol{r}_{\mathrm{N}}\cdot\boldsymbol{j}^{\mathrm{d}} \propto \boldsymbol{I}\wedge\boldsymbol{r}_{\mathrm{N}}\cdot(-\zeta\hat{\boldsymbol{j}}+\eta\hat{\boldsymbol{k}}) = \boldsymbol{I}\cdot\boldsymbol{r}_{\mathrm{N}}\wedge(-\zeta\hat{\boldsymbol{j}}+\eta\hat{\boldsymbol{k}})$$

$$\boldsymbol{r}_{\mathrm{N}}\wedge(-\zeta\hat{\boldsymbol{j}}+\eta\hat{\boldsymbol{k}}) = \begin{vmatrix}\hat{\boldsymbol{i}} & \hat{\boldsymbol{j}} & \hat{\boldsymbol{k}}\\ x_{\mathrm{N}} & y_{\mathrm{N}} & z_{\mathrm{N}}\\ 0 & -\zeta & \eta\end{vmatrix} = \hat{\boldsymbol{i}}(y_{\mathrm{N}}\eta + z_{\mathrm{N}}\zeta) - \hat{\boldsymbol{j}}\,x_{\mathrm{N}}\eta - \hat{\boldsymbol{k}}\,x_{\mathrm{N}}\zeta .$$

Consider the component of the interaction parallel to the applied field (which lies along $\hat{\boldsymbol{i}}$):

$$(\boldsymbol{I}\wedge\boldsymbol{r}_{\mathrm{N}}\cdot\boldsymbol{j}^{\mathrm{d}})_{\parallel} = (y_{\mathrm{N}}\eta + z_{\mathrm{N}}\zeta)\,I_x$$
$$= \left\{(y+\lambda R)\,\eta + z\zeta\right\}I_x$$
$$= \left\{(\eta+\lambda s)\,\eta + \zeta^2\right\}a_0\,I_x .$$

Therefore,

$$\mathcal{E}^{(1)} = (\gamma_{\mathrm{N}}j^{\dagger}I_x\mu_0/4\pi a_0^2)\int F\,\mathrm{d}\tau,$$
$$F = \left\{\gamma\,\mathrm{e}^{-2f_{\mathrm{N}}} + (1/\gamma)\,\mathrm{e}^{-2f_{\mathrm{O}}} - 2\,\mathrm{e}^{-(f_{\mathrm{O}}+f_{\mathrm{N}})}\right\}\zeta^2\left\{(\eta+\lambda s)\eta+\zeta^2\right\}/(r_{\mathrm{N}}/a_0)^3$$
$$(r_{\mathrm{N}}/a_0)^3 = \left\{\chi^2 + (\eta+\lambda s)^2 + \zeta^2\right\}^{\frac{3}{2}},$$

$$d\tau = dx\,dy\,dz = a_0^3\,d\chi\,d\eta\,d\zeta\,.$$

The interaction energy may therefore be written

$$\underline{\mathcal{E}^{(1)} = (a_0\,\mu_0\,\gamma_{\rm N}\,j^{\dagger}\,I_x/4\pi)\,\mathfrak{F}}$$

where

$$\mathfrak{F} = \int\left\{\frac{\left[\gamma\,e^{-2f_{\rm N}} + (1/\gamma)\,e^{-2f_{\rm O}} - 2\,e^{-(f_{\rm O}+f_{\rm N})}\right]\zeta^2\,[(\eta+\lambda s)\eta+\zeta^2]}{[\chi^2+(\eta+\lambda s)^2+\zeta^2]^{\frac{3}{2}}}\right\}d\chi\,d\eta\,d\zeta\,.$$

Choosing (arbitrarily) a gauge that centres $\boldsymbol{A}$ on the nitrogen nucleus (i.e. $\lambda = 0$) simplifies this to

$$\mathfrak{F} = \int\left\{\frac{\left[\gamma\,e^{-2f_{\rm N}} + (1/\gamma) - 2\,e^{-(f_{\rm O}+f_{\rm N})}\right]\zeta^2\,[\eta^2+\zeta^2]}{[\chi^2+\eta^2+\zeta^2]^{\frac{3}{2}}}\right\}d\chi\,d\eta\,d\zeta$$

$$f_{\rm N} = \tfrac{1}{2}Z_{\rm N}^*\,[\chi^2+\eta^2+\zeta^2]^{\frac{1}{2}}\,, \qquad f_{\rm O} = \tfrac{1}{2}Z_{\rm O}^*\,[\chi^2+(\eta-s)^2+\zeta^2]^{\frac{1}{2}}\,.$$

Simplify the integration by resorting to polar coordinates, with

$$\chi = \sigma\sin\theta\cos\phi\,, \quad \zeta = \sigma\sin\theta\sin\phi\,, \quad \eta = \sigma\cos\theta\,,$$

$$f_{\rm N} = \tfrac{1}{2}Z_{\rm N}^*\sigma\,, \quad f_{\rm O} = \tfrac{1}{2}Z_{\rm O}^*\,[\sigma^2\sin^2\theta + (\sigma\cos\theta - s)^2]^{\frac{1}{2}}\,,$$

$$d\chi\,d\eta\,d\zeta/[\chi^2+\eta^2+\zeta^2]^{\frac{3}{2}} = (1/\sigma)\,d\sigma\sin\theta\,d\theta\,d\phi$$

$$(\eta^2+\zeta^2)\,\zeta^2 = \sigma^4\,(\cos^2\theta+\sin^2\theta\sin^2\phi)\sin^2\theta\sin^2\phi\,.$$

The value of $\mathfrak{F}$ may then be obtained by numerical integration.

The flux density B' is such that $\mathcal{E}^{(1)} = -\gamma_{\rm N}\boldsymbol{I}\cdot\boldsymbol{B}' = -\gamma_{\rm N}I_xB'$. Therefore

$$B' = -\underline{(a_0\,\mu_0\,j^{\dagger}/4\pi)\mathfrak{F}}\,.$$

EXERCISE: Find an expression for the flux density at the oxygen nucleus. Evaluate $\mathcal{E}^{(1)}$ and B' for the nitrogen nucleus, and identify the shielding constant (evaluate $\mathfrak{F}$ numerically: write a program). Investigate the gauge dependence of the shielding constant.

14.18 $\sigma^{\rm d} = (e^2\mu_0/12\pi m_{\rm e})\,\langle(1/r)\rangle$ [eqn (14.7.19)].

$\psi(2{\rm s}) = (Z^{*5}/96\,\pi a_0^5)^{\frac{1}{2}}\,r\,e^{-Z^*r/2a_0}$ [Section 9.10; Problem 14.8],

$\psi(2{\rm p}_z) = (Z^{*5}/32\,\pi a_0^5)^{\frac{1}{2}}\,r\cos\theta\,e^{-Z^*r/2a_0}$ [Section 9.10; Problem 14.15].

$$\langle(1/r)\rangle_{2{\rm s}} = (Z^{*5}/96\,\pi a_0^5)\int_0^{2\pi}d\phi\int_0^{\pi}\sin\theta\,d\theta\int_0^{\infty}r^2\left\{r^2(1/r)\,e^{-Z^*r/a_0}\right\}dr$$

$$= (Z^{*5}/96\,\pi a_0^5)(2\pi)(2)\left\{3!/(Z^*/a_0)^4\right\} = \tfrac{1}{4}(Z^*/a_0)\,.$$

$$\langle (1/r) \rangle_{2s} = (Z^{*5}/32\pi a_0^5)\int_0^{2\pi} d\phi \int_0^{\pi} \sin\theta\, d\theta \int_0^{\infty} r^2 \left\{ r^2 (1/r) \cos^2\theta\, e^{-Z^* r/a_0} \right\} dr$$

$$= (Z^{*5}/32\pi a_0^5)(2\pi)(2/3)\left\{3!/(Z^*/a_0)^4\right\} = \tfrac{1}{4}(Z^*/a_0) .$$

Therefore, for each type of orbital,

$$\sigma^{\mathrm{d}} = (e^2\mu_0/12\pi m_{\mathrm{e}})(Z^*/4a_0) = e^2\mu_0 Z^*/48\pi m_{\mathrm{e}} a_0$$

$$= 4.44 \times 10^{-6} Z^* .$$

Consequently, $\sigma^{\mathrm{d}}/\mathrm{ppm} = \underline{4.44\, Z^*}$. For an electron in a carbon atom, for which $Z^* = 3.25$ [Table 9.1], $\underline{\sigma^{\mathrm{d}} = 14.4 \text{ ppm}}$.

EXERCISE: Calculate the contribution to σ^{d} of an electron in (a) a hydrogenic 2s-orbital, (b) Slater 3s- and 3p-orbitals.

14.19 The 2s-orbital gives zero paramagnetic contribution. The 2p-electron contributes

$$\sigma^{\mathrm{p}} = -(e^2\mu_0/12\pi m_{\mathrm{e}}^2) \sum_n{}' \left\{ l_{0n} \cdot (r^{-3} l)_{n0}/\Delta_{n0} \right\} \qquad \text{[eqn (14.7.21)]}$$

$$= -(e^2\mu_0/12\pi m_{\mathrm{e}}^2) \langle p_z | l_x | p_y \rangle \langle p_y | r^{-3} l_x | p_z \rangle / \Delta .$$

[We are assuming that the orbital mixed in is p_y, so that only l_x contributes; another p_x-orbital nearby would give an additional contribution *via* l_y.] Since $l_x p_z = -i\hbar p_y$ this expression becomes

$$\sigma^{\mathrm{p}} = -(e^2\mu_0\hbar^2/12\pi m_{\mathrm{e}}^2) \langle p_y | (1/r^3) | p_y \rangle / \Delta .$$

$$\langle (1/r^3) \rangle_{2p_y} = \langle (1/r^3) \rangle_{2p_z}$$

$$= (Z^{*5}/32\pi a_0^5)\int_0^{2\pi} d\phi \int_0^{\pi} \sin\theta\, d\theta \int_0^{\infty} r^2 \left\{ r^2 (1/r^3) \cos^2\theta\, e^{-Z^* r/a_0} \right\} dr$$

$$= (Z^{*5}/32\pi a_0^5)(2\pi)(2/3)\left\{1/(Z^*/a_0)^2\right\} = Z^{*3}/24 a_0^3 .$$

Consequently,

$$\sigma^{\mathrm{p}} = -(e^2\mu_0\hbar^2/12\pi m_{\mathrm{e}}^2)(Z^{*3}/24 a_0^3)(1/\Delta)$$

$$= -(e^2\mu_0\hbar^2/288\pi m_{\mathrm{e}}^2 a_0^3)(Z^{*3}/\Delta)$$

$$= \underline{-2.013 \times 10^{-5} Z^{*3}/(\Delta/\mathrm{eV})} .$$

For carbon, $Z^* = 3.25$ [Table 9.1], and so

$$\sigma^{\mathrm{p}} = -6.91 \times 10^{-4}/(\Delta/\mathrm{eV}) = -1.4 \times 10^{-4} \text{ when } \Delta/\mathrm{eV} = 5 .$$

Therefore, $\underline{\sigma^{\mathrm{p}} \approx -140 \text{ ppm}}$.

EXERCISE: Calculate the paramagnetic contribution to the shielding of an electron in a $3p_z$-orbital.

14.20 The magnetic perturbation transforms as a rotation. In C_{2v}, R_x, R_y, R_z transform as B_2, B_1, A_2 respectively. Therefore, since

$$A_1 \times \{B_2, B_1, A_2\} = \{B_2, B_1, A_2\},$$

it follows that for NO_2 the components g_{xx}, g_{yy}, g_{zz} depend on the admixture of $^2B_2$, $^2B_1$, $^2A_2$ terms respectively, Since

$$B_1 \times \{B_2, B_1, A_2\} = \{A_2, A_1, B_2\},$$

it follows that for ClO_2, g_{xx}, g_{yy}, g_{zz} depend on the admixture of $^2A_2$, $^2A_1$, $^2B_2$ respectively.

EXERCISE: What states contribute to g_{xx}, g_{yy}, g_{zz} in a $^2E_1'$ D_{5h} molecule?

14.21 The orbitals and energies of H_2O in the simpler scheme (neglecting O2s involvement) are

$$A_1, B_2:\quad E_\pm = \tfrac{1}{2}(\alpha_O + \alpha_H) \pm \tfrac{1}{2}\Delta, \quad \Delta^2 = (\alpha_O - \alpha_H)^2 + 4\beta^2$$

$$B_1:\quad E = \alpha_O$$

$$\left.\begin{aligned}\psi_+(A_1) &= (\mathrm{O2p}_z)\cos\theta + (1/\sqrt{2})(\mathrm{H1s_a} + \mathrm{H1s_b})\sin\theta \\ \psi_-(A_1) &= -(\mathrm{O2p}_z)\sin\theta + (1/\sqrt{2})(\mathrm{H1s_a} + \mathrm{H1s_b})\cos\theta\end{aligned}\right\}\ \tan 2\theta = -2\beta/(\alpha_H - \alpha_O).$$

$$\psi(B_1) = \mathrm{O2p}_x$$

$$\left.\begin{aligned}\psi_+(B_2) &= (\mathrm{O2p}_y)\cos\theta' + (1/\sqrt{2})(\mathrm{H1s_a} - \mathrm{H1s_b})\sin\theta' \\ \psi_-(B_2) &= -(\mathrm{O2p}_y)\sin\theta' + (1/\sqrt{2})(\mathrm{H1s_a} - \mathrm{H1s_b})\cos\theta'\end{aligned}\right\}\ \tan 2\theta' = 2\beta/(\alpha_H - \alpha_O).$$

[These were established in the solution of Problem 10.10.]

The H_2O configuration is $\mathrm{O1s}^2\,\mathrm{O2s}^2\,a_1^2 b_2^2 b_1^2$. The H_2O^+ configuration is therefore $\mathrm{O1s}^2\,\mathrm{O2s}^2\,a_1^2 b_2^2 b_1^1$ $^2B_1$. [We assume the angle to be 90° in H_2O^+; it is significantly greater, in fact.] The components g_{xx}, g_{yy}, g_{zz} depend on an admixture of $^2A_2$, $^2A_1$, $^2B_2$ respectively [Problem 14.20]. The lowest lying A_2 term arises from $\ldots a_1^2 b_2 b_1 a_1^*$, which involves excitation from a deep lying orbital; therefore we neglect it and ascribe to g_{xx} its free spin value (g_e). The lowest A_1 term arises from $\ldots a_1^2 b_2^2 a_1^*$, and the lowest B_2 term arises from $\ldots a_1^2 b_2^2 b_2^*$. The excitation energies are

$$\begin{aligned} A_1 - B_1:\quad E(a_1^*) - E(b_1) &= \{\tfrac{1}{2}(\alpha_O + \alpha_H) - \tfrac{1}{2}\Delta\} - \alpha_0 \\ &= \tfrac{1}{2}(\alpha_H - \alpha_O) - \tfrac{1}{2}\Delta. \\ B_2 - B_1:\quad E(b_2^*) - E(b_1) &= \tfrac{1}{2}(\alpha_H - \alpha_O) - \tfrac{1}{2}\Delta. \end{aligned}$$

Now note that $l_x\psi(B_1) = 0$, $l_y\psi(B_1) = -i\hbar(\mathrm{O2p}_z)$, and $l_z\psi(B_1) = i\hbar(\mathrm{O2p}_y)$.

Consequently:

$$\langle a_1^* | l_y | b_1 \rangle = -i\hbar \langle a_1^* | \mathrm{O2p}_z \rangle = i\hbar \sin\theta$$
$$\langle b_2^* | l_y | b_1 \rangle = -i\hbar \langle b_2^* | \mathrm{O2p}_y \rangle = -i\hbar \sin\theta' = i\hbar \sin\theta \qquad [\theta' = -\theta].$$

The g-values are therefore [eqn (14.9.9)]

$$g_{xx} = \underline{g_e}$$
$$g_{yy} = g_e - 2(\hbar^2\lambda \sin^2\theta)/\{\tfrac{1}{2}(\alpha_H - \alpha_O) - \tfrac{1}{2}\Delta\}$$
$$= \underline{g_e - \{4hc\zeta \sin^2\theta / [(\alpha_H - \alpha_O) - \Delta]\}} = g_{zz}\,.$$

EXERCISE: Calculate the g-values for H_2O^-. Investigate the effect of including O2s-orbital participation in H_2O^+.

14.22 The normalized forms of the d-orbitals are

$$d_{z^2} = (5/16\pi)^{\frac{1}{2}}(3\cos^2\theta - 1)\, r^2 f(r) = (5/16\pi)^{\frac{1}{2}}(3z^2 - r^2)\, f(r)$$
$$d_{xz} = (15/4\pi)^{\frac{1}{2}} \cos\theta \sin\theta \cos\phi\, r^2 f(r) = (15/4\pi)^{\frac{1}{2}}\, xz f(r)$$
$$d_{yz} = (15/4\pi)^{\frac{1}{2}} \cos\theta \sin\theta \sin\phi\, r^2 f(r) = (15/4\pi)^{\frac{1}{2}}\, yz f(r)$$

Consequently,

$$l_z d_{z^2} = 0\,,$$
$$l_y d_{z^2} = (5/16\pi)^{\frac{1}{2}}(\hbar/i)\{z(\partial/\partial x) - x(\partial/\partial z)\}(3z^2 - r^2)\, f(r)$$
$$= 6i\hbar(5/16\pi)^{\frac{1}{2}}\, xz f(r) = i\hbar\sqrt{3}\, d_{xz}\,,$$
$$l_x d_{z^2} = (5/16\pi)^{\frac{1}{2}}(\hbar/i)\{y(\partial/\partial y) - z(\partial/\partial y)\}(3z^2 - r^2)\, f(r)$$
$$= -6i\hbar(5/16\pi)^{\frac{1}{2}}\, yz f(r) = -i\hbar\sqrt{3}\, d_{yz}\,.$$

Therefore,

$$g_{zz} = \underline{g_e}\,,$$
$$g_{yy} = g_e - 2\lambda\langle d_{xz} | l_y | d_{z^2}\rangle^2/\Delta \qquad \text{[eqn (14.9.9)]}$$
$$= g_e - 6\lambda\hbar^2/\Delta = \underline{g_e - 6hc\zeta/\Delta}\,,$$
$$g_{xx} = g_e - 2\lambda\langle d_{yz} | l_x | d_{z^2}\rangle^2/\Delta = g_e - 6hc\zeta/\Delta$$

Taking $\zeta = 154\ \mathrm{cm^{-1}}$ and $\Delta/hc = 10^4\ \mathrm{cm^{-1}}$ gives

$$g_{zz} = g_e = \underline{2.002}\,, \quad g_{yy} = g_{xx} = g_e - 0.092 = \underline{1.910}\,.$$

EXERCISE: In a similar complex the d_{xy} orbital was the lowest; calculate the g-values.

14.23 $s_{1x}^2 = \frac{1}{4}(s_1^+ + s_1^-)(s_1^+ + s_1^-)$

$= \frac{1}{4}(s_1^+ s_1^- + s_1^- s_1^+)$ [because $s_i^+ s_i^+|m_s\rangle = s_i^- s_i^-|m_s\rangle = 0$ for $s_i = \frac{1}{2}$]

$= \frac{1}{4}\hbar^2$ [because $s_i^+ s_i^-|\alpha\rangle$, $s_i^+ s_i^-|\beta\rangle = 0$, $s_i^- s_i^+|\beta\rangle = \hbar^2|\beta\rangle$,

and $s_i^- s_i^+|\alpha\rangle = 0$].

$$S_x^2 = (s_{1x}+s_{2x})^2 = s_{1x}^2 + s_{2x}^2 + 2s_{1x}^2 s_{2x}^2 = \tfrac{1}{2}\hbar^2 + 2s_{1x}s_{2x}$$

$$s_{1x}s_{1y} = (1/4\mathrm{i})(s_1^+ + s_1^-)(s_1^+ - s_1^-) = (1/4\mathrm{i})(s_1^+ s_1^+ - s_1^- s_1^- + s_1^- s_1^+ - s_1^+ s_1^-)$$
$$= (1/4\mathrm{i})(s_1^- s_1^+ - s_1^+ s_1^-) = \tfrac{1}{2}\mathrm{i}\hbar s_{1z}.$$

$$S_y S_x = s_{1y}s_{2x} + s_{2y}s_{1x} - \tfrac{1}{2}\mathrm{i}\hbar(s_{1z}+s_{2z}).$$

$$S_x S_y + S_y S_x = 2(s_{1x}s_{2y} + s_{2x}s_{1y}).$$

Consequently,

$$\begin{aligned}\boldsymbol{s}_1\cdot\boldsymbol{s}_2 - 3\boldsymbol{s}_1\cdot\hat{\boldsymbol{r}}\hat{\boldsymbol{r}}\cdot\boldsymbol{s}_2 &= s_{1x}s_{2x} + s_{1y}s_{2y} + s_{1z}s_{2z}\\ &\quad - 3s_{1x}(x/r)^2 s_{2x} - 3s_{1y}(y/r)^2 s_{2y} - 3s_{1z}(z/r)^2 s_{2z}\\ &\quad - 3(s_{1x}s_{2y} + s_{1y}s_{2x})\,xy/r^2\\ &\quad - 3(s_{1y}s_{2z} + s_{1z}s_{2y})\,yz/r^2\\ &= \tfrac{1}{2}S_x^2 + \tfrac{1}{2}S_y^2 + \tfrac{1}{2}S_z^2 - \tfrac{3}{2}\hbar^2\\ &\quad - 3S_x^2(x/r)^2 - 3S_y^2(y/r)^2 - 3S_z^2(z/r)^2 + 3\cdot\tfrac{1}{2}\hbar^2(x^2+y^2+z^2)/r^2\\ &\quad - \tfrac{3}{2}(S_xS_y + S_yS_x)\,xy/r^2 - \tfrac{3}{2}(S_yS_z^2 + S_zS_y)\,yz/r^2\\ &\quad - \tfrac{3}{2}(S_yS_z + S_zS_y)\,yz/r^2\\ &= \tfrac{1}{2}\Big\{S_x^2(r^2-3x^2)/r^2 + S_y^2(r^2-3y^2)/r^2 + S_z(r^2-3z^2)/r^2\\ &\quad - (S_xS_y + S_yS_x)\,3xy/r^2 - (S_yS_z + S_zS_y)\,3yz/r^2\\ &\quad - (S_xS_z + S_zS_x)\,3xz/r^2\Big\}.\end{aligned}$$

Therefore,

$$H_{\mathrm{D}} = (g_{\mathrm{e}}^2\gamma_{\mathrm{e}}^2\mu_0/4\pi r^3)\{\boldsymbol{s}_1\cdot\boldsymbol{s}_2 - 3\boldsymbol{s}_1\cdot\hat{\boldsymbol{r}}\hat{\boldsymbol{r}}\cdot\boldsymbol{s}_2\} \qquad \text{[eqn (14.7.9)]}$$

$$= \boldsymbol{S}.\boldsymbol{D}.\boldsymbol{S} = \sum_{ij} D_{ij}S_iS_j \qquad (i,j = x,y,z),$$

with

$$D_{xx} = (g_{\mathrm{e}}^2\gamma_{\mathrm{e}}^2\mu_0/8\pi)\langle(1-3x^2/r^2)/r^3\rangle = D^{\ominus}\langle(1-3\sin^2\theta\cos^2\phi)/r^3\rangle$$

$$D_{yy} = D^{\ominus}\langle(1-3y^2/r^2)/r^3\rangle = D^{\ominus}\langle(1-3\sin^2\theta\sin^2\phi)/r^3\rangle$$

$$D_{zz} = D^{\ominus}\langle(1-3z^2/r^2)/r^3\rangle = D^{\ominus}\langle(1-3\cos^2\theta)/r^3\rangle$$

$$D_{xy} = D_{yx} = -D^{\ominus}\langle 3xy/r^5\rangle = -\tfrac{3}{2}D^{\ominus}\langle(1/r^3)\sin^2\theta\sin 2\phi\rangle$$

$$D_{yz} = D_{zy} = -D^{\ominus}\langle 3yz/r^5\rangle = -\tfrac{3}{2}D^{\ominus}\langle (1/r^3)\sin 2\theta \sin\phi\rangle$$
$$D_{xz} = D_{zx} = -D^{\ominus}\langle 3xz/r^5\rangle = -\tfrac{3}{2}D^{\ominus}\langle (1/r^3)\sin 2\theta \cos\phi\rangle,$$

with $D^{\ominus} = g_e^2\gamma_e^2\mu_0/8\pi$.

EXERCISE: Show that $\mathrm{tr}\,\boldsymbol{D}=0$, and that the diagonal elements of $\boldsymbol{D}$ may be expressed in terms of the two quantities $D=D_{zz}-\tfrac{1}{2}(D_{xx}+D_{yy})$ and $E = D_{xx}-D_{yy}$.

14.24 Write $\psi = a\psi_{2s} + b\psi_{1s}$, and choose a and b so that

$$\int \psi_{1s}\psi\, d\tau = a\int \psi_{1s}\psi_{2s}\,d\tau + b = 0,$$
$$\int \psi^2\, d\tau = \int (a\psi_{2s}+b\psi_{1s})^2\, d\tau = a^2+b^2+2ab\int \psi_{2s}\psi_{1s}\,d\tau = 1.$$

Write $S=\int \psi_{1s}\psi_{2s}\,d\tau$, then

$$b=-aS \quad\text{and}\quad a=1/(1-S^2)^{\frac{1}{2}}.$$

Consequently, the orthogonalized 2s-orbital is

$$\psi = (\psi_{2s}-S\psi_{1s})/\sqrt{(1-S^2)}, \qquad S=\int \psi_{1s}\psi_{2s}\,d\tau.$$

Then, since $\psi_{2s}(0)=0$,

$$\psi^2(0) = S^2\psi_{1s}^2(0)/(1-S^2).$$

$$\psi_{1s} = (Z_{1s}^{*3}/\pi a_0^3)^{\frac{1}{2}}\, e^{-Z_{1s}^{*}r/a_0} \quad \text{[Section 9.10, Problem 14.8]},$$
$$\psi_{1s}^2(0) = Z_{1s}^{*3}/\pi a_0^3,$$

$$\psi_{2s} = (Z_{2s}^{*5}/96\pi a_0^5)^{\frac{1}{2}}\, r\, e^{-Z_{2s}^{*3} r/a_0}.$$

$$S = 4\pi(Z_{1s}^{*3}Z_{2s}^{*5}/96\pi^2 a_0^8)^{\frac{1}{2}}\int_0 r^2\left\{r\,e^{-Z_{1s}^{*}r/2a_0}\right\}\left\{e^{-Z_{1s}^{*}r/a_0}\right\}dr$$
$$= 4\pi(Z_{1s}^{*3}Z_{2s}^{*5}/96\pi^2 a_0^8)^{\frac{1}{2}}\left\{3!\,a_0^4/(Z_{1s}^{*}+\tfrac{1}{2}Z_{2s}^{*})^4\right\}$$
$$= (6Z_{1s}^{*3}Z_{2s}^{*5})^{\frac{1}{2}}/(Z_{1s}^{*}+\tfrac{1}{2}Z_{2s}^{*})^4.$$

Consequently,

$$\psi^2(0) = \left\{\frac{(6/\pi a_0^3)\,Z_{1s}^{*6}Z_{2s}^{*5}/(Z_{1s}^{*}+\tfrac{1}{2}Z_{2s}^{*})^8}{1-6Z_{1s}^{*3}Z_{2s}^{*5}/(Z_{1s}^{*}+\tfrac{1}{2}Z_{2s}^{*})^8}\right\}$$
$$= \left\{\frac{(6/\pi a_0^3)\,Z_{1s}^{*6}Z_{2s}^{*5}}{(Z_{1s}^{*}+\tfrac{1}{2}Z_{2s}^{*})^8-6Z_{1s}^{*3}Z_{2s}^{*5}}\right\}.$$

In the case of ^{14}N, $Z_{1s}^{*}=6.70$, $Z_{2s}^{*}=3.90$ [Table 9.1], and $\psi^2(0) = 16.48/\pi a_0^3$. Therefore, with

$$H^{(\mathrm{spin})} = CI_z s_z, \quad C = (2g_e g_N \mu_B \mu_N \mu_0/3\hbar^2)\psi^2(0) \quad [\text{eqn } (14.11.12c)],$$

$$\hbar^2 C = 11.0\ (g_e g_N \mu_B \mu_N \mu_0/\pi a_0^3).$$

Since $g_N(^{14}N) = 0.403\,56$ [Table 14.1],

$$\hbar^2 C = 1.12\times 10^{-24}\ \mathrm{J}; \quad \underline{\hbar^2 C/h = 1.69\ \mathrm{GHz}}.$$

(The experimental value is 1.5 GHz.)

EXERCISE: Find an expression for the contact interaction involving an electron in an orthogonalized Slater 3s-orbital (i.e. one orthogonalized to both ψ_{1s} and the orthogonalized ψ_{2s}).

14.25 $A(\theta) = -(g_e g_N \mu_B \mu_N \mu_0/4\pi\hbar^2)\langle(1-3\cos^2\theta)/r^3\rangle$ [eqn (14.11.12b)].

$\psi(3d_{z^2}) = (Z^*/a_0)^{\frac{7}{2}}(1/2\pi\cdot 3^9)^{\frac{1}{2}}(3\cos^2\theta - 1)r^2 e^{-Z^*r/3a_0}$ [Section 9.10, Table 4.1].

$$\langle(1-3\cos^2\theta)/r^3\rangle = (-1/2\pi 3^9)(Z^*/a_0)^7\, 2\pi \times \int_0^{\pi} \sin\theta(3\cos^2\theta-1)^3\, d\theta \int_0^{\infty} r^2(r^4/r^3)\, e^{-2Z^*r/3a_0}\, dr$$

$$= -(1/3^9)(Z^*/a_0)^7 \int_{-1}^{1}(27x^6 - 27x^4 + 9x^2 - 1)\, dx \int_0^{\infty} r^3 e^{-2Z^*r/3a_0}\, dr$$

$$= -(1/3^9)(Z^*/a_0)^7(32/35)\left\{3!/(2Z^*/3a_0)^4\right\}$$

$$= -(4/2835)(Z^{*3}/a_0)^3.$$

Therefore, when the field is parallel to the z-axis,

$$A/A^{\ominus} = \underline{1.41\times 10^{-3}\, Z^{*3}}, \quad \hbar^2 A^{\ominus} = g_e g_N \mu_B \mu_N \mu_0/4\pi a_0^3.$$

When the field is perpendicular to the axis, the equivalent description is to maintain the $1-3\cos^2\theta$ form of the dipolar interaction, but to interpret d_{z^2} as d_{x^2}, the same orbital rotated through 90^0 and now lying along the x-axis. That is, we use

$$\psi(3d_{x^2}) = (Z^*/a_0)^{\frac{7}{2}}(4/5\pi\cdot 3^9)^{\frac{1}{2}}(3\cos^2\phi\sin^2\theta - 1)r^2 e^{-Z^*r/3a_0}.$$

(Note the change of the normalization factor.) Then,

$$\langle(1-3\cos^2\theta)/r^3\rangle = (4/5\pi\cdot 3^9)(Z^*/a_0)^7 \int_0^{2\pi} d\phi \int_0^{\pi} \sin\theta\, d\theta \int_0^{\infty} r^2\, dr$$

$$\times (1-3\cos^2\theta)(3\cos^2\phi\sin^2\theta - 1)^2(1/r^3)\, r^4 e^{-2Z^*r/3a_0}$$

$$= (4/5\pi\cdot 3^9)(Z^*/a_0)^7\,\pi\int_0^\pi \sin\theta(1-3\cos^2\theta)(\tfrac{45}{8}\sin^4\theta - 6\sin^2\theta + 2)\,d\theta$$

$$\times\int_0^\infty r^3\,e^{-2Z^*r/3a_0}\,dr\quad\left[\int_0^\pi \cos^2\phi\,d\phi = \pi\,,\quad \int_0^{2\pi}\cos^4\phi\,d\phi = \tfrac{5}{8}\pi\right]$$

$$= (4/5\pi\cdot 3^9)(Z^*/a_0)^7\,\pi\,(8/35)\left\{3!/(2Z^*/3a_0)^4\right\}$$

$$= (4/3^4\cdot 5^2\cdot 7)(Z^*/a_0)^3 = -\tfrac{1}{5}\langle(1-3\cos^2\theta)/r^3\rangle_{d_{z^2}}$$

Therefore, $\underline{A_\perp = -\tfrac{1}{5}A_\parallel}$.

EXERCISE: Find the expressions for the dipolar interaction between a nucleus and a d_{xy}-electron when the field is (a) along z, (b) along x.

14.26 $J \approx (2\mu_0 g_e \mu_B/3)^2 \gamma_A\gamma_B|\psi_A^2(0)||\psi_B^2(0)|c_A^2 c_B^2\,\Delta_T$ [eqn (14.12.7)].

$c_A^2 = c_B^2 \approx \tfrac{1}{2}$; $\psi_A^2(0) = \psi_B^2(0) = 1/\pi a_0^3$.

$\gamma_A = \gamma(^1H) = g(^1H)\,\mu_N$, $g(^1H) = 5.5857$ [Table 14.1]

$\gamma_B = \gamma(^2H) = g(^2H)\,\mu_N$, $g(^2H) = 0.857\,45$.

$$J \approx (2\mu_0 g_e\mu_B/3)^2 g(^1H)\,g(^2H)\,\mu_N^2\,(1/\pi a_0^3)^2\,(1/2)^2/\Delta_T$$

$$\approx (g_e\mu_0\mu_B\mu_N/3\pi a_0^3)^2\,g(^1H)\,g(^2H)/\Delta_T$$

$$\approx \left\{7.122\times 10^{-51} g(^1H)\,g(^2H)/(\Delta_T/J)\right\}J$$

$$\approx \left\{4.445\times 10^{-32} g(^1H)\,g(^2H)/(\Delta_T/eV)\right\}J$$

$$J/h = 67.1\left\{g(^1H)\,g(^2H)/(\Delta_T/eV)\right\}Hz = (321\ Hz)/(\Delta_T/eV)$$

$\approx$ $\underline{32\ Hz}$ when $\Delta_T \approx 10$ eV.

EXERCISE: Find an expression for the spin-spin coupling involving two nuclei, one of atomic number Z_1 and the other of atomic number Z_2. Express c_A and c_B in terms of α_1, α_2, and β in a Hückel type of approximation.

14.27 $H = -\gamma_H(1-\sigma_A)\,BI_{Az} - \gamma_H(1-\sigma_B)\,BI_{Bz} + (J/\hbar^2)\boldsymbol{I}_A\cdot\boldsymbol{I}_B$.

Use the basis $|\alpha_A\alpha_B\rangle$, $|\alpha_A\beta_B\rangle$, $|\beta_A\alpha_B\rangle$, $|\beta_A\beta_B\rangle$, and use $\boldsymbol{I}_A\cdot\boldsymbol{I}_B = I_{Az}I_{Bz} + \tfrac{1}{2}(I_A^+I_B^- + I_A^-I_B^+)$.

$$H|\alpha\alpha\rangle = \left\{-\tfrac{1}{2}\gamma_H(1-\sigma_A)\hbar B - \tfrac{1}{2}\gamma_H(1-\sigma_B)\hbar B + \tfrac{1}{4}J\right\}|\alpha\alpha\rangle,$$

so that $E_{\alpha\alpha} = -\hbar\gamma_H B + \tfrac{1}{2}(\sigma_A+\sigma_B)\hbar\gamma_H B + \tfrac{1}{4}J$; eigenstate $|\alpha\alpha\rangle$,

likewise $E_{\beta\beta} = \hbar\gamma_H B - \tfrac{1}{2}(\sigma_A+\sigma_B)\hbar\gamma_N B + \tfrac{1}{4}J$; eigenstate $|\beta\beta\rangle$.

$$H|\alpha\beta\rangle = -\tfrac{1}{2}\gamma_H(1-\sigma_A)\hbar B|\alpha\beta\rangle + \tfrac{1}{2}\gamma_H(1-\sigma_B)\hbar B|\alpha\beta\rangle - \tfrac{1}{4}J|\alpha\beta\rangle + \tfrac{1}{2}J|\beta\alpha\rangle ,$$

$$H|\beta\alpha\rangle = \tfrac{1}{2}\gamma_H(1-\sigma_A)\hbar B|\beta\alpha\rangle - \tfrac{1}{2}\gamma_H(1-\sigma_B)\hbar B|\beta\alpha\rangle - \tfrac{1}{4}J|\beta\alpha\rangle + \tfrac{1}{2}J|\alpha\beta\rangle .$$

The secular determinant in the basis $|\alpha\beta\rangle$, $|\beta\alpha\rangle$ is therefore

$$\begin{vmatrix} \tfrac{1}{2}(\sigma_A-\sigma_B)\hbar\gamma_H B - \tfrac{1}{4}J - E & \tfrac{1}{2}J \\ \tfrac{1}{2}J & \tfrac{1}{2}(\sigma_B-\sigma_A)\hbar\gamma_H B - \tfrac{1}{4}J - E \end{vmatrix} = 0 .$$

This has the form encountered on p.100 of the text. Therefore, with the identifications $a=\tfrac{1}{2}(\sigma_A-\sigma_B)\hbar\gamma_H B-\tfrac{1}{4}J$, $b=-\tfrac{1}{2}(\sigma_A-\sigma_B)\hbar\gamma_H B-\tfrac{1}{4}J$, $d=\tfrac{1}{2}J$,

$$E_\pm = -\tfrac{1}{4}J \pm \tfrac{1}{2}\surd\left\{(\sigma_A-\sigma_B)^2\hbar^2\gamma_H^2B^2+J^2\right\} = -\tfrac{1}{4}J \pm Q$$

$$\left.\begin{aligned} |+\rangle &= |\alpha\beta\rangle\cos\eta + |\beta\alpha\rangle\sin\eta \\ |-\rangle &= -|\alpha\beta\rangle\sin\eta + |\beta\alpha\rangle\cos\eta \end{aligned}\right\} \tan 2\eta = J/(\sigma_A-\sigma_B)\hbar\gamma_H B .$$

The transition intensities are proportional to the squares of the matrix elements of $I_{Ax}+I_{Bx} = \tfrac{1}{2}(I_A^+ + I_B^+ + I_A^- + I_B^-)$. We therefore have

$$\begin{aligned}
|\langle -|I_{Ax}+I_{Bx}|\beta\beta\rangle|^2 &= |\langle -|\alpha\beta+\beta\alpha\rangle(\tfrac{1}{2}\hbar)|^2 \\
&= \tfrac{1}{4}\hbar^2(-\sin\eta+\cos\eta)^2 = \tfrac{1}{4}\hbar^2(1-\sin 2\eta) , \\
|\langle +|I_{Ax}+I_{Bx}|\beta\beta\rangle|^2 &= |\langle +|\alpha\beta+\beta\alpha\rangle(\tfrac{1}{2}\hbar)|^2 \\
&= \tfrac{1}{4}\hbar^2(\cos\eta+\sin\eta)^2 = \tfrac{1}{4}\hbar^2(1+\sin 2\eta) , \\
|\langle -|I_{Ax}+I_{Bx}|\alpha\alpha\rangle|^2 &= \tfrac{1}{4}\hbar^2(1-\sin 2\eta) , \\
|\langle +|I_{Ax}+I_{Bx}|\alpha\alpha\rangle|^2 &= \tfrac{1}{4}\hbar^2(1+\sin 2\eta) , \\
|\langle\alpha\alpha|I_{Ax}+I_{Bx}|\beta\beta\rangle|^2 &= \tfrac{1}{4}\hbar^2|\langle\alpha\alpha|\alpha\beta+\beta\alpha\rangle|^2 = 0 , \\
|\langle +|I_{Ax}+I_{Bx}|-\rangle|^2 &= \tfrac{1}{4}\hbar^2|\langle +|-\alpha\alpha\sin\eta-\beta\beta\sin\eta+\alpha\alpha\cos\eta+\beta\beta\sin\eta\rangle|^2 \\
&= 0 .
\end{aligned}$$

Therefore, we expect four lines:

(1) $|-\rangle \leftarrow |\alpha\alpha\rangle$, intensity $\propto (1-\sin 2\eta)$,

$$\Delta E = E_- - E_{\alpha\alpha} = \gamma_H\hbar B - \tfrac{1}{2}(\sigma_A+\sigma_B)\gamma_H\hbar B - \tfrac{1}{2}J - Q ;$$

(2) $|+\rangle \leftarrow |\alpha\alpha\rangle$, intensity $\propto (1+\sin 2\eta)$,

$$\Delta E = E_+ - E_{\alpha\alpha} = \gamma_H\hbar B - \tfrac{1}{2}(\sigma_A+\sigma_B)\gamma_H\hbar B - \tfrac{1}{2}J + Q ;$$

(3) $|\beta\beta\rangle \leftarrow |+\rangle$, intensity $\propto (1+\sin 2\eta)$,

$$\Delta E = E_{\beta\beta} - E_+ = \gamma_H\hbar B - \tfrac{1}{2}(\sigma_A+\sigma_B)\gamma_H\hbar B + \tfrac{1}{2}J - Q ;$$

(4) $|\beta\beta\rangle \leftarrow |-\rangle$, intensity $\propto (1-\sin 2\eta)$,

$$\Delta E = E_{\beta\beta} - E_{-} = \gamma_{\mathbf{H}}\hbar B - \tfrac{1}{2}(\sigma_{\mathbf{A}}+\sigma_{\mathbf{B}})\gamma_{\mathbf{H}}\hbar B + \tfrac{1}{2}J + Q\,.$$

In each case, $Q = \tfrac{1}{2}\{(\sigma_{\mathbf{A}}-\sigma_{\mathbf{B}})^2\gamma_{\mathbf{H}}^2\hbar^2B^2 + J^2\}^{\frac{1}{2}}$.

(a) When $J=0$, $Q = \tfrac{1}{2}(\sigma_{\mathbf{A}}-\sigma_{\mathbf{B}})\gamma_{\mathbf{H}}\hbar B$ and $\eta = 0$. The four allowed transitions are therefore as follows:

$$\Delta E(1)/\gamma_{\mathbf{H}}\hbar = \Delta E(3)/\gamma_{\mathbf{H}}\hbar = (1-\sigma_{\mathbf{A}})B\,, \quad \text{intensity } 1$$

$$\Delta E(2)/\gamma_{\mathbf{H}}\hbar = \Delta E(4)/\gamma_{\mathbf{H}}\hbar = (1-\sigma_{\mathbf{B}})B\,, \quad \text{intensity } 1\,.$$

This is depicted in Fig. 14.8(a)

Fig. 14.8

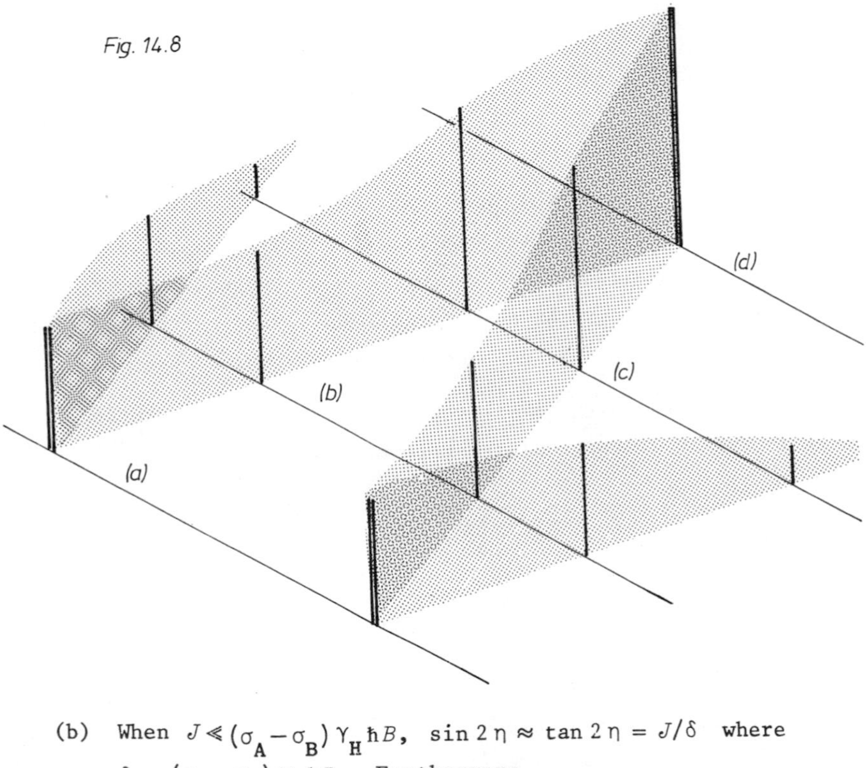

(b) When $J \ll (\sigma_{\mathbf{A}}-\sigma_{\mathbf{B}})\gamma_{\mathbf{H}}\hbar B$, $\sin 2\eta \approx \tan 2\eta = J/\delta$ where $\delta = (\sigma_{\mathbf{A}}-\sigma_{\mathbf{B}})\gamma_{\mathbf{H}}\hbar B$. Furthermore,

$$Q = \tfrac{1}{2}\delta\{1+(J/\delta)^2\}^{\frac{1}{2}} \approx \tfrac{1}{2}\delta + J^2/4\delta\,.$$

The four allowed transitions are then as follows:

$$\Delta E(1) \approx (1-\sigma_{\mathbf{A}})\gamma_{\mathbf{H}}\hbar B - \tfrac{1}{2}J - \tfrac{1}{4}(J^2/\delta)\,, \quad \text{intensity } 1-J/\delta\,,$$

$$\Delta E(2) \approx (1-\sigma_{\mathbf{B}})\gamma_{\mathbf{H}}\hbar B - \tfrac{1}{2}J + \tfrac{1}{4}(J^2/\delta)\,, \quad \text{intensity } 1+J/\delta\,,$$

$$\Delta E(3) \approx (1-\sigma_{\mathbf{A}})\gamma_{\mathbf{H}}\hbar B + \tfrac{1}{2}J - \tfrac{1}{4}(J^2/\delta)\,, \quad \text{intensity } 1+J/\delta\,,$$

$\Delta E(4) \approx (1-\sigma_B)\,\gamma_H \hbar B + \frac{1}{2}J + \frac{1}{4}(J^2/\delta)$, intensity $1-J/\delta$.

This is depicted in Fig. 14.8(b).

(c) When $J=\delta$, $Q=\delta/\sqrt{2}$ and $\tan 2\eta = 1$ so that $2\eta=\pi/4$. Then with $\sigma=\frac{1}{2}(\sigma_A+\sigma_B)$ the four transitions lie at

$\Delta E(1) \approx (1-\sigma)\,\gamma_H \hbar B - \frac{1}{2}(1+\sqrt{2})\delta$, intensity $1-1/\sqrt{2}$,

$\Delta E(2) \approx (1-\sigma)\,\gamma_H \hbar B - \frac{1}{2}(1-\sqrt{2})\delta$, intensity $1+1/\sqrt{2}$,

$\Delta E(3) \approx (1-\sigma)\,\gamma_H \hbar B + \frac{1}{2}(1-\sqrt{2})\delta$, intensity $1+1/\sqrt{2}$,

$\Delta E(4) \approx (1-\sigma)\,\gamma_H \hbar B + \frac{1}{2}(1+\sqrt{2})\delta$, intensity $1-1/\sqrt{2}$.

This is depicted in Fig. 14.8(c).

(d) When $\delta=0$, $Q=\frac{1}{2}J$ and $2\eta=\pi/2$ (so that $\sin 2\eta = 1$). Then there are only two allowed transitions $[1-\sin 2\eta = 0]$, and they lie at

$\Delta E(2) = (1-\sigma)\,\gamma_H \hbar B$, intensity 2,

$\Delta E(3) = (1-\sigma)\,\gamma_H \hbar B$, intensity 2.

This single-line spectrum is depicted in Fig. 14.8(d). Note that the condition $\delta=0$ corresponds to two equivalent protons: even though J may be non-zero, no splitting of the spectrum is observed in the case of magnetically equivalent protons.

EXERCISE: Evaluate the energy levels in the coupled representation, and establish which transitions are allowed, and their intensities.